in DETAIL Designing Circulation Areas

in **DETAIL**

Designing Circulation Areas

Stairs, ramps, lifts
Routing
Planning principles

Christian Schittich (Ed.)

Edition DETAIL – Institut für internationale
Architektur-Dokumentation GmbH & Co. KG
Munich

Editor: Christian Schittich
Editorial services: Steffi Lenzen (project management),
Sandra Leitte, Jana Rackwitz
Editorial assistants: Carola Jacob-Ritz, Michaela Linder, Eva Schönbrunner
Consultant (Planning criteria): Peter Cheret, Stuttgart

Translation German/English: Antoinette Aichele-Platen, Munich;
Alistair Gray, Dr. Yasmin Gründing
Proofreading: Stefan Widdess, Keiki Communication, Berlin

Drawings: Ralph Donhauser, Martin Hämmel, Nicola Kollmann,
Simon Kramer, Emese M. Köszegi, Dejanira Ornelas

DTP: Roswitha Siegler

Bibliographic information published by the German National Library
The German National Library lists this publication in the Deutsche Nationalbibliografie; detailed bibliographic data is available on the Internet at <http://dnb.d-nb.de>.

This book is also available in a German language edition
(ISBN: 978-3-920034-81-2).

Printed on acid-free paper produced from chlorine-free pulp (TCF ∞)

Printed in Germany
Reproduction: ludwig:media, Zell am See
Printing and binding: Aumüller Druck, Regensburg

ISBN: 978-3-920034-89-8 (Print)
ISBN: 978-3-95553-140-9 (E-Book)
ISBN: 978-3-95553-151-5 (Bundle)

MIX
Paper from responsible sources
FSC® C017373

9 8 7 6 5 4 3 2 1

Contents

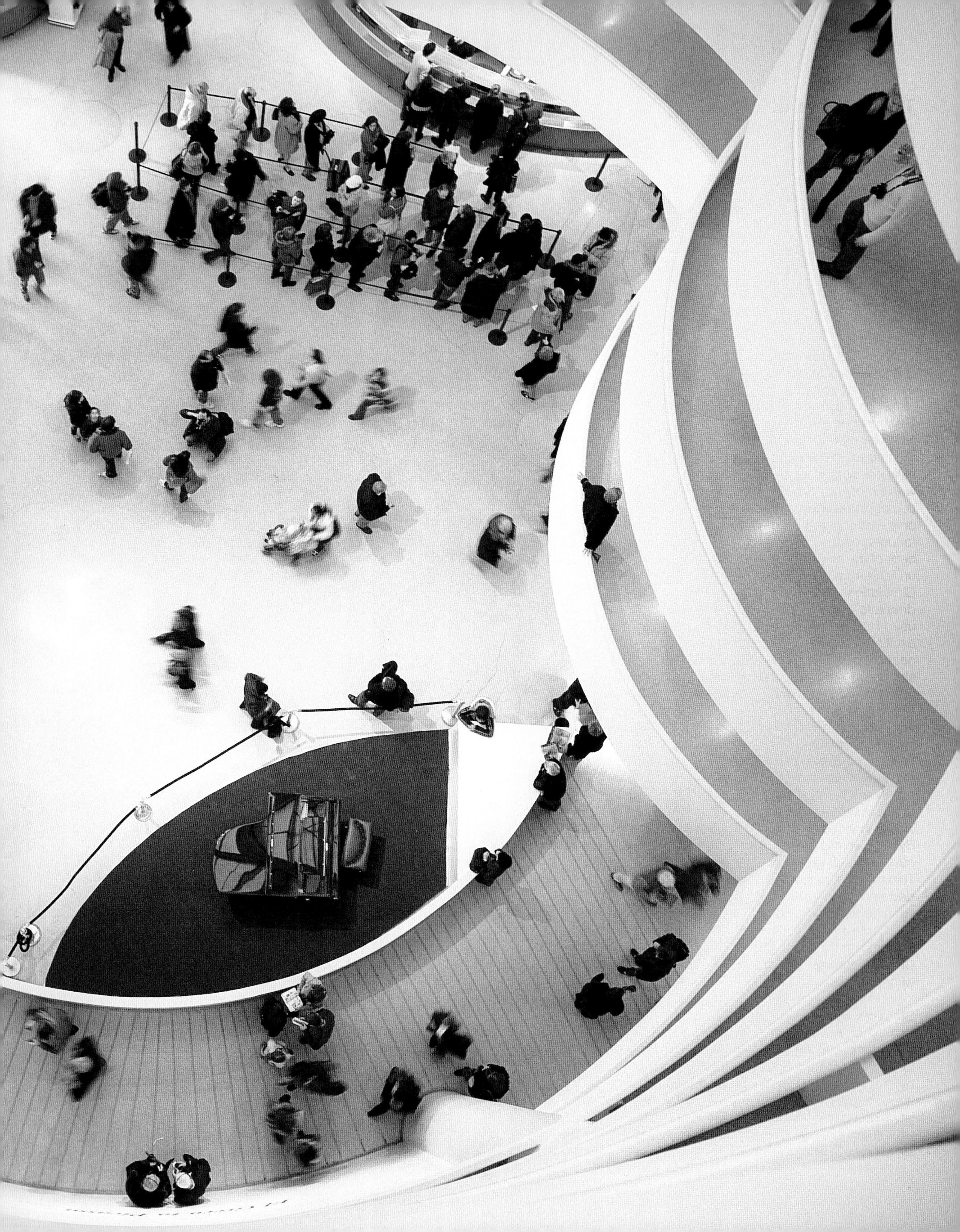

The concept of circulation

Christian Schittich

Still avant-garde, yet long an icon of modern architecture and one of the most popular exhibition venues in the art city New York today: Frank Lloyd Wright's last masterpiece, the Guggenheim Museum (fig. 1). It set off a storm of protest when it opened in October 1959. Architecture critics and leading artists were unanimous: the structure, which essentially consists of a downward-tapering concrete spiral ramp, was simply not considered suitable for a serious presentation of works of art. The architect from Wisconsin had not only – as so many others before him – emphasised and focused on the circulation, but had made it the actual concept of the museum, causing a shake-up in the traditional understanding of a museum.
Circulation – the term seems much too low-key for such a dramatic design. But then again the term, normally only used by professionals in the field, doesn't sound very exciting in general. In fact it is rather more indicative of the necessary evil of the associated rules, regulations and standards. Amazing really, considering the fact that circulation is one of the most influential components of a design. Circulation areas often serve as the calling card of a building and their individual elements are almost always used as important design features in sophisticated architecture. This applies particularly to stairs as built elements for movement within spaces. In addition to a purely functional significance – the connection of different levels – a metaphorical significance has always been attached to stairs as well, as symbols of ascent, of a transition to another world. By the Renaissance period, stairs in castles, palaces and cultural buildings had become architectural showpieces. This meant that the staircase was often a component with a decisive influence on the overall design. Great examples such as Michelangelo's stairs in the vestibule of the Laurentian Library in Florence, Balthasar Neumann's magnificent staircase in the Würzburg Residence (fig. 4, p. 16) or Charles Garnier's lavish staircase at the Paris Opera bear witness to this.

Stairs as sculptures
Unlike hardly any other structural element, stairs embody dynamism, making them ideal as enriching sculptural features within rooms. Special attention has been paid to them by architects ever since man started to build for this reason. It also explains the concentration of the essential characteristics of a particular trend in building history, sometimes even in an exaggerated form, frequently observed in stair structures. Specific trends in more recent architectural history are also reflected particularly well by the stair design – sometimes ornate or elaborate, sometimes complex and fragmented into individual components, or exhibiting a minimalist elegance. In high-tech architecture, stairs appear degraded into brackets, bolts and tensioning ropes, the overt structural design turning into a creative element (fig. 4, p. 10). Deconstructivists on the other hand use stairs to set off veritable fireworks of creativity, such as Günter Behnisch and his "Bird's Nest" in the former building of the Parliament of the Federal Republic of Germany, the "Bundestag", in Bonn (fig. 3, p. 10), where different-sized wooden sticks burst apart in all directions. In comparison, the appearance of most modern staircases is rather more simple and restrained today, although special stair

1 Solomon R. Guggenheim Museum, New York (USA) 1959, Frank Lloyd Wright
2 MUMUTH – House of Music and Music Drama at the University of Music and Performing Arts (KUG), Graz (A) 2008, UNStudio

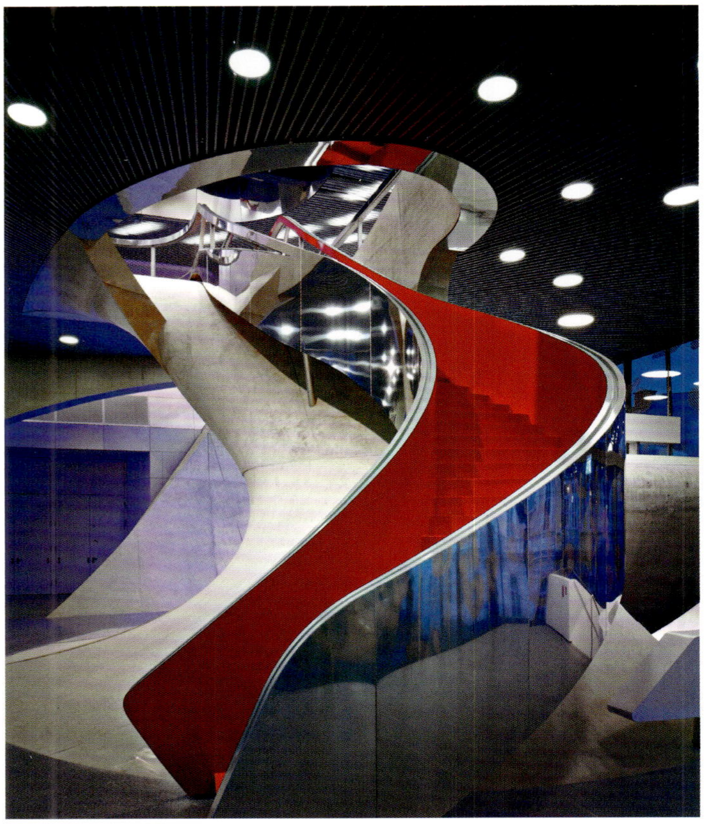
2

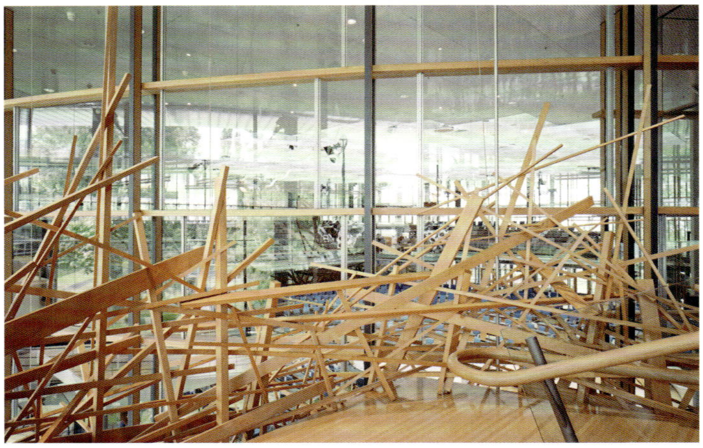

3

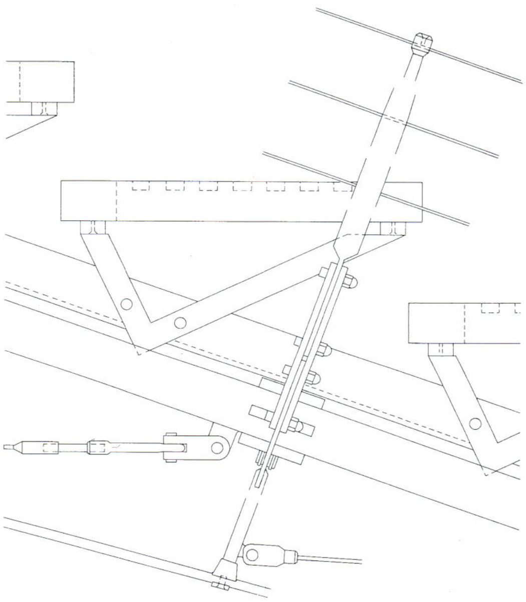

4

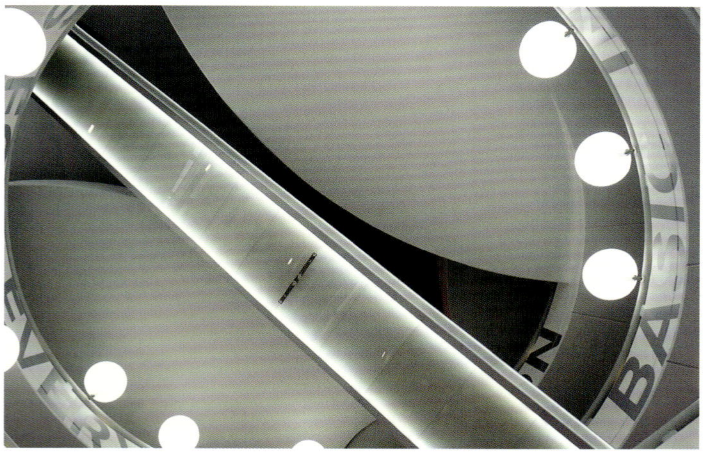

5

arrangements are also developed occasionally. These are often computer-generated free forms, capturing a furious, frozen movement or presenting themselves as organic sculptures.
A great example of this is Massimiliano Fuksas' tornado of a staircase in the Armani store in Manhattan (see p. 150ff.): the four-storey stair sculpture dominates the whole interior of the shop. As an intentional attraction, the exceptionally powerful and geometrically barely comprehensible shape becomes part of the corporate identity of the fashion label, symbolising values such as boundlessness and dynamism. Similarly organic, and also in New York, is a dramatic staircase in the Cooper Union Academic Building (see p. 93ff.) designed by Morphosis as the central communication centre of the university over a total of nine floors. In the House of Music and Music Drama at the University of Music and Performing Arts Graz designed by UNStudio (fig. 2), it is once more an impressively fluent staircase that forms the backbone of the entire design concept. The architects organised the internal circulation on the basis of a spiralling constructive element made of reinforced concrete emerging out of the rear walls of the music rooms. This connects and welds together the various storeys with a "twist", tempting and leading visitors upwards. As in most UNStudio projects, circulation and structure merge seamlessly. This also applies to the much smaller "Haus am Weinberg" (a single family house in a vineyard setting) in Stuttgart (see p. 66ff.). Here too, Ben van Berkel describes the central staircase as a "twist", transporting people upstairs as well as integrating the surrounding area outside, thereby creating an artificial landscape that merges with the natural landscape.
Structurally less way out, but equally impressive and clever, are the stair sculptures by the Danish architecture studio 3XN. Circulation almost always plays a central role in their designs. This is especially true for the high school building in Ørestad, which breaks with all conventions of school construction design, because there are no classrooms, no communication barriers and of course no corridors (see p. 98ff.). The whole school consists of a single large room, in which everything takes place in parallel – an architectural implementation of the requirements specified by the Danish school reform demanding a venue for flexible, open and independent learning. A central cascading stair system becomes the defining element of a vertical learning landscape composed of circular platforms, stairs and open spaces.

Jakob + MacFarlane also turn the circulation of their Cité de la Mode et du Design in Les Docks in Paris into an essential design element (fig. 6) – but on the outside, in the form of bright green structures attached to the facade. The solution is an echo of the principle so impressively demonstrated in the same city almost 40 years earlier by Renzo Piano and Richard Rogers with the Centre Pompidou (fig. 7). The escalator tubes colonising the facade give the building a special mechanical aesthetic. Around that time the escalator – until then to be found mainly in the rather functional surroundings of train stations and department stores – was first able to gain place in superior architecture.

Stunning escalators

A good decade after the opening of Frank Lloyd Wright's Guggenheim Museum, Karl Schwanzer made use of the same circulation principle in his "bowl" design of the BMW Museum in Munich, and enriched it with a new – spectacular – element (fig. 5). While Wright transports visitors upwards with lifts to let them wander downwards again past the works of art, Schwanzer's – much smaller – automobile museum requires them to walk upwards before taking an impressively positioned escalator back to where they started. The escalator cuts through the hemispherical space like an arrow and demonstrates effectively, why escalators that can be designed leaner and without landings, are currently so popular among architects, when aiming to create impressive circulation spaces in museums or shopping centres. Rem Koolhaas leads visitors into the former Coal Washing Plant in Essen's Zollverein Colliery now converted to the Ruhr Museum, via a bright orange glowing escalator. Zaha Hadid uses escalators in the Phaeno Science Center in Wolfsburg and Delugan Meissl integrate them in their design of the Porsche Museum in Stuttgart (see p. 146ff.). Like in many modern museums, the circulation and exhibition design concepts merge in the Porsche Museum. This is particularly appropriate for a car museum. The exhibition design by HG Merz is based on the structure of a road. The route through the museum landscape follows a permanent architectural movement in the shape of an ascending spiral composed of different levels and a series of flat ramps and stairs.

Ingeniously designed pathways

An orchestrated circulation through the building is an integral part of the museum construction concept. It can however also be an interesting design approach for other construction projects – even in a normally "high security" embassy build-

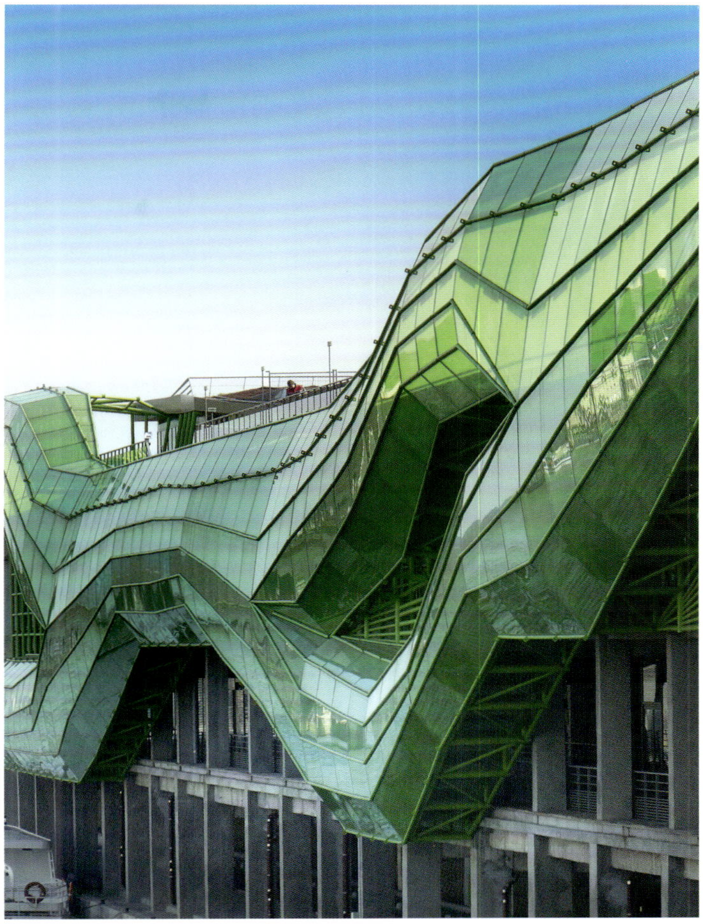

6

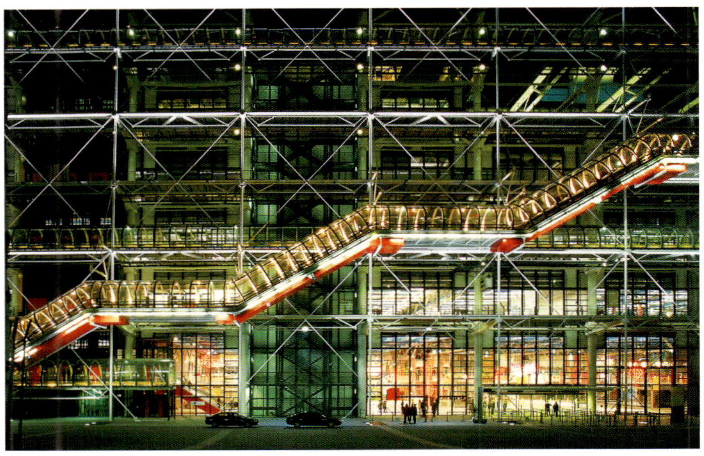

7

3 Former Bundestag building of the Federal Republic of Germany, Bonn (D) 1993, Günter Behnisch
4 Vertical section of a stair step, scale 1:10, apartment, London (GB) 1991, John Young
5 BMW Museum, Munich (D) 1973/2008, Karl Schwanzer, Atelier Brückner
6 Cité de la Mode et du Design, Paris (F) 2008, Jakob + MacFarlane
7 Centre Pompidou, Paris (F) 1977, Renzo Piano and Richard Rogers

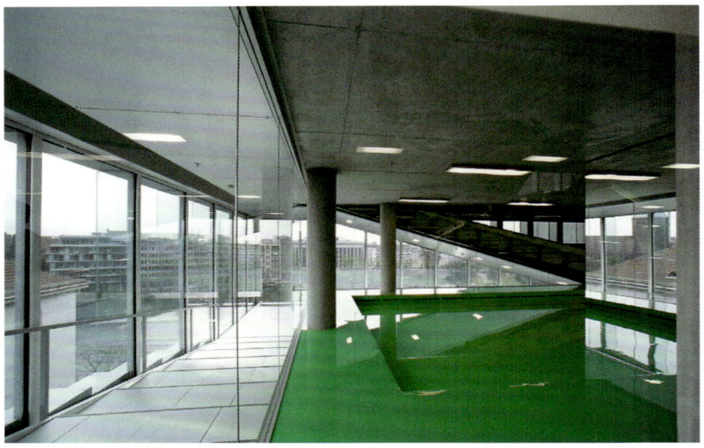
8

ing, as demonstrated by Rem Koolhaas in his design of the Netherlands Embassy in Berlin. He cuts through the glass cube with a pathway system made up of successive stairs, ramps and corridors (fig. 8). Instead of distinct floors, there are countless functionally related levels. Changing in width, the route zigzags through the ten storeys of the building. Sometimes narrow and sometimes wide, occasionally opening up to a common area, mostly located in the building interior, but also in a bay window section of the façade – a variety of views and perspectives within and outside the building are offered. Koolhaas aims to bring the urban circulation design inside the building in this way, as well as attempting to overcome the customary stiffness of diplomatic conventions.
He uses a similar principle in the Casa da Música in Porto completed soon afterwards (fig. 11). Here too, the circulation route is a complicated and very varied path through the concert hall, one that opens up to spaces finished with different materials – conveying a great diversity of sensory impressions.

Total space as circulation
At the Rolex Learning Center of the EPFL Lausanne, Kazuyo Sejima and Ruye Nishizawa of SANAA practically transfer the topography of hills and valleys inside, modelling a kind of ideal educational landscape – a space without boundaries at the disposal of students and other users, but one in which each must find his or her own way (fig. 9). The differentiation between circulation and usable areas disappears almost completely in the process. Hardly any dividing walls can be found on the area of almost 17,000 m^2 accommodating a library, workplaces, various cafeterias, a restaurant, a book

8 Royal Netherlands Embassy, Berlin (D) 2004, OMA
9 Rolex Learning Centre at the EPFL, Lausanne (CH) 2010, Kazuyo Sejima + Ruye Nishizawa / SANAA
10 Chikatsu Asuka Museum, Osaka (J) 1994, Tadao Ando
11 Casa da Música, Porto (P) 2004, OMA

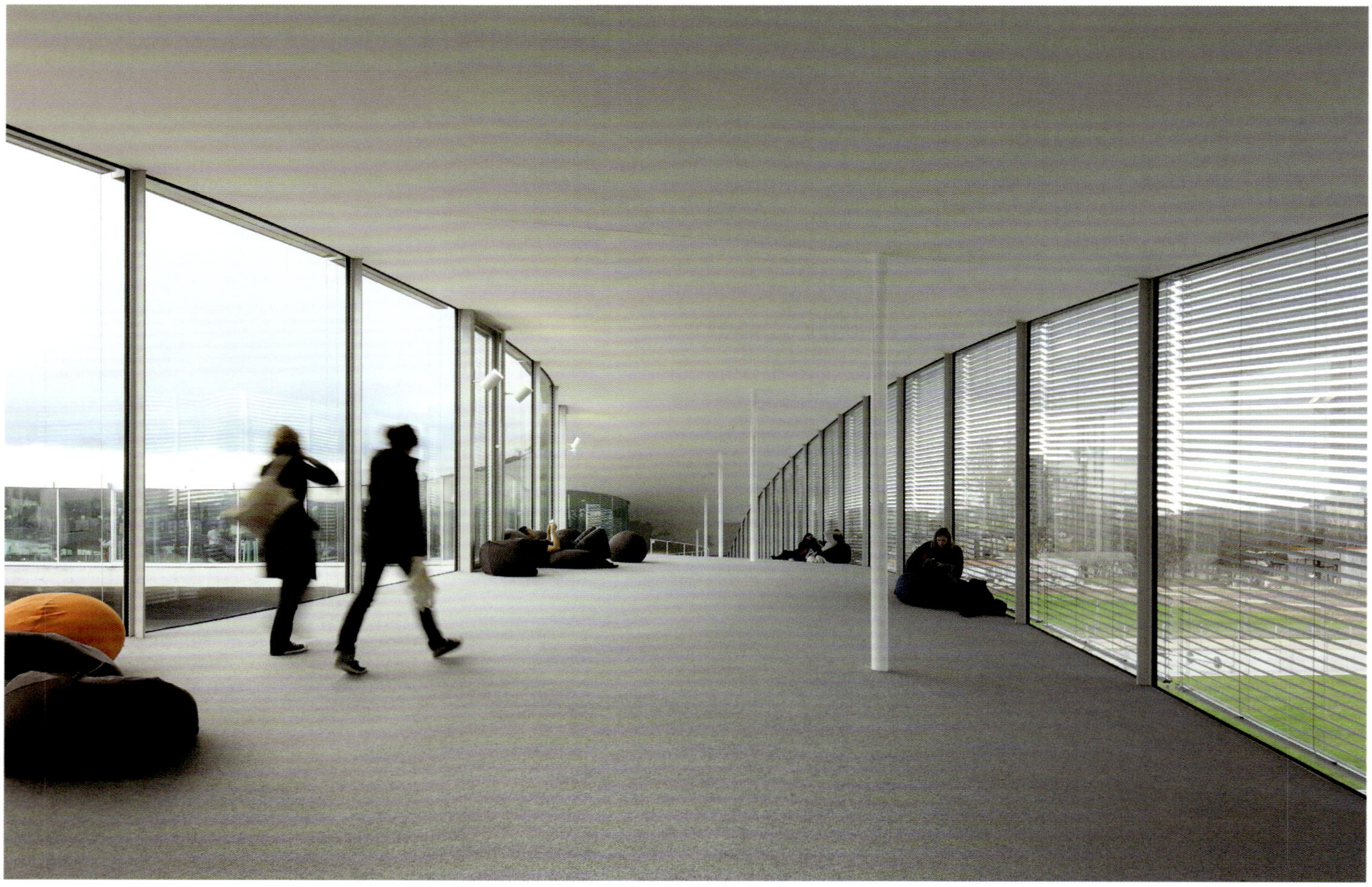
9

shop, a multifunctional auditorium and offices – the spatial flow is seamless. The structure is an ideal architectural realisation of the values of the institute: transparency, interrelation and innovation. With its unconventional concept and organic design language, the Rolex Learning Centre primarily aims to promote spontaneous encounters and thereby an exchange of ideas – a principle that is becoming increasingly important in working life, in office buildings and research institutes. The openness of the library landscape in Lausanne can of course rarely be implemented to such an extent in these facilities. But rather than limiting the function of foyer, corridor or staircase areas to transit zones, designs are generally aiming to invite to linger and thereby promote by-chance encounters and informal communication.

Building sculptures as a walk-on landscape

While the topography of the Rolex Learning Centre is clearly shifted inwards, the complete outer structure of the new Opera House in Oslo, designed by Snøhetta like a giant rock sitting in a bay, is presented as a walk-on landscape (see p. 140ff.). Almost the entire roof area is a public space. Access is not limited to prescribed paths – people can walk all over the roofscape. From the water level right up to the panorama plateau at the top, people can use the area as their own, enjoy the views of the city or the fjord, and look inside the building through a massive glass skylight.

About 20 years earlier, Tadao Ando designed the whole roof of his Chikatsu Asuka Museum near Osaka as an accessible area offering a view of the surrounding tumuli and countryside (fig. 10). An open-sky walk up a gigantic flight of steps, is the dramatic and scenic finale of a carefully designed route around and through the museum. As with all the cultural buildings designed by the Japanese architect, and in line with Japanese tradition, this route begins long before actually entering the building. The path is rarely straight; instead a meditative progression and a conscious approach towards the structure is sought, by offering a variety of preparatory perspectives on the way there. This is followed by the path through the interior, which is also designed to be a stimulating experience: spatial situations alternate between large and small, linear and curved, and especially between light and dark. Like the other examples of architecture described earlier, Tadao Ando's Chikatsu Asuka Museum illustrates above all that circulation should be much more than a pure necessity. It can be the concept on which a design is based, provide spaces offering a specific experience, serve as a communication zone or an area to linger in, or in some cases, even just be an end in itself.

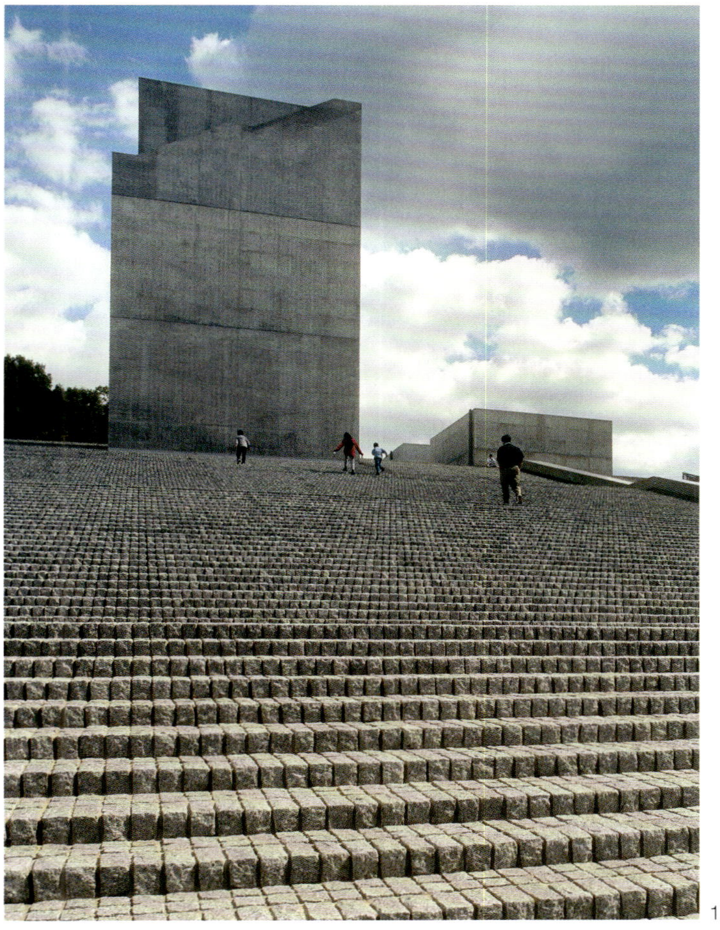

10

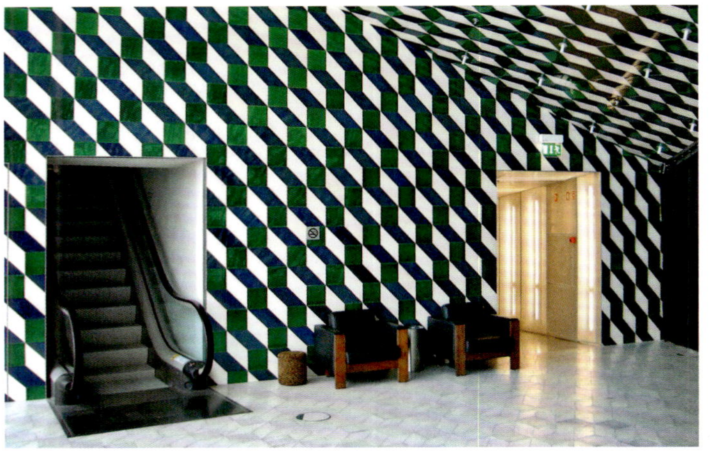

11

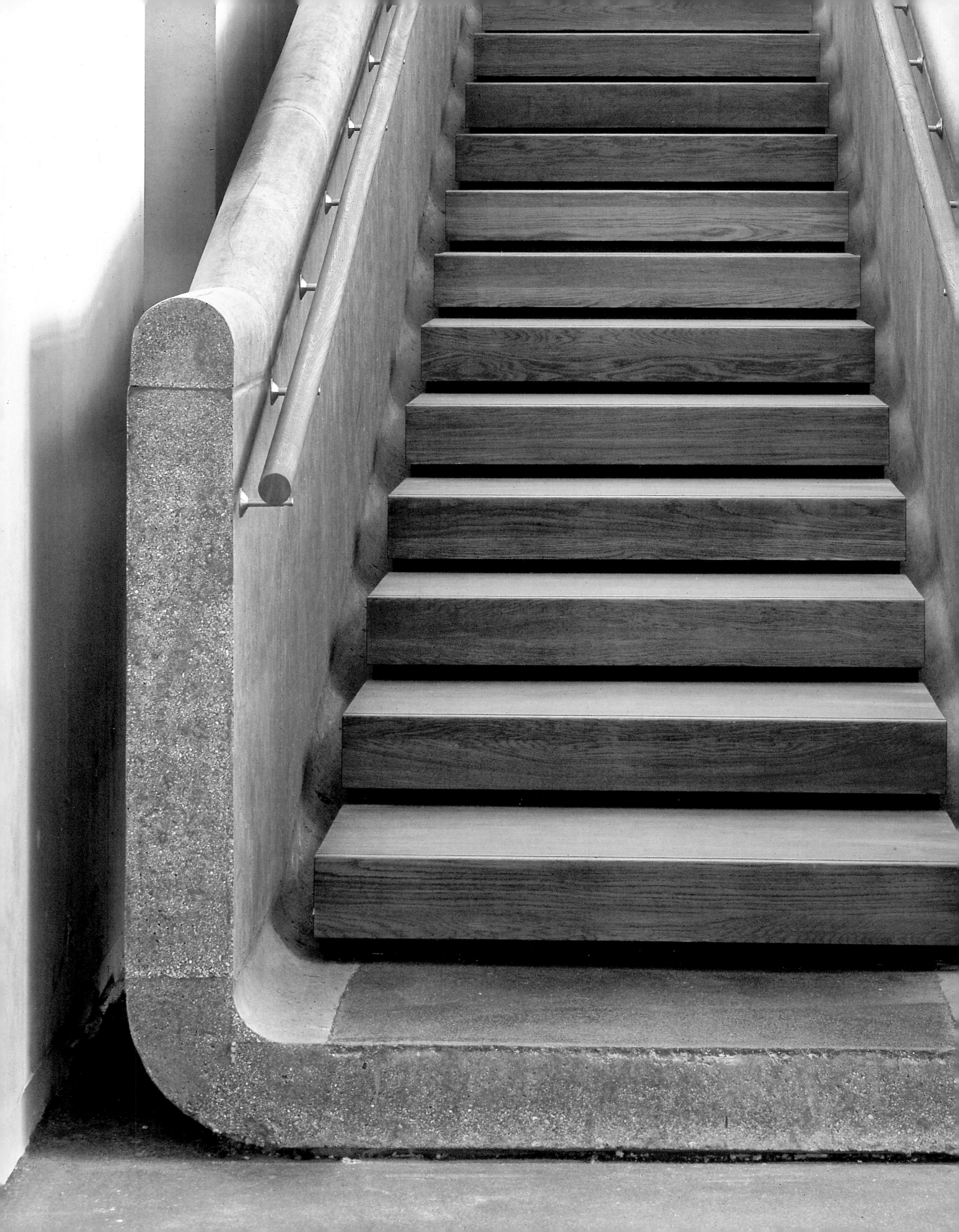

Circulation areas – places for interaction or an El Dorado for building regulations?

Arno Lederer

The original title of this publication is "Erschließung". In an architectural context, this German term is used to describe "an opening up or making accessible" in the sense of site development, as well as in the sense of circulation design. Searching the specialist literature on architectural history and theory for information about circulation design is alas not very fruitful. From Leon Battista Alberti's "On the Art of Building in Ten Books" [1] to the "Hatje-Lexikon der Architektur des 20. Jahrhunderts" [2], an encyclopaedia of 20th century architecture: circulation design – so important in teaching and practice – apparently has no significance in architectural literature. Even the 24-volume encyclopaedia "Brockhaus" merely states that the term refers to the totality of measures specified for development of a plot pursuant to the Federal Building Code (BauGB).

Attempts to find out more through the English or French equivalents of the German term are just as futile. The words offered by the two languages refer to the exploitation of resources, the provision of infrastructural services (of an economical or technical nature) to landscape areas, building zones or urban districts. This seems odd, since the circulation design of a building constitutes a primary design element for architects. The ground plan of a house is essentially determined by two types of spaces: those for spending time in (intended for resting) and those leading to these (intended for movement).

To differentiate between the two types and their fundamentally differing functions, they can be described as: spaces for movement/spaces for resting, or transitional spaces/destination spaces. On closer inspection, this is not entirely correct. There are numerous spaces that, though intended for circulation, invite to linger in, and conversely, a degree of movement takes place in every space that people stay in. The large entrance hall of a railway station, the lobby of an office building or a museum foyer – these circulation areas are all places that people also stay in, just as a living room additionally offers enough space for movement. The term circulation primarily refers only to the areas making the rooms themselves accessible. We can however sense that this is a purely technical designation in an attempt to express the simplified conception that movement and rest may be split into two separate functions.

The concept of being able to functionalise spaces according to whether they are primarily used for staying in or for circulation, calls to mind a degree of Taylorism, which aims to optimise work flow. In 1911, the American mechanical engineer Frederick W. Taylor, one of the founders of work science, wrote a book about the principles of scientific management [3], which also had a big impact on housing layout planning. Christine Frederick, a supporter of these ideas, for example studied the distances that a housewife has to cover unnecessarily in the kitchen, because of functional interrelations. Her book on rational housekeeping, entitled "Efficiency Studies in Home Management" [4], was published in 1913. It served as a basis for the development of compact ground plans, not only with regard to kitchens, but also applied to whole houses and apartments. Relief from time-consuming housework was however not the only issue here. Another effect was the economisation of the building measures themselves by a reduction in area, achieved with the help of rationalisation. Last but not least, this was also ideal for meeting the housing shortage at the time, since such optimised ground plans led to smaller area requirements and hence to lower building costs.

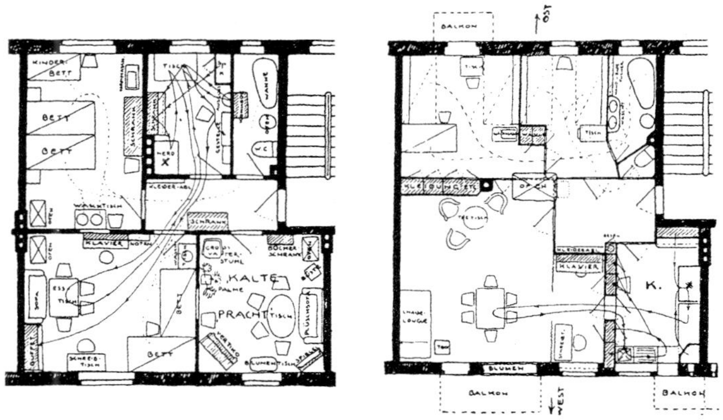

2

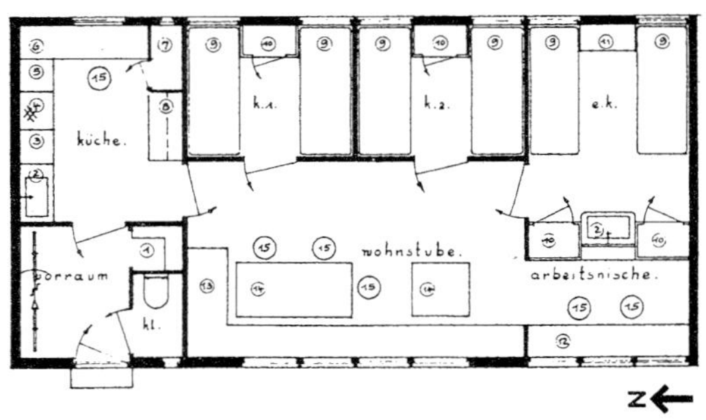

3

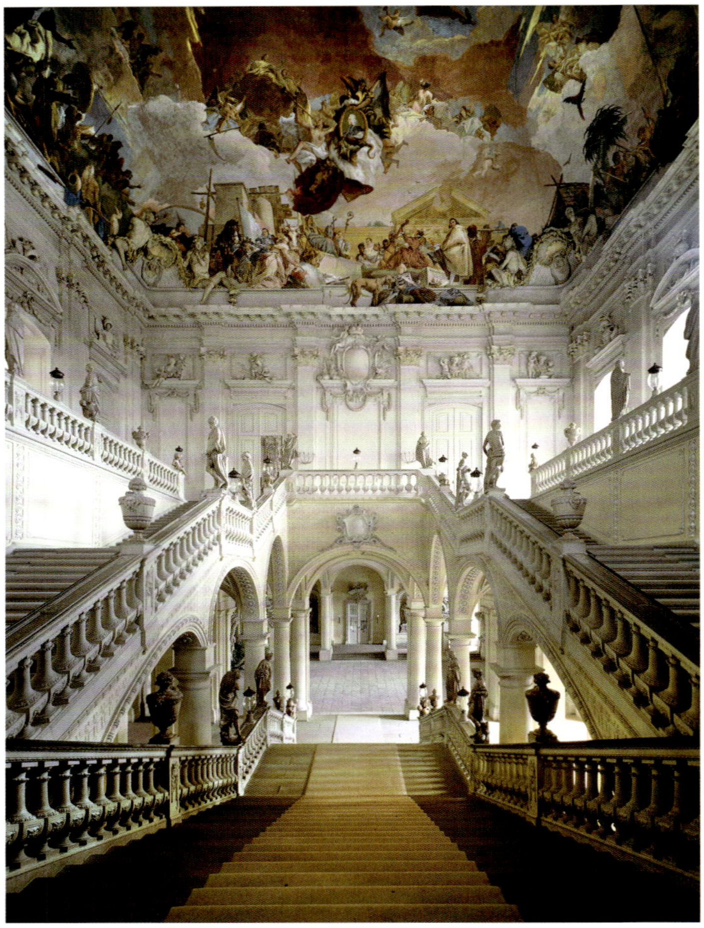

4

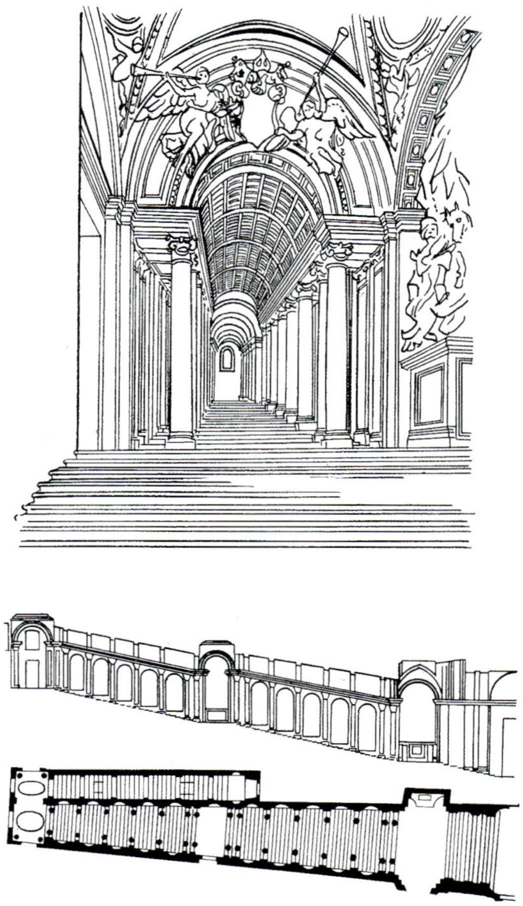

5

The approaches taken by architects in the 1920s to reduce the required floor space in housing construction, can be illustrated effectively by means of two examples [5]. Firstly, the drawings by Bruno Taut that represented a functional improvement of the typical ground plans at the time, by inclusion of walklines in the drawings (fig. 2, p. 15). Secondly, and rather more extreme, Otto Haesler's "Kabinengrundriss" (Cabin Ground Plan, fig. 3, p. 15) from 1926 that "simply did without a central corridor altogether. [...] Under the economic conditions prevailing at the time, this doing away with any kind of circulation area was convincingly logical." [6] The reduction of circulation areas in favour of larger living spaces, was one of the major design topics then, and it appears as though the architects were engaged in a veritable competition about the most favourable relation between circulation area and living space.

Quality of circulation areas

Housing construction in the 1920s is exemplary for the reduction of areas for corridors and stairs to the sole function of providing a shortest possible connection route. The idea that these areas should only serve to get from one room to another as quickly as possible, is not only favourable to economisation. The unambiguous assignment of purpose for instance also allows safety requirements to be satisfied much more easily than for rooms with combined utilisations. But this development characterised by functionalism is often associated with an impoverishment in terms of design, because the spatial experience itself no longer plays a role. The spaces merely represent the shortest distance between two points – nothing else is expected from a functionally compliant circulation. Assessing how economical a building is, therefore often involves a consideration of the ratio of usable area to auxiliary area. This approach makes it impossible for the design quality of circulation areas to compensate for unsatisfactory figures on account of excessively long distances consumed by stairs and corridors.

Wouldn't we though much rather speak of stair spaces, of entrance halls and bright corridors, of vestibules or of Le Corbusier's "promenade architecturale"? Just think of the Spanish Steps in Rome or the Scala Regia in the Vatican (fig. 5), the double-helix staircase at the Château de Chambord on the Loire, or the three-flight stairway in the Residence in Würzburg (fig. 4) – all stairs that could on no account be built in Germany today, because of the existing rules and regulations. Generous pathways through buildings, that have always been places for encountering others, are only possible with significant technical and constructional expenditure, such as sprinkler systems for ensuring fire protection. Further aspects such as barrier-free access and prescribed escape routes from buildings additionally help to limit architectural diversity.

Nobody would seriously challenge the implementation of regulations permitting disabled persons to use buildings and rooms without the help of others. Irrespective of the regulations and a willingness to provide barrier-free access, the question arises, especially in association with existing buildings, how the spatial disadvantage very often associated with this can be compensated and how these requirements can be fulfilled in the available space. A subsequent provision of barrier-free circulation is often not even possible here for reasons of space, or only with immense constructional, technical and financial expenditures.

The costs arising from the required compliance with the regulations are certainly also problematic. Although requirements, e.g. with regard to fire safety and barrier-free construction, have been increased, the available budgets generally have not – the specifications have to be fulfilled within a "regular" budget. The only way that this can usually be achieved is to design the circulation areas in compliance with the minimum area and volume specifications.
The linguistic change that has taken place in the sense that stairs, halls or corridors are now referred to as circulation elements, may have been accompanied by a negative influence on the design quality of these spaces. And yet, numerous buildings in the world of industry and finance erected in the past years, do in fact have opulent entrance halls (fig. 6). Although many of these spaces are in fact not really essential for circulation, with the employees seldom accessing their offices through the publicly visible entrance, but rather through a very conservative circulation – stairs, elevator and corridor – from the underground parking space straight to the required office floor. Why then did the architect's clients want to have a generous entrance hall?
Whether single-family home or palatial premises of financial institutions: entrance and stairs, or more appropriately, the spaces that we perceive first when we enter, function as both calling cards and social places. On the one hand, the proprietor can document his or her social standing, by prestigious, impressively large areas, or conversely, by classy reserve in conjunction with excellent design. On the other hand, entrances and stairs always serve as places of social interaction, since the ability to meet others is directly associated with movement. The pathways through a house there-

1 Youth hostel, Possenhofen (D) 2002, Hierl Architekten
2 Floor plans of typical and improved flats including walklines, Bruno Taut, 1924
3 Cabin ground plan, 6-bed type, Otto Haesler, 1926
4 Staircase, Würzburg Residence (D) 1720–1744, Balthasar Neumann
5 Scala Regia, Vatican (V) 1663–1666, Gian Lorenzo Bernini
6 Foyer, adidas Laces, Herzogenaurach (D) 2011, kadawittfeldarchitektur

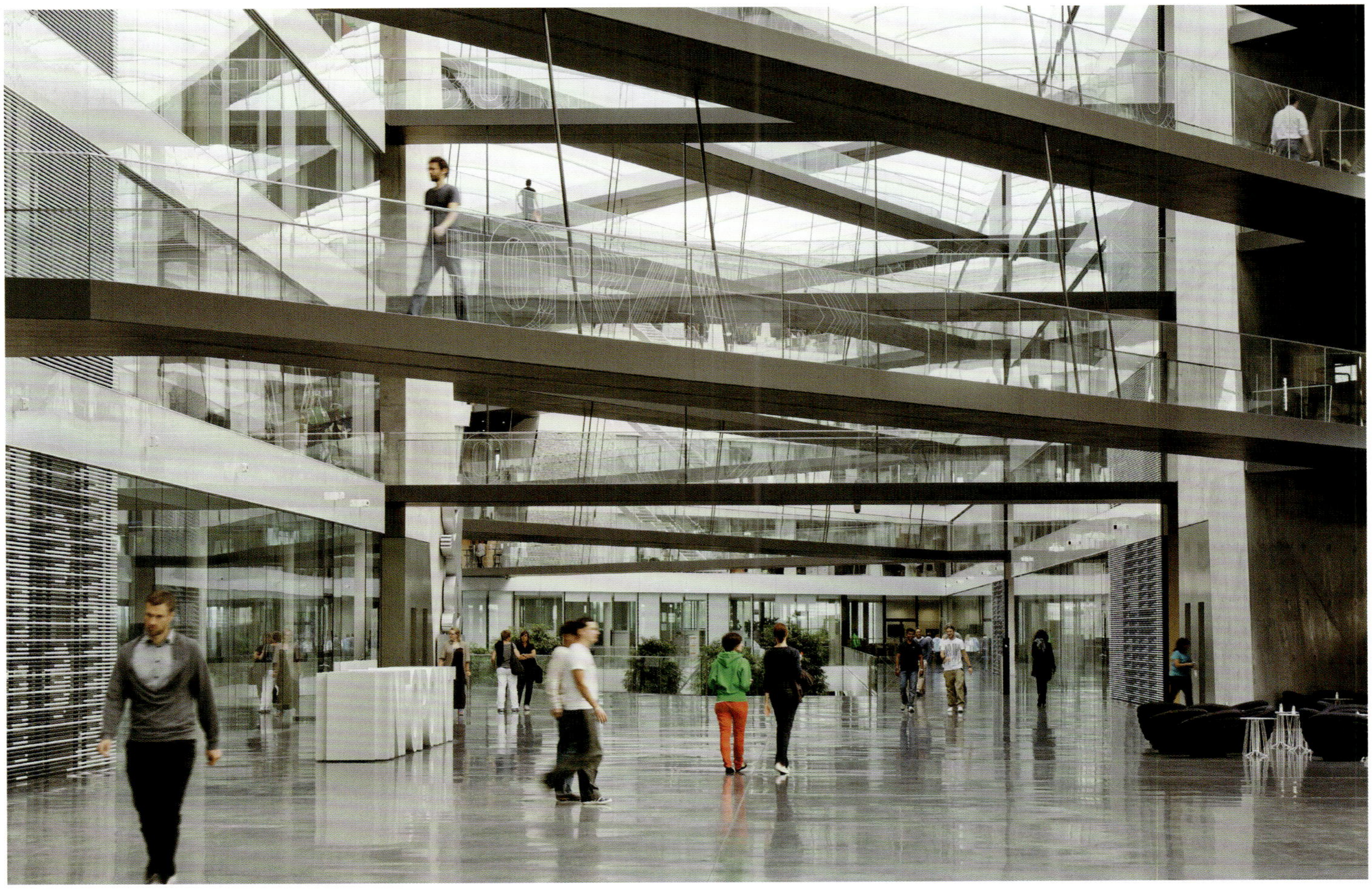

6

fore not only satisfy the purpose of connecting rooms, but also the desire for social exchange. Roads, promenades or boulevards in cities not only serve to safeguard quick accessibility of urban locations, but also fulfil the important function of being places for strolling about or going for walks in. Similarly, it should be possible for entrance, stairs or corridor to fulfil the same dual function in a building. Not without reason are these spaces often the settings for various types of scenes of social encounter in films, the theatre or in paintings.

Safety versus design

Doing justice to the dual function thereby attributed to circulation systems – i.e. walking and standing – necessarily leads to conflicts. A common phenomenon familiar to everyone can be observed at home. Popular places for greeting others and engaging in social exchange are the doorways of corridors, living rooms or kitchens, although everyone knows that this leads to standing in the way and blocking the accessibility of the particular rooms. Considering this issue in terms of circulation systems in public buildings, safety experts see a problem in the associated risk of accidents that occur when stairs are not only used as circulation routes, but also as places to meet and linger.

An interesting question in this regard would be whether tightening the safety regulations automatically leads to a prevention of accidents. Could a perfection of accident prevention not also lead to a reduction in the extent of circumspection and caution practised by individuals? Isn't it amazing how stairs in other countries with much less stringent safety regulations are used by young and old with an impressive degree of safety? We stand in awe of stairs that have been used for

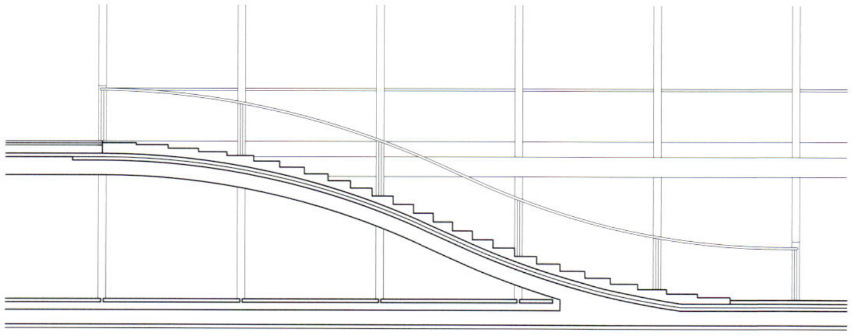

7

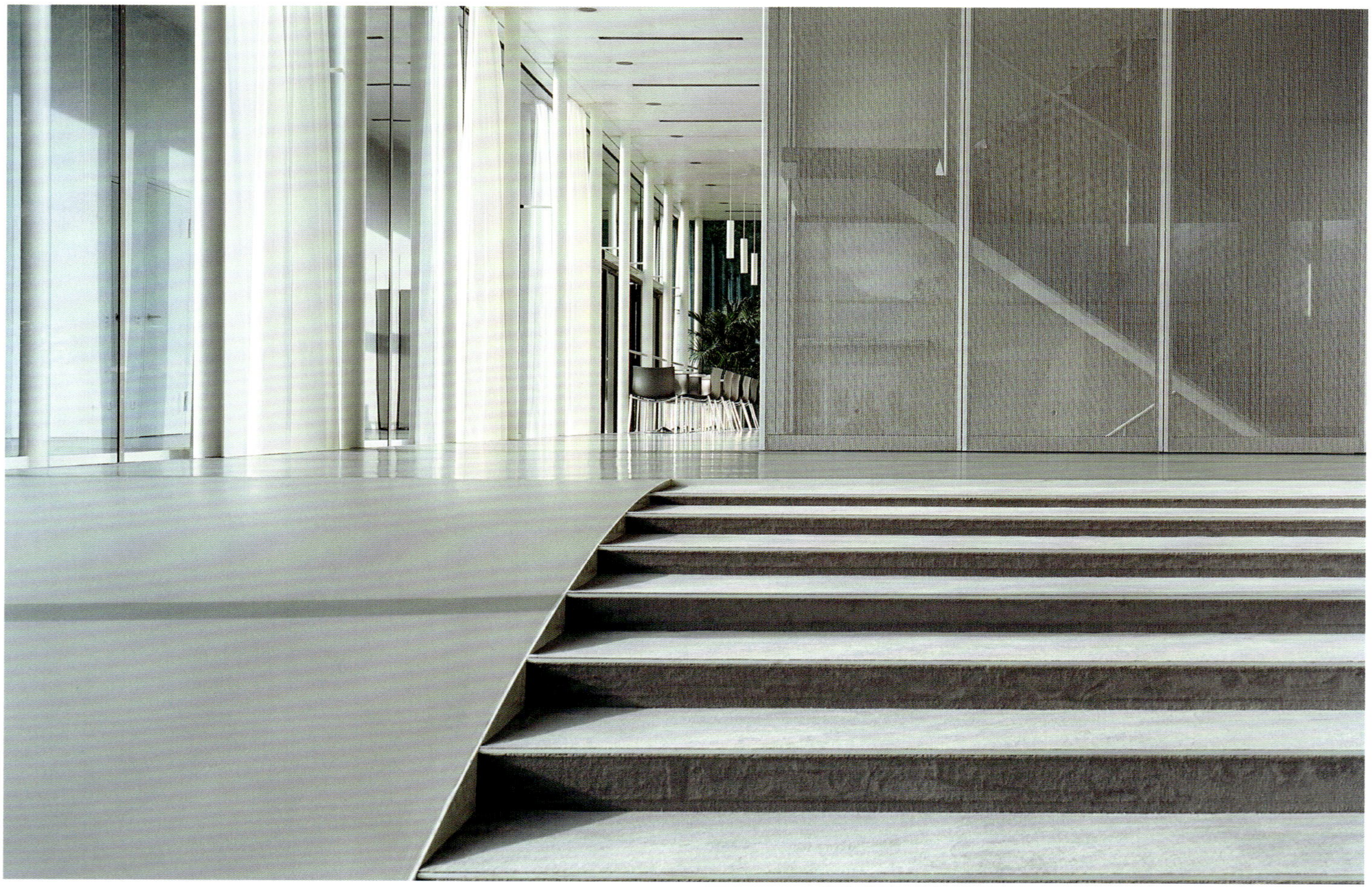

8

centuries without protective railing or a precise and consistent ratio of riser and tread. These aspects are also studied by heritage conservationist Friedrich Mielke, who founded the Centre for Staircase Research for this purpose [7].
With the fulfilment of the regulations, the responsibility is in a sense transferred from the user to the construction – and analogously, a little of the personal responsibility of every individual to the state? Far from being a specific problem related to circulation, this point affects the entire building sector – from accident prevention to social interaction between users.
The fulfilment of safety regulations has top priority in the hierarchy of the building requirements specified for circulation systems. Design issues appear to play a secondary role. And yet, the significance of the aesthetic element is much greater than may be assumed at first glance. This is naturally more relevant to stairs than corridors or entrance areas. The act of climbing stairs is a challenge to the human musculoskeletal system. Moving the body from one level to another requires an expenditure of energy, as well as increased attention to master the motion sequence. Climbing stairs is a strenuous activity that most people prefer to avoid. It's not unusual for people to wait minutes for a lift to arrive, rather than use the stairs, even if they only want to access a floor below. Not even a packed lift cubicle can tempt most people to use the mostly empty staircases. This type of behaviour is in complete contrast to the opinion voiced by the majority of doctors: that climbing stairs every day is good physical exercise and generally preferable to taking a lift. For this reason, it may be an idea to mistrust the recommendations made by the designers and manufacturers of lift systems, which are based on minimum waiting times that users can be expected to put up with, and instead make sure that the waiting times are much longer than predicted by the calculations. In addition, rather than positioning lift shafts in a central location, inviting stairs should be placed conveniently in the entrance area.

Exemplary stairs
Stairs should therefore be made in way that makes them popular with users. There are stairs that are so beautiful that one automatically wants to climb them without even thinking about it. The ziggurat in Ur, Michelangelo's stairs to the Capitoline Hill in Rome, or the helical staircase by Oscar Niemeyer at the Ministry of Foreign Affairs in Brasília (fig. 10, p. 21): it's a sheer pleasure to climb these stairs. When, in the course of the conversion of his house ("Haus am Frauenplan") in Weimar, Johann Wolfgang von Goethe was thinking about installing new stairs, he formulated the specification that the stairs should be such that one wouldn't tire of going up and down them. This staircase is exemplary in many respects, because it essentially fulfils the various functions that we expect from an ideal circulation element (fig. 9). To start with, the stairs are extremely comfortable to use, making them also suitable for older people. With a riser height of 10 cm and a tread depth of 35 cm, the stairs do not however correspond to the standard formula riser + tread = 64 cm. The height of each step is far below today's conventional measure. But whoever uses the three-flight staircase leading to the main storey, immediately senses how it satisfies the human desire for comfort. The breadth and slope of the staircase makes it very suitable as a communicative circulation element and that's exactly what it was supposed to be: a

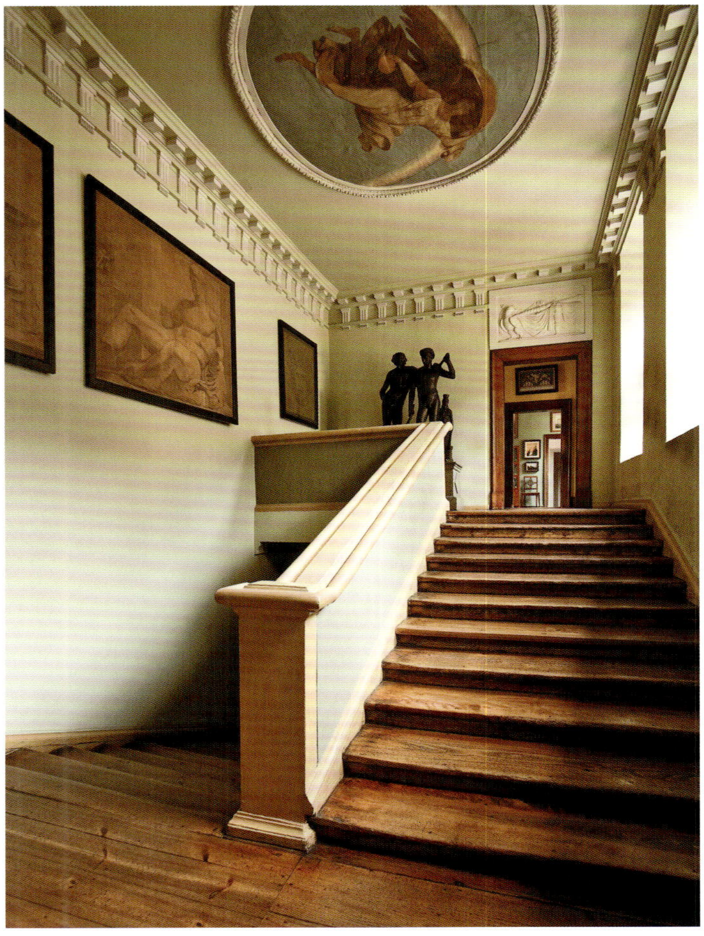
9

7 Longitudinal section, staircase with inconsistent ratio of riser and tread, scale 1:150, office building, Klaus (A) 2004, Oskar Leo Kaufmann
8 Staircase, office building, Klaus, Oskar Leo Kaufmann
9 Staircase, Haus am Frauenplan, Weimar (D) 1792

place to meet and greet others. For comparison: Balthasar Neumann's three-flight stairway in the Würzburg Residence has a riser height of 11.5 cm and a tread depth of 37 cm. In front of the staircase in Goethe's house, there is a corridor connecting it to the actual entrance. The location of the entrance in relation to the thoroughfare in the inner courtyard is ideal from a circulation design aspect, offering direct access from the vehicle, i.e. the carriage. The almost imperceptible ascent of the staircase is additionally supported by uplifting decoration elements: niches with sculptures and busts on the landings, as well as a ceiling fresco. The way Goethe managed to connect the three-flight staircase to adjacent spiral stairs via a landing bears witness to his astounding spatial imagination.
The staircase unmistakeably also served for self-representation of the celebrity poet. The Italian word "scala" meaning both stairs and scale, gives a fitting indication of the close association between vertical circulation and the nature – the scale – of a house. Since staircases are always a difficult constructional component for architects to master, reflecting the ability to visualise in three dimensions as well as demonstrating technical and design skills, they are a good measure of the architectural competence of the designer. In spite of all the limitations mentioned above, and despite the reduction of its function to the term "vertical circulation", the staircase tells us something about the qualification of its creator. It is therefore not reprehensible to also consider a staircase as an object in which the architect represents him- or herself. The hobby architect Goethe has moreover left us a sketch that reproduces a very accurate three-dimensional impression of the planned staircase, which can certainly be called professional in this regard.
The previously mentioned staircase in the Ministry of Foreign Affairs in Brasília should also be considered in this context (fig. 10). It stands as a pars pro toto for the work of Oscar Niemeyer and is much publicised, not without reason. It should however be noted that everything that creates the identity of this floating spiral would not be permitted to be built as such in Germany today. Take the big air space for example that would be considered as a deficiency on account of lacking fire compartments. The upper floor and the flight of stairs have no fall protection, there are too many steps without an intermediate landing, the spiralling would be impermissible, as well as the over-dimensional width without any railing in the middle and the outer edge with nosings etc. All this would not be possible with the currently applicable DIN standards, plus the requirements of building legislation, fire protection and accident prevention. Anyone climbing Niemeyer's staircase is however delighted by the ease with which the next floor is reached. And safety reservations concerning the lack of any railings disappear after stepping on the first step: no one wants to tread on the edges of the steps protruding over the sides anyway – they replace the necessary guard rail.

Inviting corridors and halls

Corridors and connecting passages are subject to fewer regulations. Restrictions are mainly posed by fire protection requirements and anyone who has ever been involved in school construction knows exactly what this entails. The architect is asked to produce a design that adheres to the economic principle on which most school construction programmes are based. This specifies that the ratio of the usable area to auxiliary area should be 60–40 % to ensure an adequate financing efficiency. Visions of bright, sun-lit corridors – preferably achieved with layouts that have rooms on only one side of the corridor – that are also attractive areas to spend time in, soon disappear into thin air. Modern pedagogy has long wanted to furnish corridors for work in small groups, but this is simply not possible in many cases for fire safety reasons. Alone where to put the coat rack can already trigger extensive discussions with regard to preventive fire protection.
Back in the 1960s, the City of Stuttgart published a special series of books concerning new constructions, where it was proudly stated that only the classroom sizes and equipment would be prescribed for the design of school buildings, with complete freedom in the design of the layout and structure otherwise [8]. Most of the published projects have generously designed circulation systems with land improvements outside, broad corridors that are inviting to stay in and flowing room sequences. The implementation of modern educational concepts concerning corridors and in-between spaces was already an issue then, but one that gradually had to give way to financial stipulations and fire protection regulations. The creation of similar spatial constellations today would involve a correspondingly higher technical and financial expenditure, which would in turn necessitate an adaptation of the financing to the increased requirements. A comparison of current circulation systems with those of the 1950s and 1960s clearly shows a regression that architecture has suffered in this respect.
Not only have the areas of the circulation systems, but also their material qualities deteriorated in the course of this development. Seating benches made of solid wood alongside glazed facades, carefully designed natural stone floors and a lovingly sketched fountain under a skylight featuring glass bricks characterise a perfectly normal school in Stuttgart designed by the architect Helmut Erdle in the 1950s. A material standard that was a matter of course then. Similar qualities are rarely found in school constructions today, certainly providing standard-compliant floor coverings, wall finishings and other furnishings, yet signalising a minimum of financial expenditure – even though, as mentioned earlier, entrance halls, corridors and stairs are the calling card of a building. There are many buildings with beautiful halls and air spaces which today's regulations would not allow to be used any more, but which are considered worthy of protection. The subject of circulation is in dire need of attention, development and research. The solutions offered nowadays rarely agree with the needs, the fulfilment of which certainly was an architectural task until a few decades ago.

Circulation as route and place

Circulation, as comprehended by architects, does not commence behind the front door. As observed earlier, a route is not a place and a place is not a route. The modern idea of splitting up human activities and spatial requirements into functions is contradictory to our behaviour, feeling and perception. It is right to think about how needs and necessities can be expressed in terms of functions. But are we also aware of the mixtures and superimpositions of functions occurring in every space that we create? Functions are components, which, together with other things, lead to a whole, the space. The whole cannot however be reduced to the components again. Nobody likes to use a circulation system that only serves the mechanical purpose of securing the accessibility of spaces. Architecture that is not attractive to use doesn't make any sense.

References:
[1] Alberti, Leon Battista: On the Art of Building in Ten Books. Cambridge, Mass. 1988
[2] Lampugnani, Vittorio Magnago (pub.): Hatje-Lexikon der Architektur des 20. Jahrhunderts. Ostfildern-Ruit 1983/1998
[3] Taylor, Frederick Winslow: The Principles of Scientific Management. London 1911, reprint New York 2006
[4] Frederick, Christine: The New Housekeeping. Efficiency Studies in Home Management. Garden City, NY 1913
[5] Faller, Peter: Der Wohngrundriss. Munich 2002, p. 17f.
[6] Ibid.
[7] www.scalalogie.de
[8] Ley, Helmut: Bauherr Stadt Stuttgart. Ein Leistungsbericht der Stadt Stuttgart und des schwäbischen Baugewerbes. Band 3. Stuttgart 1963

10 Helical staircase, Palácio Itamaraty, Brasília (BR) 1970, Oscar Niemeyer

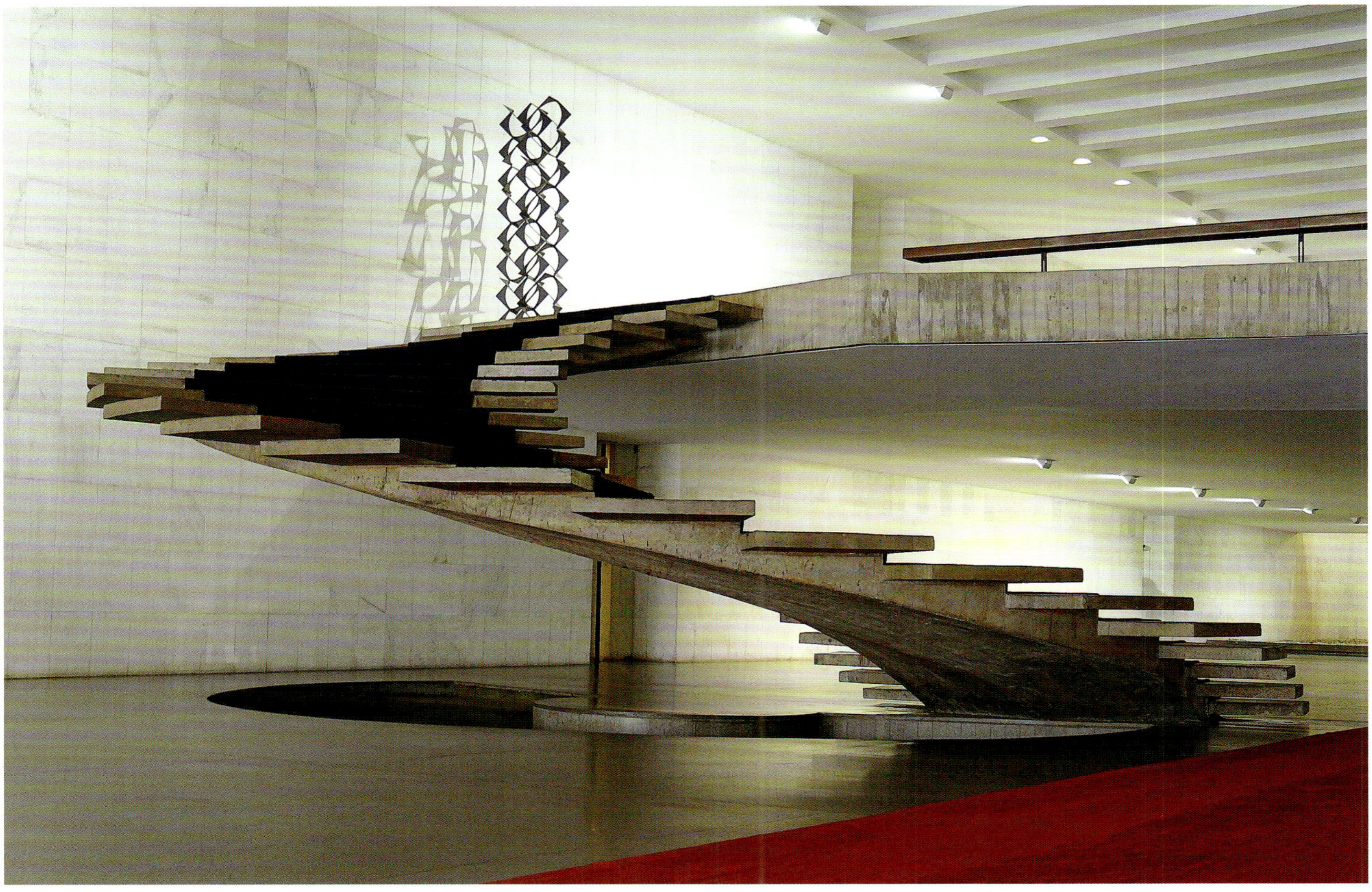

10

Planning criteria

In the building industry, in addition to the infrastructural development of land and the technical infrastructure of a building (service engineering), access means, in particular, the access routes and spaces which make it possible for people to reach areas of use such as offices, apartments, rooms etc.
The functionality of a building depends on its accessibility, which makes a coherent access concept a key element in successful planning. The requirements in terms of access routes vary considerably according to the nature of the building project and depending on the building typology, the site and the number of users, in particular due to the requirements of the technical regulations and standards relating to preventive and structural fire protection. There are no standardised international regulations, and even within Germany requirements may differ between different federal states.

Horizontal and vertical access
Access routes in buildings can be divided into vertical elements which connect the individual floors of a building and horizontal elements which determine the accessibility of areas of use on one level. The vertical elements include stairs, ramps, lifts and escalators. Horizontal elements are much less clearly definable – these include vestibules, corridors, access balconies, moving walkways, but open areas or whole sequences of rooms can also take on an access function. Access spaces primarily fulfil functional purposes, but generally speaking they are also part of the spatial and aesthetic concept when it comes to creating quality of experience or spaces which encourage communication.
The planning of the access to a building is based on the nature of the building project and the number and type of potential users. Obviously, requirements differ depending on whether this involves the organisation of a single-family home, an apartment block, an office building, a school or a factory etc.
In apartment building construction, as a rule a distinction is made between single, double, triple and multiple access plans, depending on the number of residential units on each floor which can be reached from a single landing. Where access is provided by means of access balconies, an access route leads on from the stair landing, providing access to the individual apartments (figs. 1 and 3).
As a rule, much larger units are accessed in office and administrative buildings than in apartment buildings. In most cases access is provided via central cores containing stairwells and lifts, which can be arranged and combined according to different principles (fig. 4).

Fire protection
Since some of the access routes serve as escape and rescue routes in the event of a fire, a whole series of building code regulations relating to preventive fire protection apply to the planning of access areas. In order to allow practical fire protection requirements to be formulated, the Model Building Code (MBO) classifies buildings into different building classes according to their size and height.
In Germany, requirements in terms of preventive fire protection are regulated in the individual regional building codes (LBO), which are based on MBO 2002. The aims are:
- to prevent the outbreak of fire and the spread of fire and smoke
- to make it possible to rescue people and animals
- to make it possible to fight fires effectively through smoke extraction and extinguishing measures
- to prevent damage to property

For this purpose, requirements regarding the combustibility of building materials and the fire resistance duration of the building components are stipulated in the regional building codes.
Building material classes according to DIN 4102-1 (A 1 and A 2 non-combustible and B 1–3 combustible) and DIN EN 13 501 allow the fire risk of building materials to be judged in order to predict in advance the stability of building components in the event of a fire (fig. 5). The fire behaviour of building components is described through their fire resistance duration, that is to say the period of time during which a building component remains functionally serviceable.
In addition, a distinction is made between:
- planning measures relating to preventive structural fire protection (dimensioning of escape routes as well as pedestrian and vehicle access routes for the fire service, planning of vertical and horizontal fire compartments, structural measures etc.)
- technical measures (fire alarm systems, fire extinguishers, hydrants, sprinkler systems, smoke and heat extraction systems, fire doors, fire shutters etc.)
- structural measures (building construction, choice of suitable building materials, protective measures for endangered building components etc.)

Emergency stairs and emergency corridors
The design of emergency stairs, which must be located in emergency stairwells, is strictly regulated since in the event of a fire these serve as secure escape routes and therefore

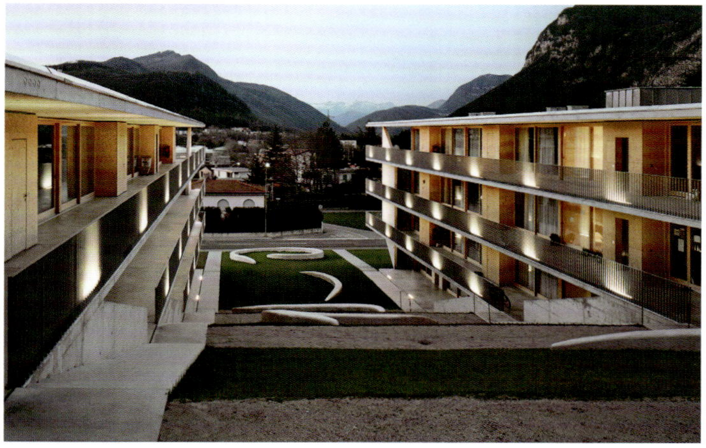
1

Emergency stairs	Minimum dimension for usable tread width
In apartment buildings with no more than two flats and within flats	80 cm
In all other buildings (LBO)	100 cm
In high-rise buildings (HochhVO)	125 cm
In places of public assembly: per 150 persons reliant on the escape route (max. width 250 cm / tread), designed according to how many people can be seated in the room (VStättVO)	100 cm
In shops (VkVO; max. width 250 cm / tread)	200 cm
Non-emergency stairs	
In all buildings	50 cm

2

have to fulfil strict fire protection requirements (fig. 2). Fundamentally, the design of stairwells depends on the nature of the use of the building as well as the number and type of users. The number of stairwells required also varies depending on the building's use. In special categories of building (e.g. schools, kindergartens, old people's homes), rescue via facade openings on the upper floors is not permitted, so as a rule these buildings require a second emergency stairwell. More or less strict regulations apply depending on the size and use of the building. As a rule, emergency stairs must be provided in buildings containing more than two apartments and more than three full storeys, as well as in all public buildings, office buildings and commercial buildings.
Emergency stairwells are stairwells which represent an isolated fire compartment and, as an escape and rescue route, lead safely into the open. Their walls must fulfil the specifications for fire walls and the stairs must be of the F 90 quality standard, or F 30 in lower buildings. Doors and glazing in emergency stairwells must be of corresponding quality.
Emergency stairs must be accessible from every point in the building at a distance of no more than 35 m.
Special regulations, which are contained in the High-rise Buildings Ordinance (HochhVO), apply to high-rise buildings, i.e. buildings in which the top storey lies more than 22 m above ground level.
From occupied spaces, emergency corridors lead either to an emergency stairwell or to an exit into the open. Emergency corridors must have a minimum width of 1.25 m. If they contain stairs, a sequence of fewer than three steps is not permitted. Emergency corridors are to be divided by means of unlockable, smoke-tight and self-closing doors into smoke compartments which may not be longer than 30 m.

Stairs

In addition to the emergency stairs as defined in the building codes, a building can contain any number of non-emergency stairs. These are mostly open, aesthetically designed stairs

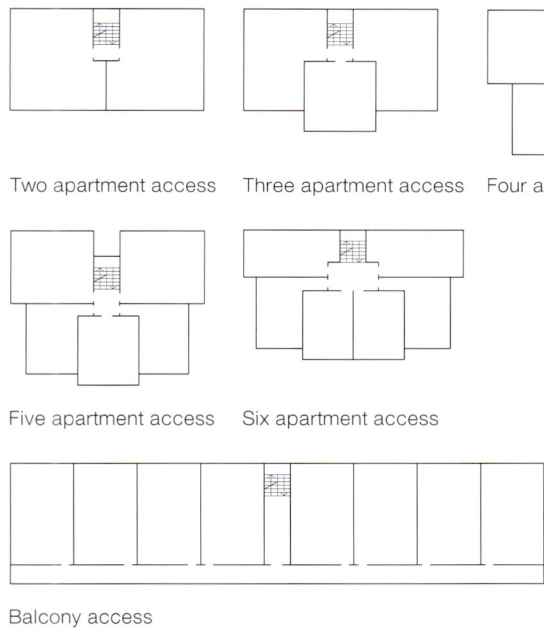

3

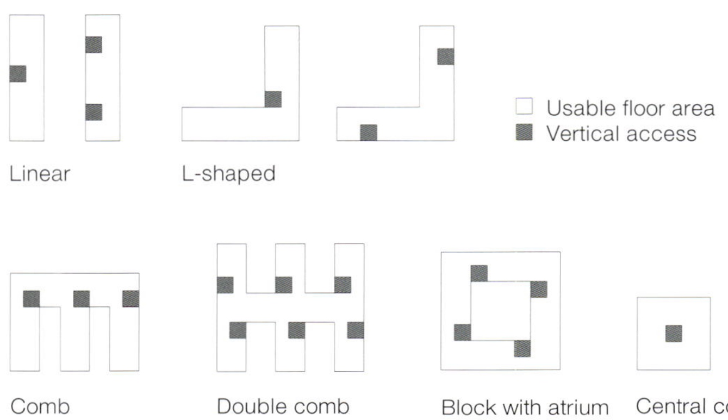

4

Building material class			Designation
A	A1	Non-combustible building materials	Contains no combustible building materials
	A2		Essentially consists of non-combustible building materials
B	B1	Combustible building materials	Flame-retardant
	B2		Normally flammable
	B3		Easily flammable

5

1 Balcony access, student accommodation, Mendrisio (CH) 2006, Könz-Molo and Barchi Architekten
2 Minimum dimensions for the usable tread width in emergency and non-emergency stairs
3 Different access concepts in apartment buildings
4 Different access concepts in office buildings
5 Building material classes according to DIN 4102-1 and DIN EN 13501

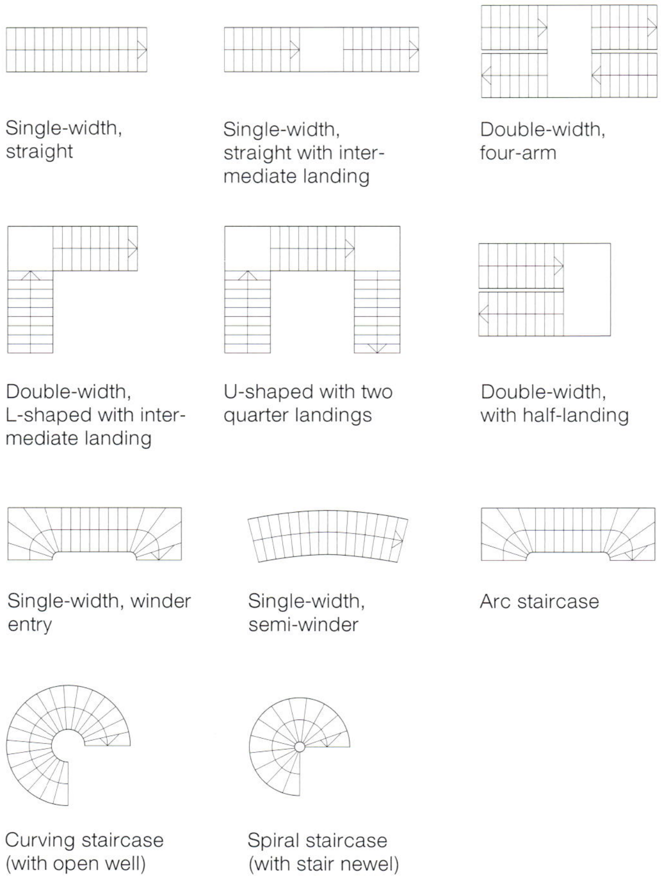

6

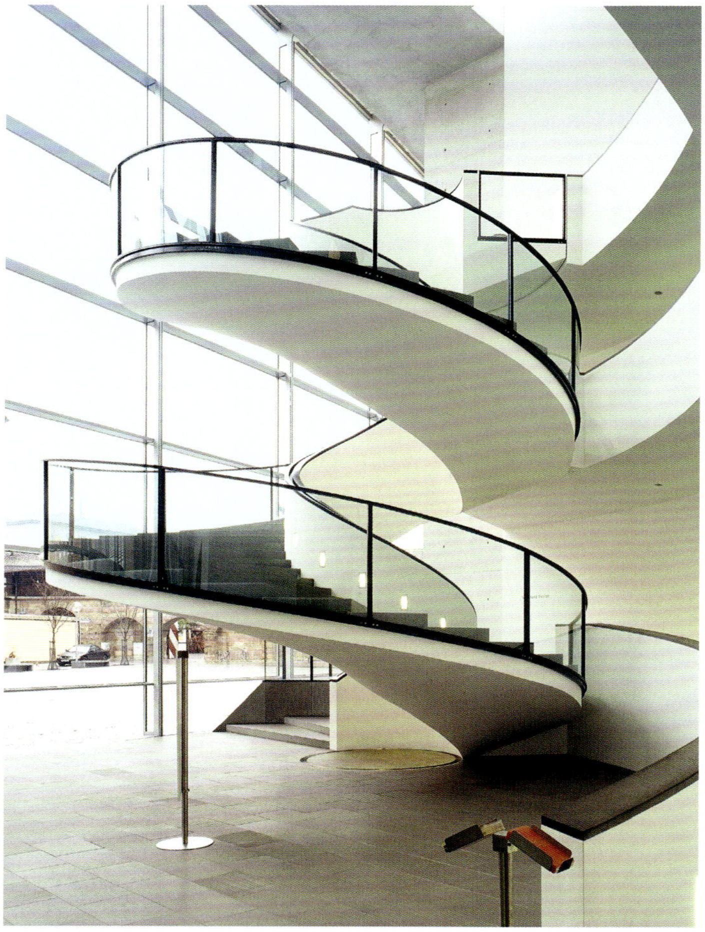

7

which are not subject to the fire regulations applicable to emergency stairs. However, stairs which are not classified as emergency stairs according to building regulations are also subject to the requirements set forth in DIN 18065, the building codes of the federal states or other ordinances. These relate to the type of stairs, rise, step design, handrails, the material used for the floor covering etc. For example, safety regulations are very strict in hospitals, schools, kindergartens and old people's homes.

DIN 18065 makes a distinction between stairs in "buildings in general" and in "residential buildings with up to two flats and within flats". There are straight, angled and winding stairs, ones with or without landing, single, double or multiple-width (fig. 6). The walkline is the central axis of the walking area of a staircase. It must run steadily and without any kinks. The tread width of stairs in residential buildings with no more than two residential units is at least 80 cm, otherwise at least 1 m. The minimum width is specified depending on the use of the building, i.e. greater tread widths may well be necessary in schools, assembly rooms, department stores etc. in which there is a lot of traffic (fig. 2).

A comfortable rise is achieved with a step height of 16 to 19 cm, and if there are more than 18 steps a landing should be included. Landings must have a depth corresponding at least to the tread width.

As a rule, the rise is calculated from the height between floors divided by the assumed step height (16–19 cm).

Step dimensioning rule S: $2r + t = 59$ to 65 cm (ideal 63 cm), r = rise, t = tread width

Comfort rule: $t - r = 12$ cm

The tread depth should be at least 21 cm and at most 37 cm, the rise at least 14 cm (fig. 10).

Balustrades

Stair balustrades provide security against falling and are intended to make it easier to climb the stairs. They must be installed at prescribed minimum heights and may not restrict the minimum tread width of the stairs. All stairs with more than three steps require a balustrade. As a rule, measured above the front edge of the step, stair balustrades must be at least 90 cm high, and at least 110 cm high in the case of stairs with a drop of more than 12 m and on the inner side of spiral stairs. Special requirements apply in buildings in which children can be expected to be present as well as in old people's homes. The number of necessary balustrades (on one side, on both sides, additional central balustrade) depends on the width of the stairs.

6 Diagrammatic plan views of different stair types
7 Spiral staircase, Neues Museum, Nuremberg (D) 1999, Staab Architekten
8 Industrial facility in Munich (D) 2012, bogevischs buero architekten & stadtplaner
9 Angles of inclination for ramps, open steps, residential stairs and ladders and examplary dimensions (in cm) for rise (r) and tread width (t)
10 Determination of the rise (r) and tread width (t) of different designs of steps:
   a Straight riser without undercut
   b Step with undercut (u)
   c Riser with sloping face
11 Minimum dimensions (in cm) for stairs

However, the distances between vertical and horizontal infill elements as well as the necessary balustrade heights and fire safety requirements can vary and are governed by the individual regional building codes, which are based on the DIN standards.
Sound insulation requirements also vary widely depending on the building type. As a rule, a flight of stairs must be acoustically isolated from floors, landings and stairwell walls.

Ramps

Ramps are used in place of stairs in order to overcome differences in level in an accessible way. Ramps for pedestrians and wheelchair users should have a maximum gradient of 6 % and should feature a landing of 150 cm in length after every 6 m (fig. 14, p. 26). A ramp may not have any transverse slope, the clear width should be at least 120 cm and an 85 cm high handrail should be installed at the side and continued for 30 cm beyond the beginning and end of the ramp. Normally, a level area with a minimum diameter of 150 cm should be provided before and after the ramp as a turning space for wheelchair users. Since, due to these requirements, ramps take up a great deal of space, they are in most cases very specifically planned from an architectural viewpoint, frequently as a visual feature.

Lifts

In Germany, the regional building codes (LBO) and the Equal Treatment of Disabled Persons Act (BGG) stipulate that public buildings must be of accessible design. Also, each building above a certain height must be provided with an adequate number of lifts for access, at least one of which should, as a rule, be suitable for carrying heavy loads, stretchers and wheelchairs.
Lifts are distinguished according to the type of drive (electric traction drive lifts and hydraulic lifts) and the type of use (passenger, goods, bed lifts etc.). DIN 15306, 15309 and DIN EN 81 specify the requirements for lifts, irrespective of make, and their technical performance (figs. 12 and 13).
Lifts should be easily accessible and if possible positioned at points in the floor plan where the greatest volume of traffic is to be expected. Waiting zones in front of the lift may not lie in circulation routes, the depth of the waiting zone in front of or between opposing passenger lifts should be between 3 and 4 m. As a rule, the number and size of lifts needed is determined by means of a traffic calculation or transportation concept which takes into consideration the basic data such as the number of persons per day, workplaces per floor, times

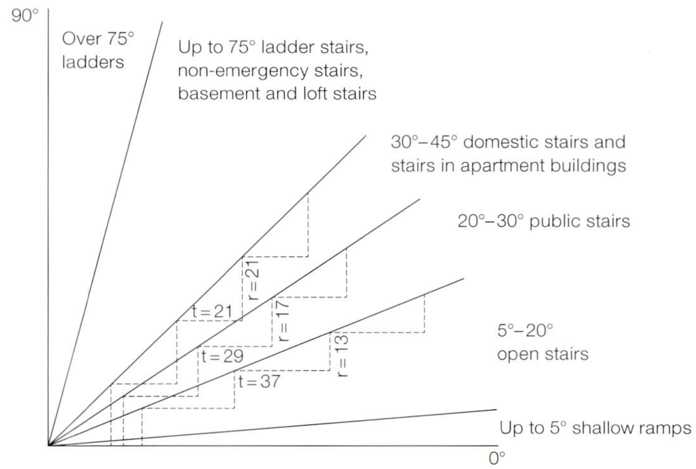

9

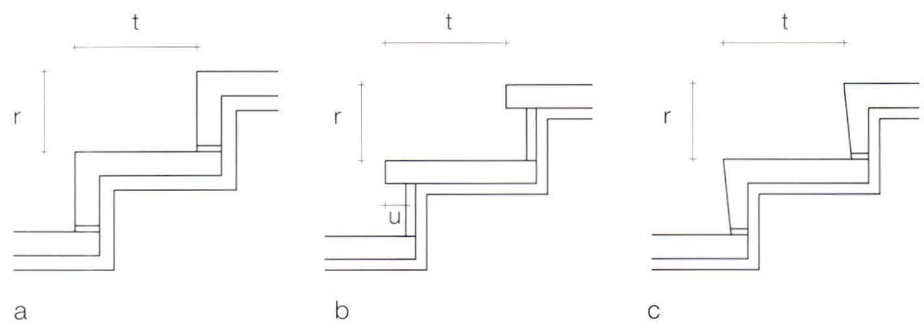

10

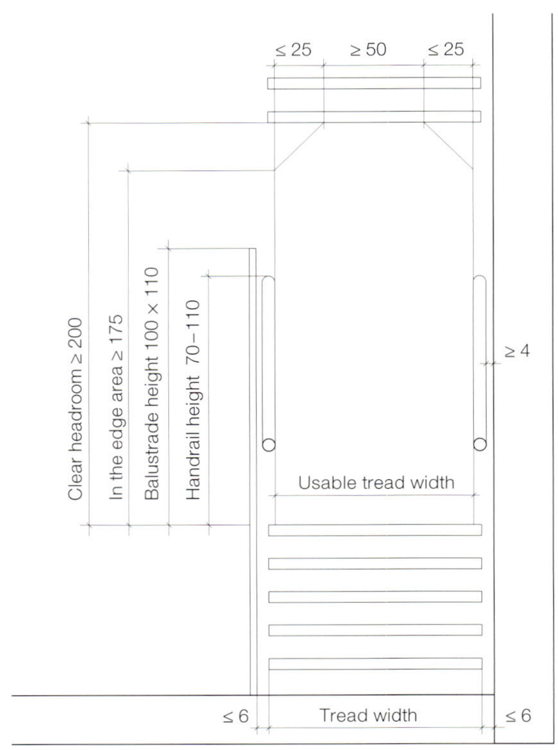

11

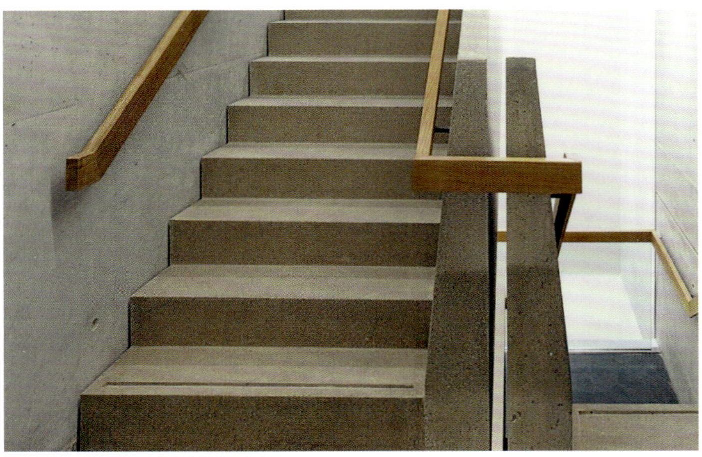
8

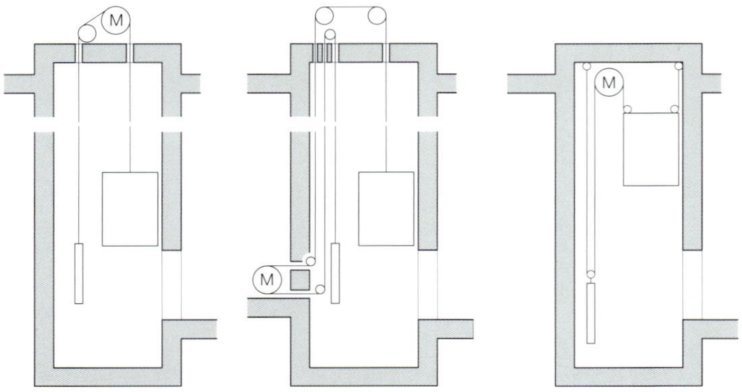

12

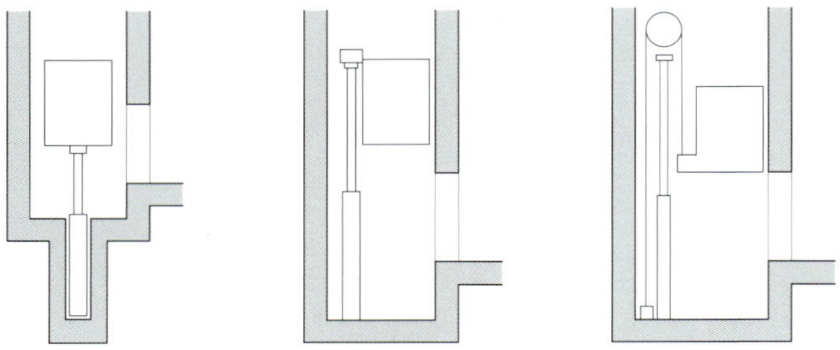

13

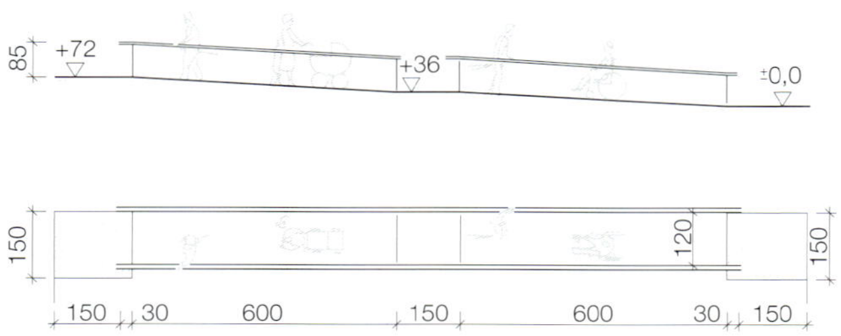

14

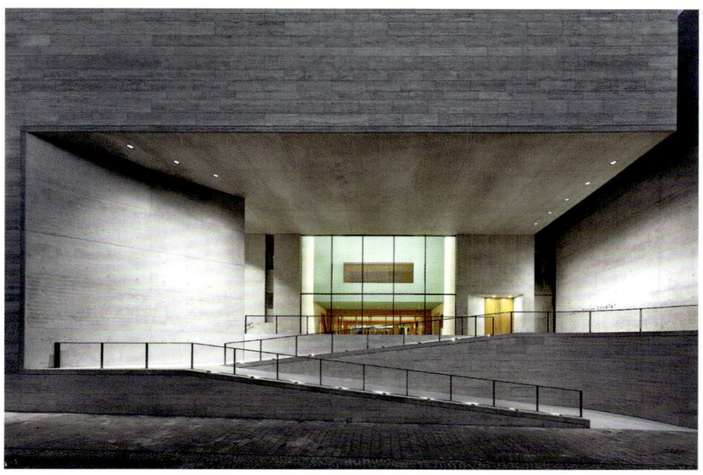

15

of use, location of the canteen, underground car park, lecture theatres, meeting rooms and conference rooms etc. On the basis of the determined nominal load, the associated characteristic data (shaft dimensions, car dimensions) can be taken from the DIN standard.
The construction design of lift shafts is subject to numerous requirements which, however, can in some cases vary depending on the installation situation and lift manufacturer. As a rule, minimum structural requirements must be met in terms of fire safety and sound insulation, health and safety and accident prevention, in some cases the prevention of water contamination (in the case of hydraulic lifts), shaft, shaft head and shaft pit, housing of technical equipment (lift machinery, counterweight, switchgear cabinet etc.) as well as the provision of sufficient space for maintenance and repairs.
The nominal speed specifies the speed of the lift car for which the lift is designed; it is stated in m/s. The maximum permissible load or load-bearing capacity of a lift is referred to as the nominal load. A sign in the lift car provides information on the load capacity of passenger lifts.
The necessity for fire service lifts is determined according to the provisions of the regional building codes and the Model Guideline on High-rise Buildings (MHHR).

Escalators

Escalators are frequently used where large numbers of people are regularly moving around, for example in department stores or at public transport stations. Their angle of inclination is between 27° and 35° depending on space requirements, their transportation capacity depends on the angle of inclination.
The following rules apply to the pre-dimensioning of the length L in the floor plan:
L = 1.75 × ceiling height (at 30° inclination)
L = 1.45 × ceiling height (at 35° inclination)
In addition there is a minimum free space of approx. 2.50 m before and after the escalator.
The speed of travel varies between approx. 0.5 and 0.65 m/s depending on the building's use, generally higher in transport facilities and lower in department stores. The maximum permitted speed is 0.75 m/s.

Floor coverings

Floor coverings are a key design component of a room and the way they are perceived has an influence on the overall

12 Different positions of the drive mechanism in friction drive lifts
   a Above the shaft
   b Next to the shaft
   c In the shaft, below the ceiling of the top floor
13 Drive types in hydraulic lifts
   a Direct hydraulic drive: hydraulic cylinder beneath the lift car
   b Direct hydraulic drive with lateral piston or next to the lift car
   c Indirect hydraulic drive: piston next to the lift car, which is moved by a cable with a pulley on the piston
14 Ramp with necessary dimensions according to DIN 18040-1, dimensioning with max. 6% gradient and intermediate landings after max. 6 m length
15 Entrance area, Museum Georg Schäfer, Schweinfurt (D) 2000, Staab Architekten
16 Evaluation of anti-slip properties for floors in public facilities, determined using the standardised "inclined surface" method according to DIN 51130 and BGR 181. The higher the R-value, the more slip-resistant the floor covering.
17 School extension, Marburg (D) 2010, Hess Talhof Kusmierz

spatial effect and thus the users' feeling of well-being. Their surface structure and material properties also influence the room's acoustics, footfall noise and relative atmospheric humidity. Floor coverings must be able to withstand mechanical, thermal and chemical stress. The surface properties of floor coverings are sensed haptically through the feet and are perceived as being pleasant or unpleasant. Both the objectively measurable characteristics and the subjective sensory qualities should therefore always play a role in the selection of floor coverings.
The individual properties such as thickness, weight, manufacturing method, anti-slip qualities, safety and combustibility etc. of different floor coverings depend on the material and vary greatly. As a rule, the individual manufacturers provide information on special technical regulations and special recommendations regarding the laying of floor coverings. Regulations on anti-slip properties and safety apply to floors in public facilities. The anti-slip properties of surfaces are classified into R-groups (R 9–R 13) and, depending on the nature of the space, recommended for different uses (fig. 16). Stricter requirements apply to sloping surfaces and wet barefoot areas.

Rating group	**Acceptance angle on inclined surface [°]**	**Room type/use**
R 9	6°–10°	E.g. indoor floor coverings in general public areas (office)
R 10	10°–19°	Public toilets, changing rooms and washrooms
R 11	19°–27°	Shop entrances and stairs as well as in kitchens used for communal catering, e.g. in care homes or nurseries
R 12	27°–35°	Hospital kitchens and large kitchens in which more than 100 meals are prepared daily
R 13	> 35°	Floor coverings in almost any working area in hairdressers, all manufacturing facilities, slaughterhouses etc.

16

Relevant guidelines:
- DIN 4102 Fire behaviour of building materials and building components
- DIN 4109 Sound insulation in construction
- DIN 4174 Ceiling heights and stair pitches
- DIN 18040-1 Construction of accessible buildings: publicly accessible buildings
- DIN 18040-2 Construction of accessible buildings: dwellings
- DIN 18065 Stairs in buildings
- E DIN 18 070 Public transport spaces and open spaces (in preparation)
- DIN EN 115 Safety of escalators and moving walkways
- DIN 18 365 (VOB) General technical regulations for the installation of floor coverings
- DIN 51 130 Testing of floor coverings
- Law on the equal treatment of disabled persons (Equal Treatment of Disabled Persons Act – BGG)
- Ordinance on places of public assembly (VStättVO)
- Regional building codes of the individual federal states (LBO)
- High-rise buildings ordinance (HochhVO)
- Model Guideline on High-rise Buildings (MHHR)
- School construction guidelines (SchulbauR)
- DIN EN 81 Safety rules for the construction and installation of lifts
- Guideline on lifts (AufzR)
- DIN 15306 Passenger lifts for residential buildings
- DIN 15309 Passenger lifts for non-residential buildings and bed lifts

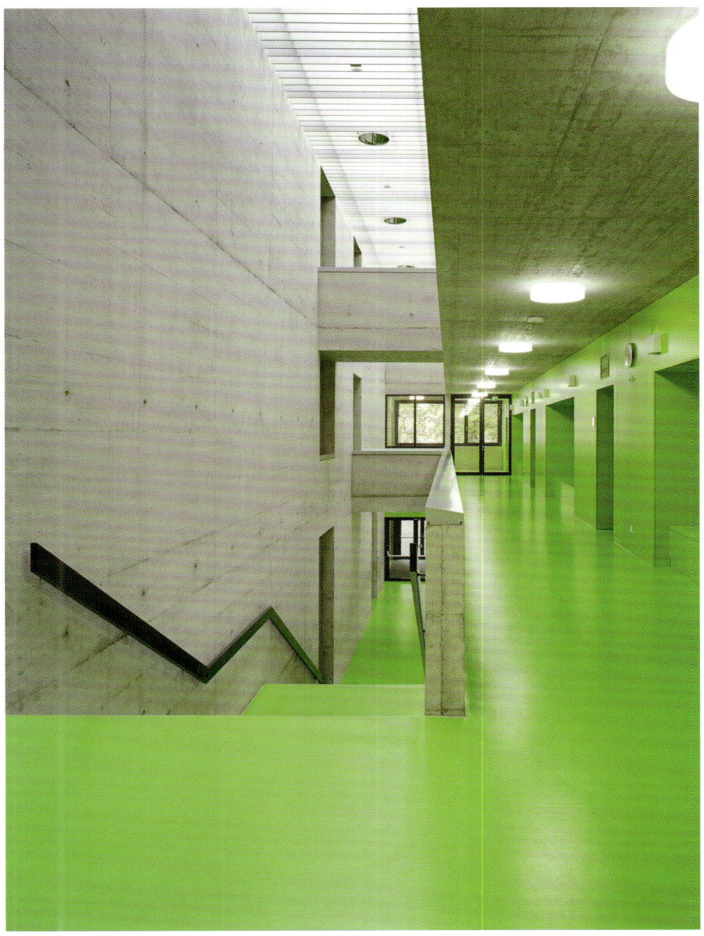

17

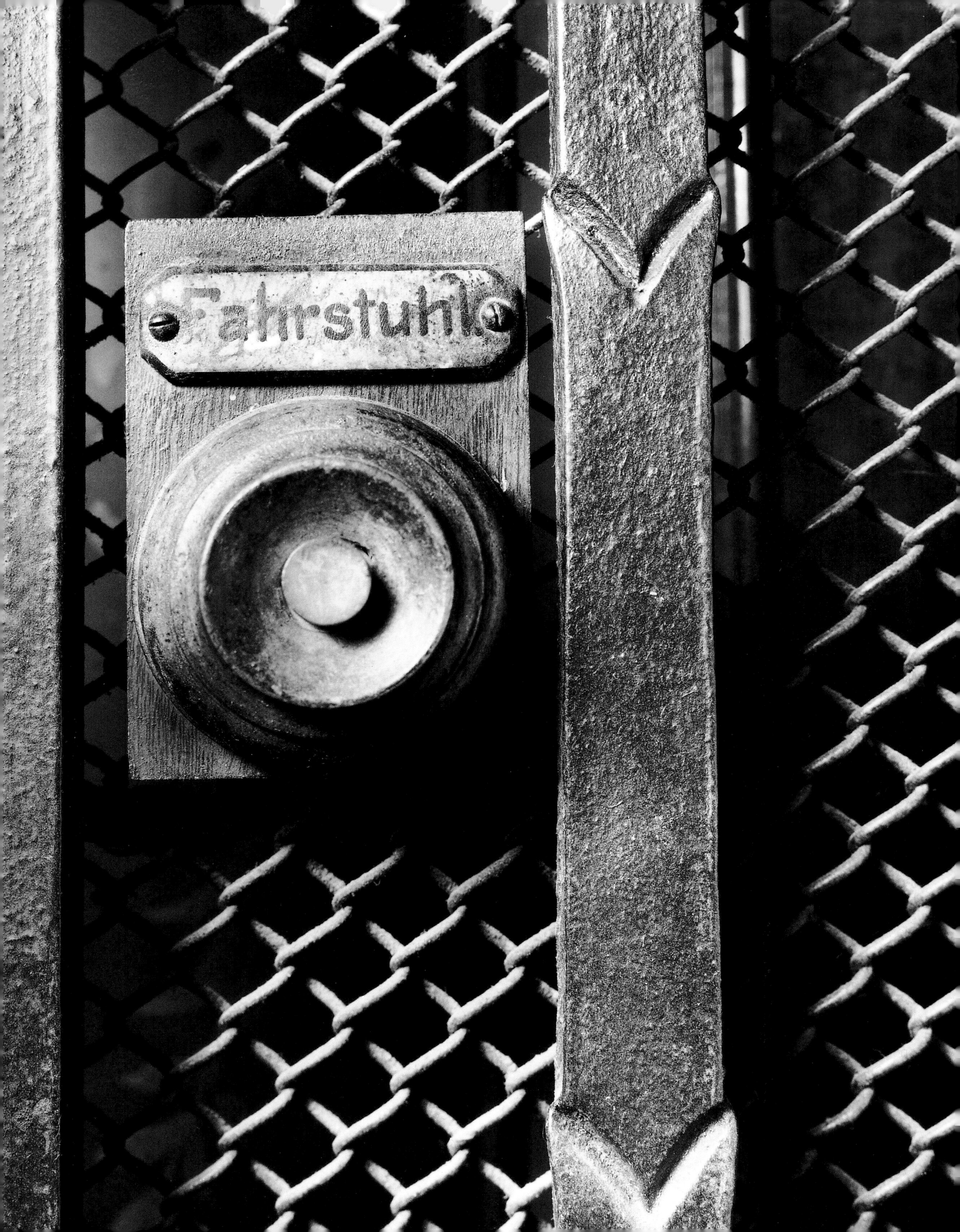
Fahrstuhl

A cultural history of elevators and lifts

Jeannot Simmen

The dream of a vertical ascent – Jacob's vision in the bible of a ladder between heaven and earth – is realized technologically in modern times. Taking an elevator or lift is our modern, rather more profane and earthbound version of taking a "stairway to heaven": gravity is overcome and less physical effort is required.
Technology defeats the force of gravity, from simple ladder to high-speed lift. Lifts glide seemingly effortlessly to the observation deck on the 124th floor of the Burj Khalifa in Dubai, with a height of 828 m currently the tallest building in the world (fig. 2). The lifts zoom skywards at speeds of 10 m/s (36 km/h). The maximum lifting height of a passenger elevator is 504 m. A goods elevator holds the record for the highest lift stop in the world at a height of 638 m. What today is considered as "the biggest" in Arab countries, was formerly realised in US metropolises, and described as traumatic by a European author.

Modern convenience technology vs. fatigue
"Amerika", synonymous for the hopes and dreams of a new life, is the title of an unfinished novel by the German writer Franz Kafka (1883–1924) [1], in which he portrays the strangeness of a faraway world. At the age of sixteen, the protagonist Karl Roßmann is sent to America by his poor parents [2]. The young hero stands in awe before the tall houses in New York, where one could look down to the street from the balcony using opera glasses. Roßmann finds a job in the city as a hotel lift boy. Already after the first week, Karl realises that he is fully up to the job, quickly mastering the short low bows expected of lift boys and collecting his tips in mid-air.
His friend Delamarche lives in a New York high-rise without an elevator. He visits him, and climbing up the stairs his friend keeps telling him that they will be right up in a moment. But it just doesn't happen: again and again another flight of stairs continues in a hardly perceptible different direction. Karl even stands still once, not from fatigue, but out of sheer helplessness in view of the length of the stairway. Delamarche tells him that the fact that the apartment is so high up also has its advantages: one hardly goes out, spending the whole day in a dressing gown – very cosy really. And of course no visitors come that far. [3]
Kafka's description of this feeling of helplessness with regard to the length of the stairs, is in fact a precise description of a new disease of civilization: psycho-physical exhaustion. Rather than being caused by a mechanical or visible injury, this is an invisible nervous debility. Neurasthenia – a weakness or exhaustion of the bodily functions – turns out to be a protective mechanism of the body against the new demands of society. Kafka's "Amerika" describes the European insecurity in the face of the then brand new skyscrapers. New York's vertical dimension threatens to be a traumatic strain on the physical body.
In the absence of an elevator, height implies fatigue and exhaustion.
In New York, the high-rise buildings equipped with the latest "anti-fatigue technology" were initially called "elevator buildings". In Europe, the American skyscrapers were referred to as "tower houses" until 1930 and described in the tradition of church towers, as solitary monuments similar to cathedrals: the appearance of an office tower may be highly artistic, but it should only be allowed as an exception – such as for singular emphasis in the overall appearance of the city – for example as town hall or the sole central commercial building [4].
The lift is used to counteract the rush and fatigue mania in big cities [5]. Around 1880, the elevator became part of the facilities generally found in all comfortably furnished hotels in larger cities [6]. The presence of elevator systems had an impact on the design of the hotel reception area – the ensemble of entrance and staircase (fig. 8, p. 31).

1 Birdcage elevator bell, Berlin-Charlottenburg (D)
2 Floor plan with three "wings" and central core accommodating the 57 lifts in total, Burj Khalifa, Dubai (UAE) 2010, Skidmore, Owings & Merrill

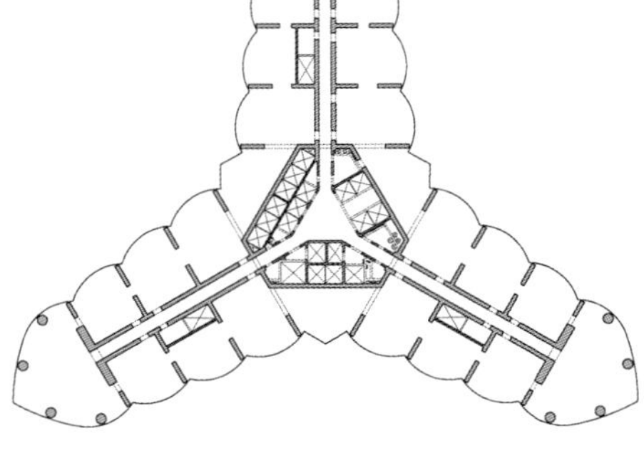

2

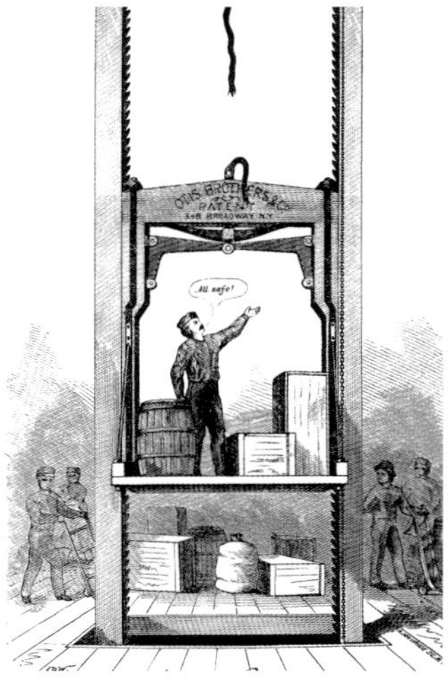
3

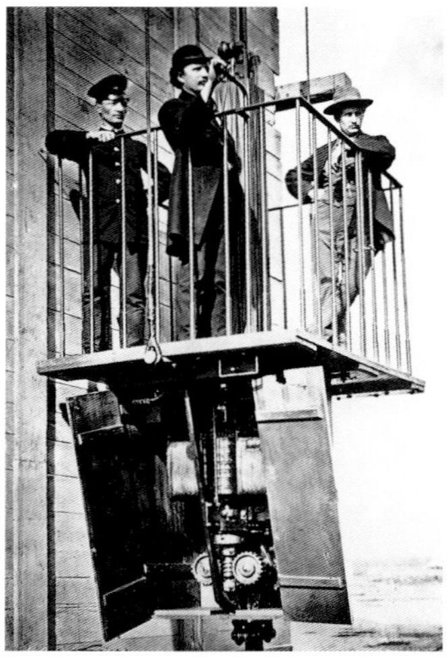
4

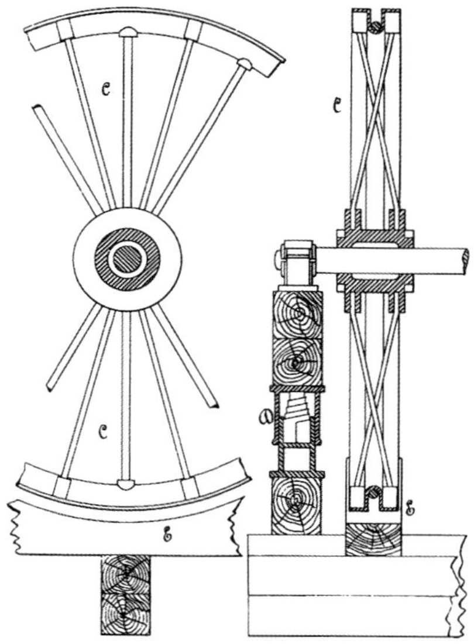
5

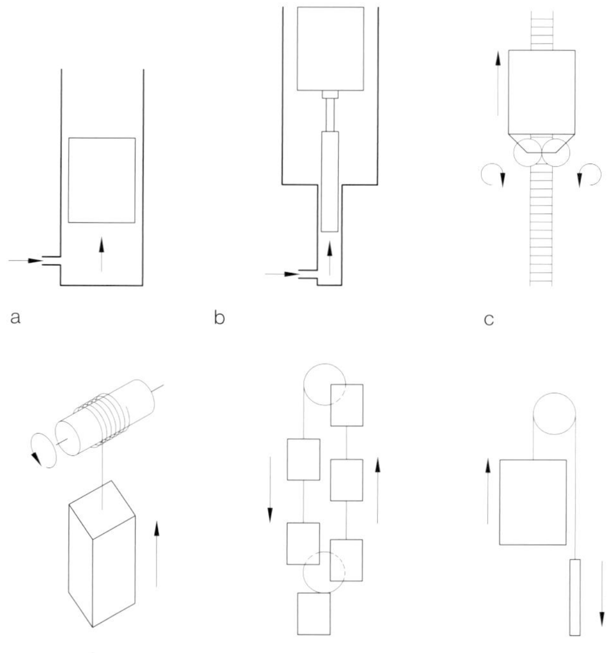

6 d e f

Moreover, the installation of lifts in upper middle-class residential buildings revived the conflict between the École des Beaux-Arts and the Polytechnic, between architect and engineer. The old-established staircase had to put up with a new contender: a technical device initially perceived as alien, mechanical and even ugly.
Hoist, lift cage, ascending room, elevator or express lift – vertical mechanical transportation systems went through various stages of technical development. The systems are however all clearly definable using the term “guided load”: the total movement is along a “track” composed of rigid or flexible guides whose position is determined by fixed points. The term “guided load” excludes other handling equipment such as crane, cable car or ski lift, while the funicular or inclined plane railway is a kind of hybrid system.

Three inventions towards development of the modern lift

From a historical point of view, there were three inventions that transformed the hoisting cage or freight elevator to the modern passenger lift. These three inventions – all in the second half of the 19th century – were essential for convenient and profitable access to the skyscraper.

Automatic emergency brake

After an elevator accident in his employer’s firm, Elisha Graves Otis was asked to build a safe device for vertical transportation. In 1853, he developed a special safety brake that made a lightning-like, human reaction in the event of a cabin crash unnecessary. A self-triggering mechanism was to bring the elevator to a halt in the event of an accidental fall. The automatic emergency brake consisted of a clamping mechanism, capable of immediately stopping the unsecured, ropeless platform from speeding downwards.
As star and stuntman of the experiment, Otis personally demonstrated his invention very effectively to the public in the New York Crystal Palace. A colleague cut the hoisting rope with an axe: the heavily laden unsecured 15-metre-high platform starting to hurtle down in free fall was stopped automatically by a tensioned spring and engaging ratchet bars. Rem Koolhaas called this spectacular demonstration the “triumph of a non-event” [7]. “All safe, gentlemen. All safe,” Otis is said to have commented after the averted plunge into the depths, as he sat triumphantly on the rescued platform high above the heads of the gasping audience (fig. 3).
This automatically-functioning emergency brake transformed the freight elevator into a safe passenger elevator. Since 1878, the system has been perfected by a centrifugal governor, which initiates braking when specific speed limits are exceeded. Formerly attached directly on the cab and now mounted stationary in the upper part of the shaft, the governor is connected to the cab through a rope that moves with the cab. At speeds beyond tolerance, the rope is clamped, triggering the braking mechanism. Today, a gentle sliding brake system has also replaced the abrupt stopping typical of the automatic snap mechanism.

Electric drive system

The second great invention on the way to the modern lift was the electric drive. Utilisation of steam- or gas-powered motors limited elevator operation to factory yards and

steam systems requiring a machine attendant to constantly monitor the pressure. Another drive mechanism was the water-based hydraulic system, especially common in Paris, where the inventor Léon Edoux had built hydraulic systems for the World Exhibitions in 1867 and 1889. But hydraulic systems were also dependent on pressurized water, which required steam-driven pumps at the time.
All these problems were solved by electric operation – the new systems were maintenance-free and flexible with regard to location. The electric motor was smaller and lighter compared to steam systems. As in other areas, the use of electricity for power transmission changed how familiar energy sources like wood, coal or oil were used.
In 1880, Werner von Siemens and the electrical engineer Johann Georg Halske presented a novel kind of perpendicular transport system at the Pfalzgau Exhibition in Mannheim (fig. 4). The device had an automotive platform, with the engine attached under the viewing platform. What was irritating about this lift that climbed up an exterior wall, was that the track system composed of a ladder went right through the hoisting platform.
Although the climbing lift was modified by a number of technical developments and improvements in the course of the years that followed, its speed remained limited. The wall ladder shaft was a rather dangerous track: one broken tooth would have resulted in the cabin dropping down to the next rung with full force and from there with acceleration downwards. The electric dynamo-motor solved the drive problem, but what was still lacking was a safe and efficient transmission system.

Friction as drive – an invention from mining

Longer vertical distances have been overcome in the mining world in the past than in high-rise construction. The cable drums however grew in size with increasing working depths and reached dimensions of several metres in diameter. The sequential unwinding of the rows of cable on the drum subjects the machine shafts and bearings to an uneven strain, which increases the risk of cable breakage.
In 1877, the mining engineer Carl Friedrich Koepe converted the drum machine to a new type of drive system for a 234-metre-deep shaft – equivalent to 70 floors in multistorey construction. Instead of using a cable drum, Koepe put the cable loosely over a disc; the cable was located in a dovetailed groove and solely driven by friction (fig. 5). The fear of the cable slipping was big, because it meant a threefold danger for the miners: should a worker survive the then enormous height of fall of up to 2,000 m down the shaft, he could drown in the pit water, or be killed by the damaged shaft casing or the falling hoisting cable. But contrary to all fears, these traction sheaves offered a high level of security, and the length of the cable in fact even turned out to increase friction and prevent cable slippage.
Security in case of cable breakage can be increased by multi-cable operation, with several cables over parallel discs. Each cable is thinner and therefore more flexible. The cable is deflected only once, reducing cable wear. Forces are simple to calculate because the cable always runs axially. The danger of the cable becoming slack is eliminated. An even distribution of tensile forces protects the cable, drum and motor; the difference in cable weight is constantly balanced by a cable hanging loosely below.

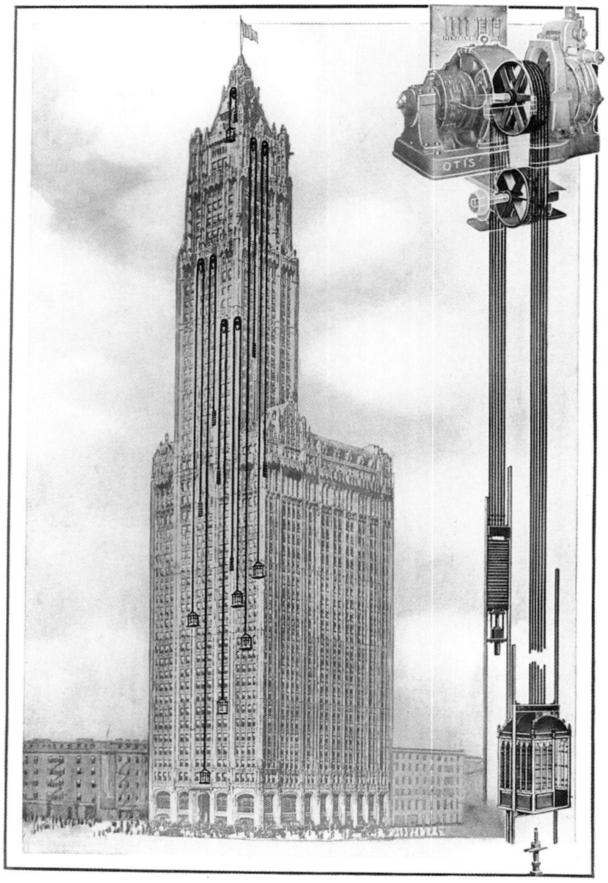

7

3 Elisha Graves Otis during his free fall test at the Crystal Palace, New York (USA) 1854
4 The first electrically operated passenger lift, a climbing lift developed by Siemens & Halske at an industrial exhibition, Mannheim (D) 1880
5 Drawings from the patent specification of Carl Friedrich Koepe's invention of the traction sheave drive in 1877: the cable is placed directly over a wheel (left) and is located in a groove (right).
6 Schematic representation of different types of drives:
   a Air (pneumatic)
   b Water / oil (hydraulic)
   c Cog / ladder (climbing elevator)
   d Drum
   e Paternoster
   f Traction sheave
7 Woolworth Building, New York (USA) 1913, 57 floors, with the most modern traction sheave elevators, until 1930 the tallest building in the world
8 An "ascenseur" in Paris (F) around 1920: integration of the elevator system in the entrance area with clear geometric horizontal and vertical lines

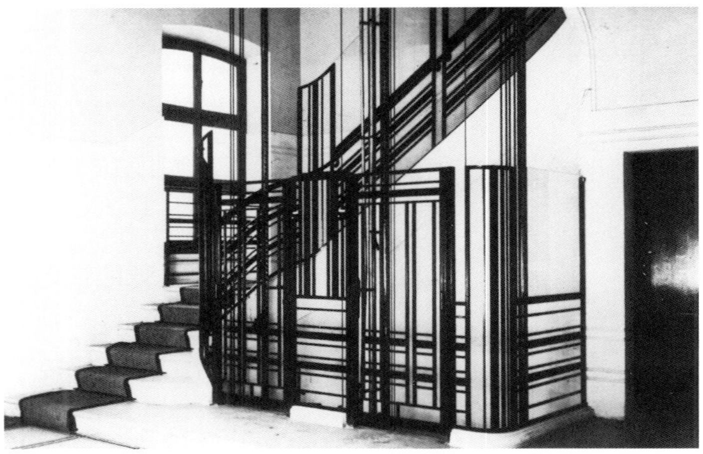

8

9
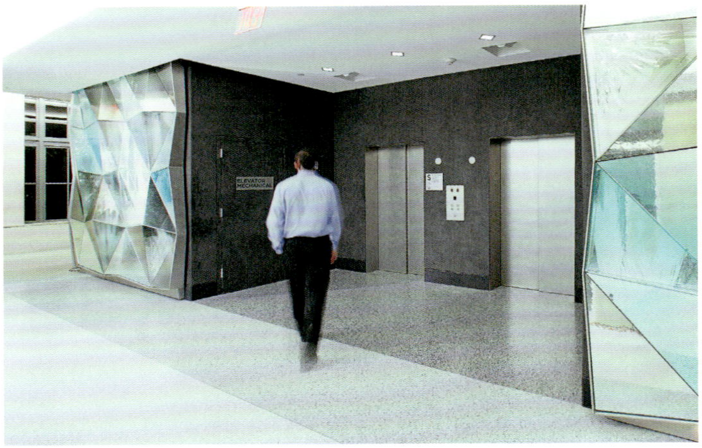

9 Elevators, “Top of the Rock” observation deck, Rockefeller Center, New York (USA)
10 Panoramic elevator, Lloyd’s Building, London (GB) 1986, Richard Rogers Partnership
11 New York Times Building, New York (USA) 2007, Renzo Piano Building Workshop, FXFowle Architects, Interior Design: Gensler (based on guidelines by RPBW)
a Cafeteria with sky lobby on 13th floor
b Floor plan 13th floor
12 Schematic elevator concept of a modern high-rise building

10
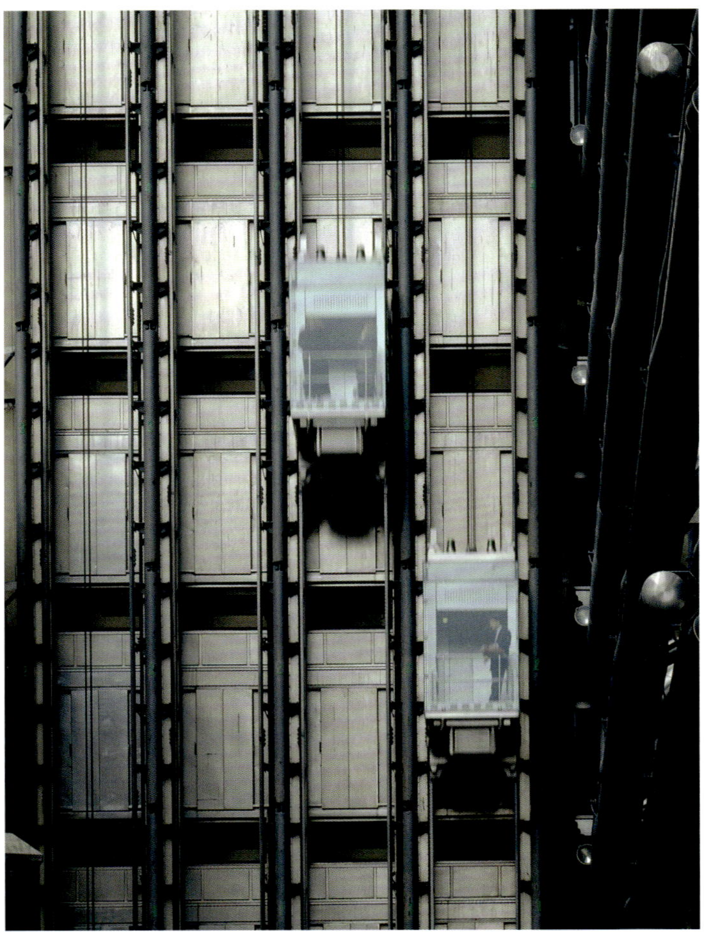

No skyscrapers without the modern lift
It is the association of these three inventions that made the modern, safe and comfortable elevator possible. A significant advantage of the combination of safety device, electric propulsion and traction drive transmission is the small space requirement and light weight of the machine ensemble. This modern set-up was installed in the Woolworth Building in New York in 1913, which was the tallest skyscraper with the fastest lift systems until 1930 (fig. 7, p. 31). The 29 elevator systems travelled at a speed of about 3.50 m/s and a quick lift up to a height of 207 m was available. The more efficient the lift technology, the taller the constructions that can be built and used. Building height is dependent on lift technology.
Unlike a journey by ship or rail, a lift ride doesn’t usually offer a visual experience on account of the change in location. The decisive parameter is not the spatial variation, but the negatively experienced time. An elevator ride in a closed car through a concrete shaft confronts the passenger with a particularly uniform experience of time. The human two-minute limit in daily lift journeys is comparable to the absolute speed of light in modern physics: a waiting time of about 20 seconds and a travelling time of 100 seconds are considered as bearable during a normal working day. Tolerance is increased by floor stops and successively entering passengers, and is generally higher in residential high-rise buildings than in commercial ones. Tourists are considered to be the most patient lift users. They do not feel that the time spent in elevators, especially in attractions such as the Eiffel Tower or other TV towers, is wasted.
In addition to the waiting time parameter, the space required for lift systems is another limiting factor. The space required for elevator shafts increases disproportionately in relation to the usable area gained. In buildings with conventional elevator systems, the percentage rises from 7 % to up to 20 % for an increase from 70 to 100 floors. There are three possibilities for keeping the space requirement for lift systems low:

- Double cabs allow a 30 % increase in the transport capacity. Models can be found in the mining sector. Visitors can access the elevator from two levels and after the ride they get off on different floors.
- The express/shuttle elevator offers a simple solution for buildings with over 80 storeys. The building is divided into independent sections: express elevators go directly to the transfer stations, so-called sky lobbies. From there, shuttle lifts take passengers to the individual floors (fig. 12). Sky lobbies can be turned into shopping or food areas by addition of cafés and restaurants, or supplemented with entertainment or sports facilities such as cinemas or gyms (fig. 11).
- Facade lifts or panoramic elevators make the elevator shaft in the core of the building unnecessary, but may minimize the window area (fig. 10). Panoramic lift systems can be architectural prestige objects, especially in atria and lobbies of office buildings or hotels, and have long been popular as dramatic film motifs.

Today the elevator is a technically established, safe mode of transport; getting stuck in a lift is a possible, but rare event thanks to safety regulations (an availability of about 99 % is quoted by manufacturers). High capacity utilization

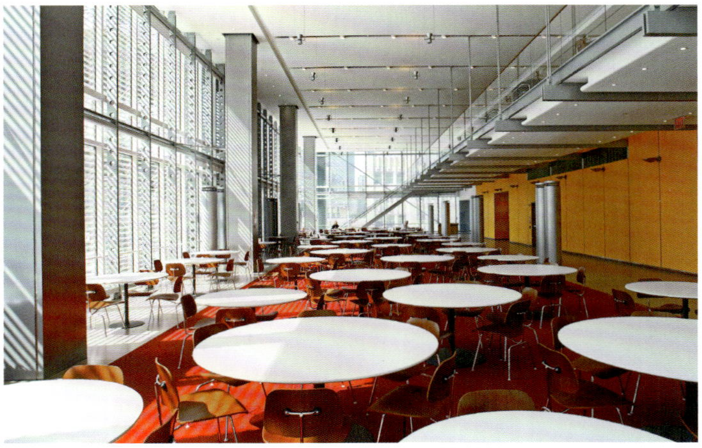

a

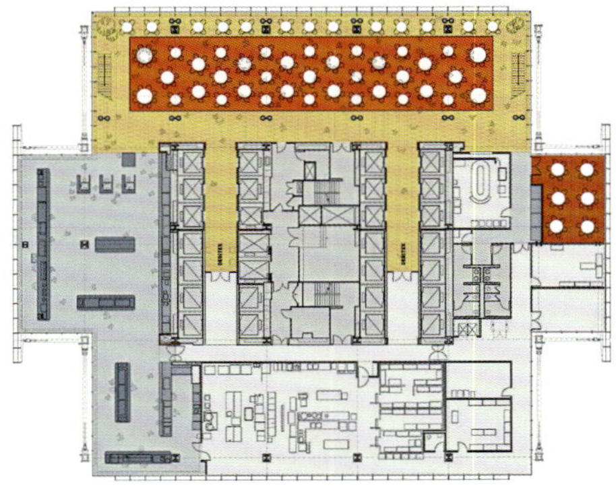

b

11

is achieved by computer control; empty and single trips are avoided. Energy efficiency is the innovative area at the moment: energy savings of up to 70% are possible compared to old systems. Various manufacturers offer elevator systems with power recovery during descending journeys, with the ultimate aim of not requiring any electricity from the public power grid at all in the future.
The mechanical ensemble of drive and transmission is set to remain in the established combination of electricity and traction sheave, with the emergency brake functioning automatically on a mechanical basis. The lift is the first automotive device that starts at the press of a button, brakes with millimetre precision, opens and closes doors automatically – all without an operator.
This automatically accessed modernity changes our perception, especially in cities: we drive our car straight into the underground car park and then take the elevator directly to our work place or to the mall. Buildings are experienced less in terms of what they look like from the outside, but more through their circulation, starting from the basement. The architectural image, the facade and a representative entrance area are in fact secondary. The quality of the lift journey experience needs to be improved through virtuality, as the elevator car is turned into a communicative space [8].

In a section of his novel "Amerika", Franz Kafka refers to a wonderful earthly-heavenly theatre of deliverance that lies beyond all the automated modernity of the new world. Karl, the former lift boy, sees a poster: The Great Theatre of Oklahoma is calling you! It's only calling today, only once! Whoever misses this opportunity now, will miss it forever! Whoever thinks of his future should join us! Everyone is welcome! That was the European promise of modernity, the salvation through art and culture. Kafka wrote "Amerika" in 1912, the year that the Titanic, believed to be unsinkable, struck an iceberg and sank.
Some 100 years later, modernity is globalized electronically through the World Wide Web, reality is extended sustainably through simultaneity and immateriality, metaphysically real through cyberspace. Beyond the vertical dimension physically insurmountable by us, we create for ourselves a weightless beaming into other spheres, and experience an all-dimensional simultaneousness beyond local coordinates. This virtual utopia and sensuously real promise makes us feel almighty – the sinking of the Titanic was yesterday.

References:
[1] Franz Kafka published the first chapter "The Stoker" in 1913. Max Brod edited and published the text posthumously in 1927 under the title "Amerika". The title "The Man Who Disappeared" ("Der Verschollene") appears in a letter by Kafka to Felice Bauer dated 11 November 1912.
[2] Franz Kafka: Der Verschollene. Ed. by Jost Schillemeit, Frankfurt a.M. 1983, p. 7
[3] Ibid., p. 289f.
[4] Werner Hegemann: Amerikanische Architektur & Stadtbaukunst. Berlin 1925, p. 11
[5] Ibid.
[6] M. Busse: "Über eine neue Sicherheitsvorrichtung für Fahrstühle" In: Zeitschrift des Vereins Deutscher Ingenieure. Berlin 1879, p. 422
[7] Rem Koolhaas: Delirious New York. New York 1978, p. 19
[8] Gestaltung der Vertikal-Wahrnehmung in der Liftkabine: Wettbewerb Universität Essen und Universität der Künste Berlin; Jeannot Simmen (Ed.): Up and Down. Essen/Berlin 1999

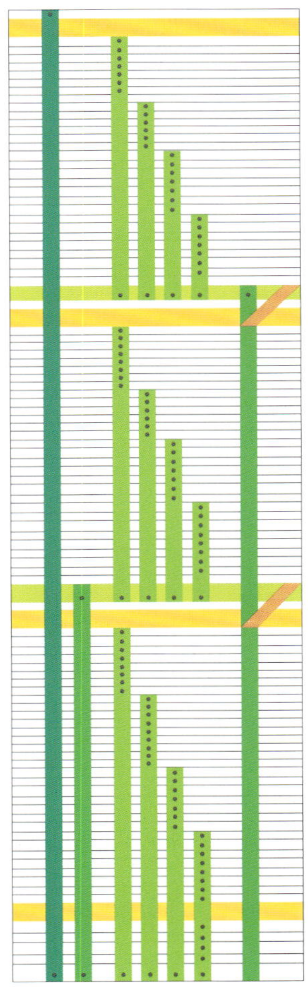

- Express elevator to the top floor
- Express elevator to the sky lobby
- Local elevator
- Sky lobby
- Mechanical floor
- Escalator
- · Floor stop

12

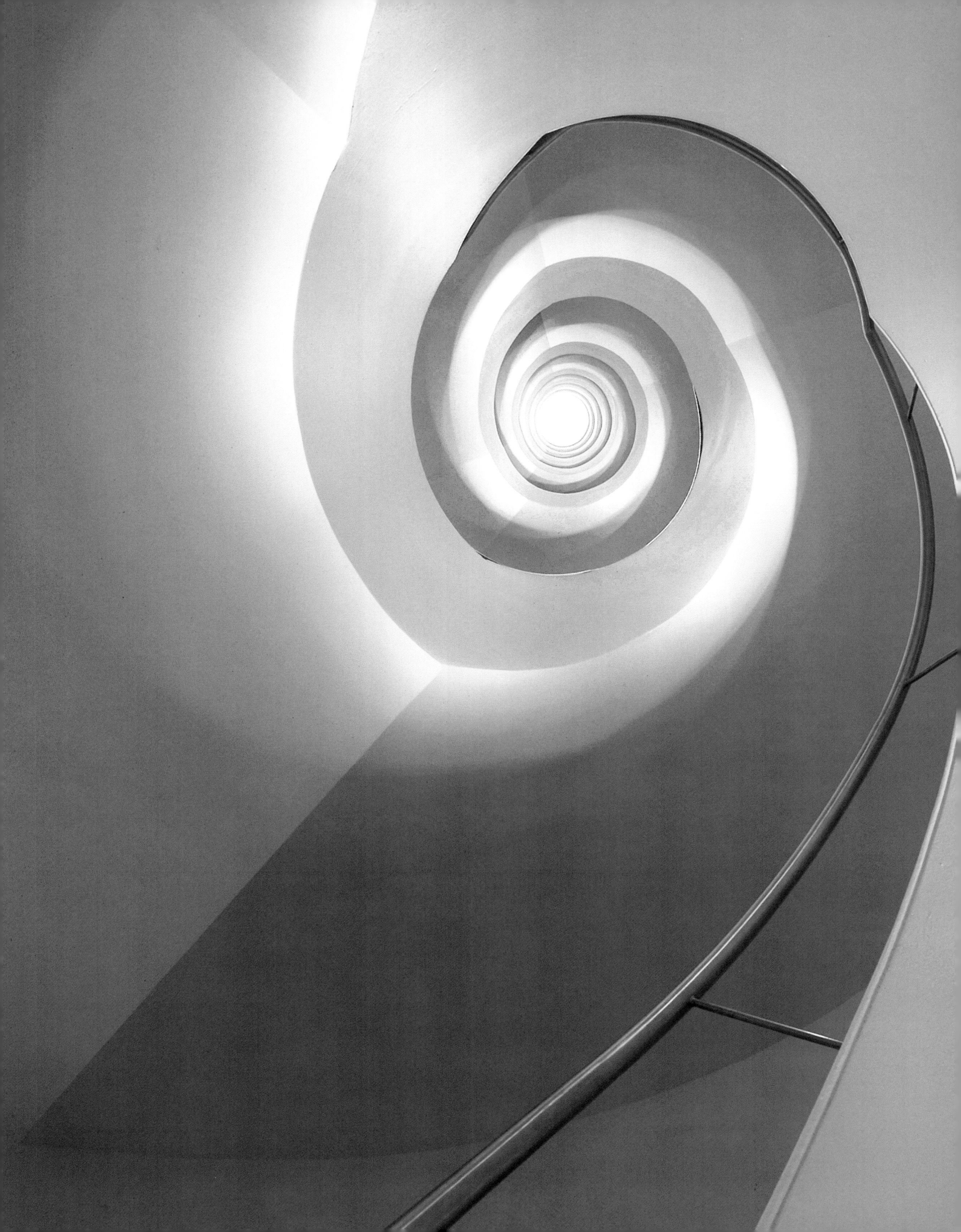

Light design for orientation and direction

Thomas Schielke

How can lighting be used for better orientation in situations involving stairs and pathways? And can light also be used to create a calm or contrast-rich atmosphere? The three principles formulated by the American lighting designer Richard Kelly – "focal glow", "ambient illuminance" and "play of brilliants" – are helpful for project analysis and design. "Focal glow" is the basic general illumination required to be able to recognise the surroundings. "Ambient illuminance" creates perceptual hierarchies through contrasts in the brightness of accent lighting. Targeted light is particularly suitable for accentuating specific zones, entrances or wall exhibits in circulation areas. This results in guiding the direction of view and supports spatial orientation. "Play of brilliants" encompasses numerous lighting effects used for their own sake, for creating a special atmosphere or for decorative purposes, but which do not primarily have a practical function, such as a light art object or coloured lighting that mainly serves to influence the colour climate of a room.

In the case of new constructions, many effects can be specified at the design stage, such as integration of daylight to illuminate areas like foyers or corridors via windows or skylights, incorporation of interesting views, as well as minimisation of energy requirements attributable to electric lighting. The significant drop in the luminous intensity of daylight indoors, however makes it necessary to provide additional electric illumination of circulation areas inside buildings during the day.

Illuminated surfaces can be used to create spatial patterns for orientation, such as wall washing indicating the way through corridors, or accentuations emphasising entrance or crossing situations. In the shape of a line in the floor, light can identify a threshold, draw attention to critical situations such as steps or lead through rooms. Staggered light lines can be used to indicate stairs from afar. A better comprehension of spaces and circulation routes can be achieved by means of a differentiated lighting design for the architecture.

In addition to orientation, light makes an important contribution to the creation of an appropriate atmosphere. Light-coloured materials and bright illumination can make a narrow corridor appear less narrow. Intensive contrasts of dark and light on the other hand, can conjure up a mysterious setting for dramatic effects. Dynamic changes in colour can even turn circulation areas into scenographic design stages. The depth of a room can be increased or decreased through the type of lighting used. Narrow cones of light can be used to create private zones in large circulation areas, for instance, to differentiate sitting areas from the surroundings. Lighting concepts that not only focus on the pathway, but also include the surroundings in the design are particularly worth paying attention to.

Exterior circulation design

Different circulation areas have different lighting requirements. The following overview identifies the most important aspects of light planning and presents the various different types of lighting.

Pathways can be marked using a series of pinpoint lights. A similar effect is achieved by an illuminated row of trees along a path. An illuminated wall can also be used for orientation. By including the surroundings in the lighting design in this way, the spatial experience becomes more pronounced. Alternatively, the path itself can be illuminated with side bollard lighting, with the light directed towards the path. A calm atmosphere can be achieved in this way, with an even distribution of brightness and soft-contoured light cones. The luminance at the light-emitting surface of the luminaire should be low to avoid glare effects. Soft-contoured cones of light avoid sharp contrasts with the dark surroundings, and facilitate easier adaptation of the eye at night. Instead of bollard lights, orientation lights embedded in the ground can also be used to emphasise the direction with high-contrast light lines or dots. Special crossing points can additionally be accentuated by means of a different lighting type or colour temperature, or by implementation of an increased luminous intensity.

As for the pedestrian pathways before, the surroundings can also be included in the lighting concepts developed for roadways, to achieve a greater sense of three-dimensionality. Intersections, curves or pedestrian crossings require more attention for greater safety, which can for example be realised using a higher luminous intensity. Shorter roadways can be illuminated using roadside bollards. Shielded bollard lights prevent drivers from being blinded by glare effects, since the eyes of drivers are often closer to the ground than those of pedestrians, a point that has to be taken into account by the lighting specifications. Mast lights with higher light sources permit larger intervals between luminaires and are suitable for longer distances.

Stairs visible from afar signalise deviations in a vertical direction. An intense contrast in tread and riser brightness emphasises the step-wise gradation in height with a delicate light pattern. Light lines along a handrail or banister tend to

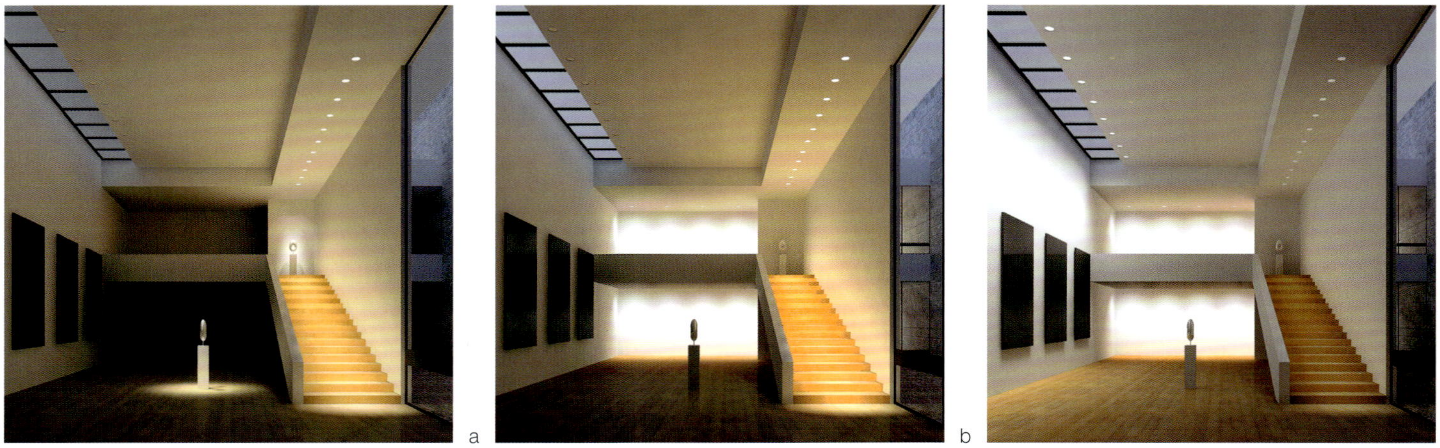

2 a b c

place the visual focus on the length and breadth of the stairs, and can be used to make a generous gesture. Good illumination at the points where the stairs begin and end accentuates the critical zones during walking. Easy recognition of the steps is achieved by an intense contrast between riser and tread in conjunction with adequate brightness. Luminaires can be fitted in the steps, in the handrail, on the wall or placed as free-standing bollard lights. Shielded luminaires prevent glare effects that could originate from the upper steps while climbing the stairs.
A more brightly illuminated entrance area compared to the remaining facade facilitates quicker orientation as well as legibility of the house number or graphic elements. A bright welcome mat on the floor creates a friendly entrance for visitors (fig. 4). During the day, this is the place where the eye adapts from bright daylight conditions outside to comparatively dim lighting conditions inside. At night, the welcome mat is an orientation aid. Shielded luminaires in interior spaces prevent glare effects on entering the building. As opposed to the horizontal illumination of floor areas, vertical illumination components of downlights or wall washers ensure effective lighting of door surfaces on the one hand, and facilitate good recognisability of faces during greeting on the other. Laterally placed luminaires for illumination at face height will not have host and visitor appear as silhouettes, in contrast to a situation where the lighting shines on the backs of the persons.

Interior circulation design

Making a prestigious or specifically representative impression in the foyer area is not only limited to utilisation of the right furniture and materials, but can also be achieved with an appropriate lighting concept. A light and spacious impression can be created by means of extensive wall washing. Accent lighting on the other hand, emphasises individual objects and can be used to create private zones in large spaces through islands of light. The targeted light of accent lighting is also suitable for creating effects of lustre. Grazing light allows emphasis of the textures of wall materials. Decorative luminaires can additionally function as eye-catchers. A calm overall lighting design, accentuating architectural elements such as walls or columns, helps to clearly structure various functional areas such as reception desk, lobby or elevator, and facilitate orientation. Differentiations in the luminous intensity allow perceptual hierarchies to be developed in order to accentuate important zones. As transition areas between exterior and interior space, foyers reconcile the high luminous intensities outside and the reduced brightness inside buildings during the day. At night, illuminated foyer walls allow a glimpse of the building interior from the outside, in contrast to the daytime situation, when reflections on the glazed foyer facades are common due to the higher luminance outside. Narrow light cones and shielded luminaires are particularly recommendable for tall foyer spaces in order to avoid glare effects.
A more friendly impression can be created in corridors by means of daylight through windows or overhead lights. A window at the end of a corridor makes a connection to the exterior space and emphasises the direction of the corridor (fig. 7, p. 39). Artificial lighting can be used to create a pleasant atmosphere in the evenings and in areas without any daylight illumination. A brightly illuminated end wall with accentuation of a picture or an object can be an attractive eye-catcher in interior spaces, and contribute to spatial orientation (fig. 2). Since the walls of corridors constitute dominant elements in their spatial perception, they offer a large potential for creating a bright overall impression. A calm and clear appearance can be achieved with homogeneous wall washing. In contrast to this, accent lighting can counteract the monotony of longer corridors and divide them into shorter units, by for instance accentuating pictures on the walls with spotlights, or by placing light cones so as to create a rhythmic pattern on the floor. Grazing light can also be used to produce diverse light patterns on the walls. The luminaire itself can of course also become an attractive eye-catcher, and help to structure the room by a sequential arrangement for example, provided that the luminance does not give rise to any unpleasant glare effects. Materials with light colours generally help to avoid a gloomy atmosphere in corridors. Since most corridors are not used constantly, the utilisation of ensors and a time control system is recommendable to reduce the illumination (and hence energy consumption) to a minimum, except when actually required.
Transitions from one room to another can be made more significant by the accentuation of doors with narrow light cones. Similar to the entrance situation, a light cone pointing to the ground can create a welcoming impression and point out a door from a distance, even if the door, hidden by the reveal, is not yet visible from that perspective. To avoid glare effects after stepping through the door, the luminaires in the next room should be shielded and spotlights should not be directed towards the door.

Traversal of a space in a vertical direction can be turned into a dramatic experience by a contrast-rich stair lighting design (fig. 3). Light lines in the handrail, luminous joints between the steps, accent lighting on pictures, or vertical light sculptures in the centre of the stairwell, can transform a staircase into an interesting and attractive place. Unusual features also include illuminated frosted glass steps, or luminous steps with dynamic images introducing a narrative element. On the other side of the spectrum of spectacular designs are clearly structured concepts for creating a calm spatial impression. Shielded luminaires or indirect lighting through wall washing in the landing areas minimise glare effects, especially when ascending the stairs. The luminaires can be fitted to the ceiling or walls, or integrated in the handrail or steps. In the case of step-integrated lighting, it should be taken into account that every single step needs to be illuminated to ensure clear orientation and avoid accidents. Broad-beam illumination with adequate luminous intensity is recommended for landings and stairs to facilitate general orientation as well as a calm and clear spatial impression. Lighting that shines down from the upper landing creates a distinct contrast between riser and tread for safe walking. An exclusively diffuse illumination leads to less contrasts and makes it more difficult to identify the steps as well as the transitions between stairs and landing. An energy-saving light control system is also recommendable for less-frequented stairwells. Stairwells used as escape routes require installation of emergency and safety lighting in compliance with the relevant standards.

The dynamic element of escalators and moving walkways can be emphasised by placement of light lines in the handrail or on the ceiling, and/or by means of luminous balustrades. To prevent the attention of passers-by from being exclusively drawn to the escalator or the moving walkway, the spatial surroundings should be incorporated in the lighting concept. Accent lighting identifies the change from firm ground to moving elements for increased safety. In very long moving walkways, transitions may additionally be indicated by flashing start and end points. Escalator luminaires can be fitted to the underside of the escalator above, or integrated in the balustrade (fig. 6, p. 38).

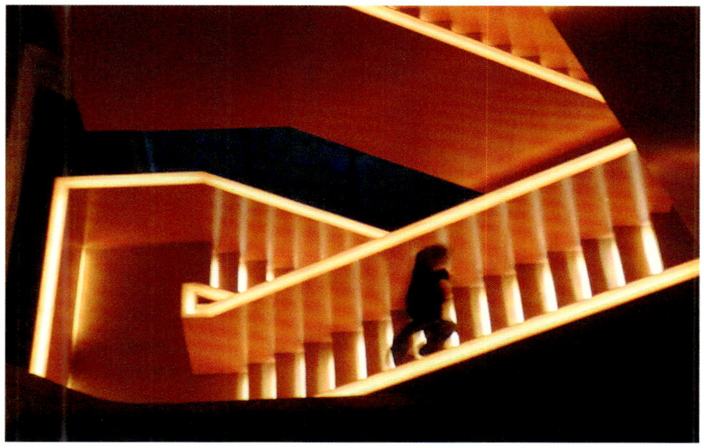

3

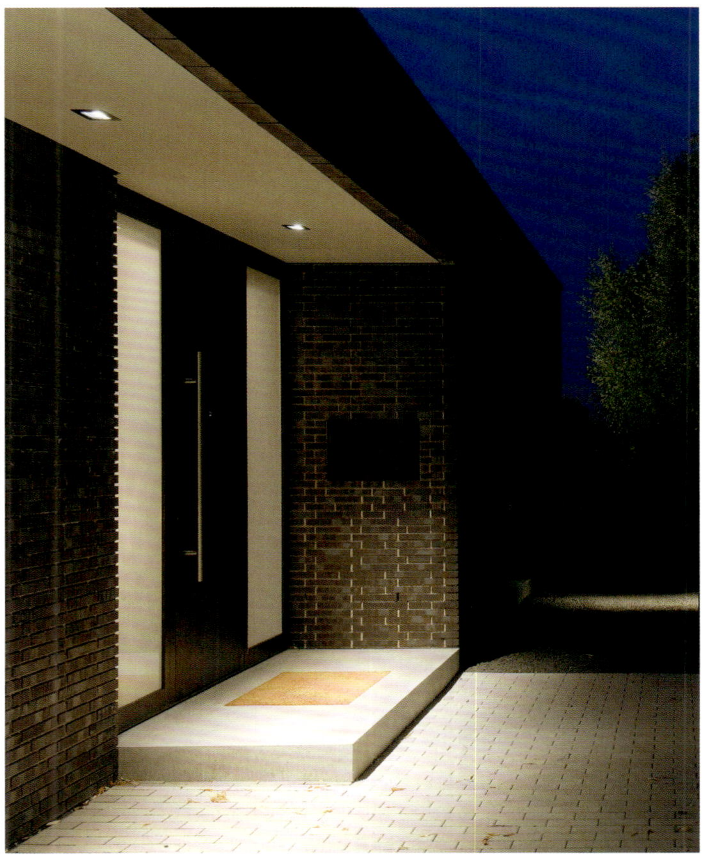

4

1 Uplights in the basement create a light line on the underside of the stairs for orientation. Daylight enters the stairwell through a round opening in the roof.

2 Three different lighting scenarios and their spatial effects:
  a Functional zones such as circulation areas can be emphasised by lighting. Accentuation of the columns creates an additional eye-catcher.
  b Illumination of the entire rear wall shifts the focus of attention to the back area and visually combines the lower and upper floors in addition to the illuminated stairs.
  c Illumination of the left and rear wall, as well as the accentuation of the columns, increases the spatial experience and emphasises the art objects. The dimmed illumination of the stairs signalises that the way to the upper floor is of secondary importance.

3 Reminiscent of embers, the orange sides as well as the additional lighting under the stairs and in the handrail clearly differentiate the staircase from the listed building fabric. Coal Washing Plant "Kohlenwäsche", Essen (D) 2009, OMA, Böll & Krabel, lighting design: Licht Kunst Licht

4 The bright welcome mat on the floor together with the luminous panes of glass on either side of the door create a friendly entrance situation for visitors. House P, Dortmund (D) 2012, Heiderich Architekten

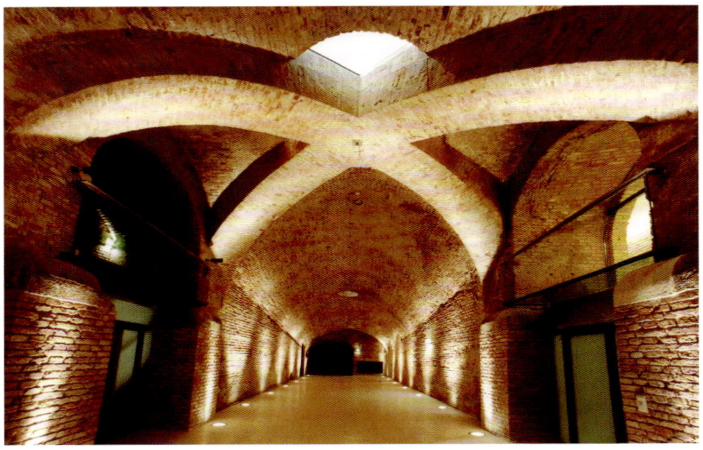

5

5 The recessed floor lights right next to the walls achieve a striking grazing light effect accentuating the texture of the masonry. Hotel Palais Coburg, Vienna (A) 2003, lighting planning: Elmar A. Wimmer
6 Luminaires in the balustrades of the excalators provide good lighting for the stairs and connect the storeys visually. A string of recessed ceiling lights adds additional focus. Selfridges, Birmingham (GB) 2003, Future Systems
7 A window at the end of a corridor enhances the depth effect, offers a view, allows daylight to enter and dramatizes the illuminated space.
8 Comparison of the properties of LEDs, fluorescent lamps and metal-halide lamps
9 Different luminaires and lighting types

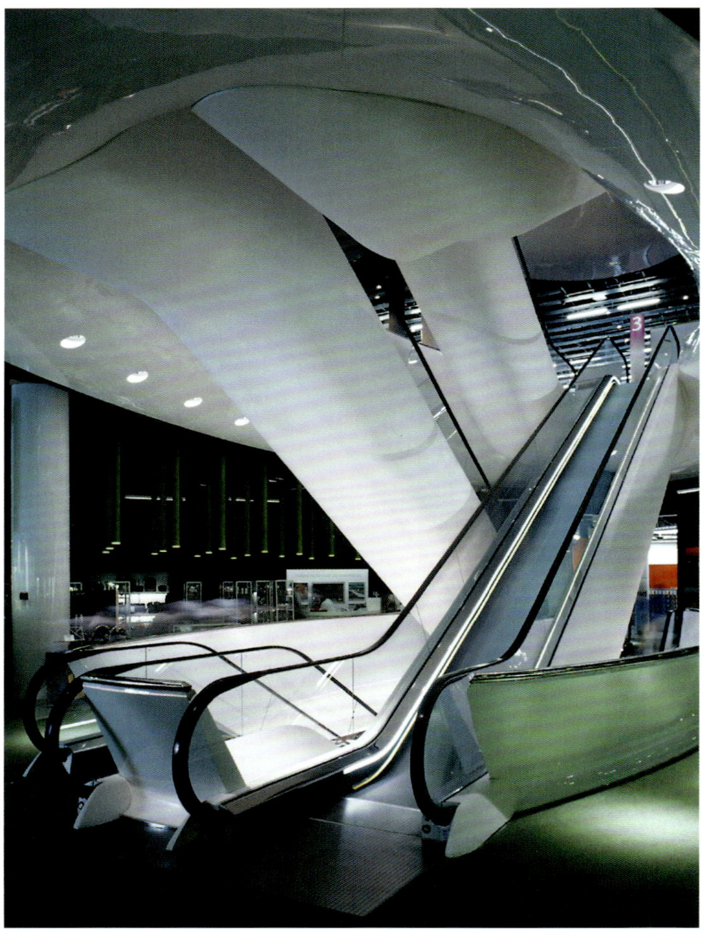

6

In contrast to a diffuse luminous ceiling, which creates a calm and soft atmosphere, brilliant lighting effects are produced by targeted light emanating from point light sources in an elevator. The vertical proportion of the lighting ensures an even illumination of the faces. An appropriate luminous intensity and colour temperature is recommended for a smooth transition between foyer and elevator. Shielded luminaires outside the elevator are favourable for providing pleasant lighting conditions to passengers stepping out of the elevator. If there are reflective surfaces in front of the elevator, possible glare effects due to reflections should be taken into account. Effects that communicate the dynamic character of the elevator can be created using a light control system. Glass ceilings and walls offer the option to include the elevator shaft in the lighting concept for a more intensive spatial experience.

Illuminants – criteria and types

The decision for an illuminant type depends on the desired lighting effects and the framework conditions of the project. Out of a large number of available lamps, only a limited number will fulfil the specified requirements. Modelling and lustre are for instance effects achieved by targeted light. These effects need point sources, the light of which is often additionally bundled using lenses and reflectors. The light colour of a lamp influences the colour appearance of illuminated objects. Warm white light colours correspond to the colour temperature of incandescent lamps and emphasise warm, e.g. red and yellow colours, while daylight white light colours accentuate cold colours like blue and green, and correspond to the daylight outside. Neutral white light colours offer a compromise colour temperature solution. The colour rendering index describes the quality in which material colours are reproduced.
LEDs are characterised by a very long service life (about 50,000 hours), good luminous efficacy (80–100 lm/W) and dimmability, as well as small size. Light colours at various colour temperatures are available. The small size of LEDs makes it possible to develop compact component-integrated light solutions. Since luminous flux and service life of LEDs changes with heat, it is recommended (particularly in individual solutions) to ensure an adequate thermal management, to ule out a decrease in luminous flux as well as a shorter service life.
Fluorescent lamps are characterised by high luminous efficacy and long service life. The light colours available for this type of lamp are warm, neutral or daylight white; they can also be dimmed with an appropriate controller. Two different structural designs are available: linear fluorescent lamps and compact fluorescent lamps with a diffuse lighting effect. Fluorescent lamps are suitable for homogeneous and economical basic illumination.
Metal-halide discharge lamps are very efficient with a service life of 9,000 hours and a luminous efficacy of up to 90 lm/W. They also have high luminous flux values. Modern metal-halide lamps with ceramic burners offer good colour stability and rendering as well as an integrated UV protection. Metal-halide lamps are available in the light colours warm white, neutral white and daylight white, and can only be dimmed to a limited extent or not at all (fig. 8).

Selecting luminaires

The spectrum of recessed ceiling luminaires ranges from classical downlights with a broad symmetrical light distribu-

tion for basic illumination, through downlight wall washers and lens wall washers for vertical illumination, to directional spotlights for flexible accentuation (fig. 9). The light output ratio is an indication of efficiency. When comparing luminaires, this figure should be supplemented by two others: thevalue of the possible luminous intensity for an area in relation to the energy expended, and a visual comfort analysis.
A high cut-off angle reduces glare effects. In the case of ceiling-integrated illumination with downlights, the lighting technology can be integrated in the architecture elegantly. In contrast to this, the technical details exposed by additive lighting systems such as spotlights on tracks or pendant luminaires, give these systems a greater expressive potential in addition to the lighting effect.
The flexibility of power track spotlights predestines them for illumination of images on walls. They are available with various beam angles for accent lighting. Changing presentation requirements can be met by a corresponding adaptation of the place of installation and orientation. Projection light sources can be combined with insertable templates made of metal or glass, referred to as gobos, to create various light cone contours or light patterns for decorative effects.
A further group of luminaires that can be used with power tracks are floodlights. These are characterised by a broad symmetrical light distribution with a soft gradient, for illuminating long rows of images or large objects. Wall washers on the other hand, with their asymmetrical light distribution, only illuminate vertical surfaces. The result is a highly uniform luminous intensity on the wall.
Further illumination options include utilisation of light structures and orientation luminaires, as well as the integration of lighting in architecture, such as cove lighting, diffuse luminous ceilings or coloured luminous walls. When using recessed floor lighting, for wall washing or grazing light for instance, glare effects on passers-by should be avoided by adequately shielding and appropriately positioning the luminaires. Various protection classes are relevant for exterior installation of luminaires. The rating IP65 for instance indicates that a luminaire is dust-tight and protected against water jets, while the higher rating IP68 means that the luminaire is additionally protected against immersion in water.
Light control systems offer high flexibility, both during installation and operation, for adaptation of the lighting to varying requirements. Specific lighting scenarios depending on the time of day can be defined. A light control system can automatically respond to the presence of daylight through sensors and regulate the interior illumination efficiently. Incorporation of coloured lighting also requires a light control system for programming.
Apart from the aspect of providing pleasant and appropriate illumination in circulation areas, the quantitative requirements of emergency lighting also have to be fulfilled. Safety illumination ensures safe evacuation of a building by placement of relevant markings and ensuring adequate visibility on the escape routes. This includes, for example, illuminated or back-lit emergency signs and a minimum luminous intensity of 1 lux on the central axis of emergency routes. More detailed information about safety lighting is provided in DIN EN 1838.

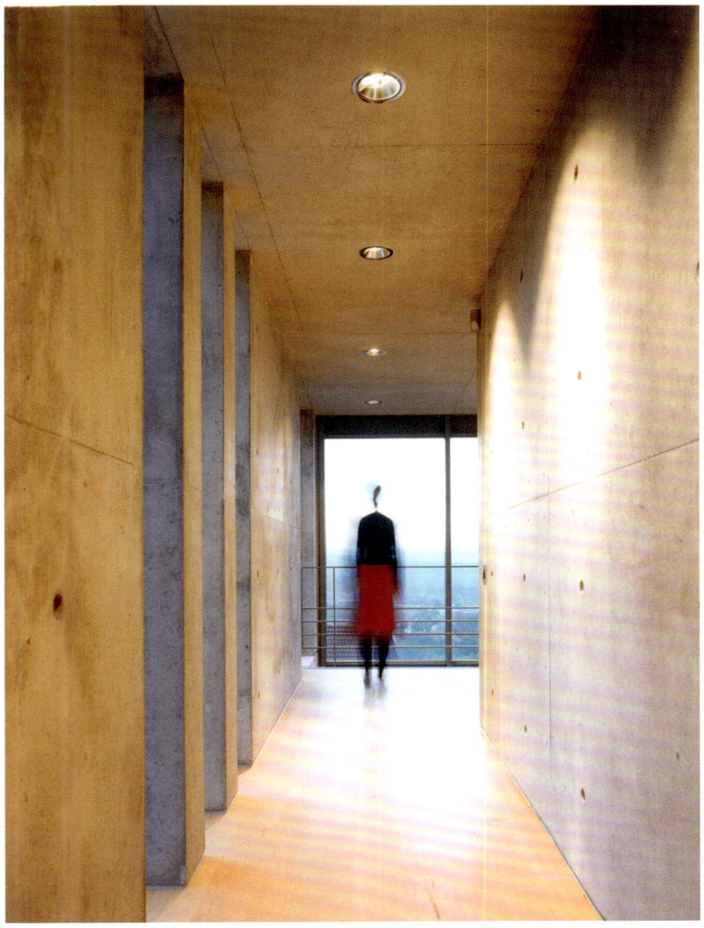
7

	LED	Fluorescent lamps	Metal-halide lamps
Max. luminous efficacy (lm/W)	80–100	94	114
Colour rendering index R_a	80–90	80–89	81–90
Service life t (h)	50,000	5,000–15,000	12,000–20,000
Dimmer behaviour	possible	possible	not possible
Start-up behaviour	possible	possible	not possible

8

Luminaires	Lighting types
Recessed ceiling lights	• General illumination • Wall washing • Accentuation
Spotlights	• Accentuation • Flooding • Wall washing • Projection
Light structures	• General illumination • Wall washing
Recessed floor lights	• Wall washing • Accentuation • Grazing light • Orientation illumination

9

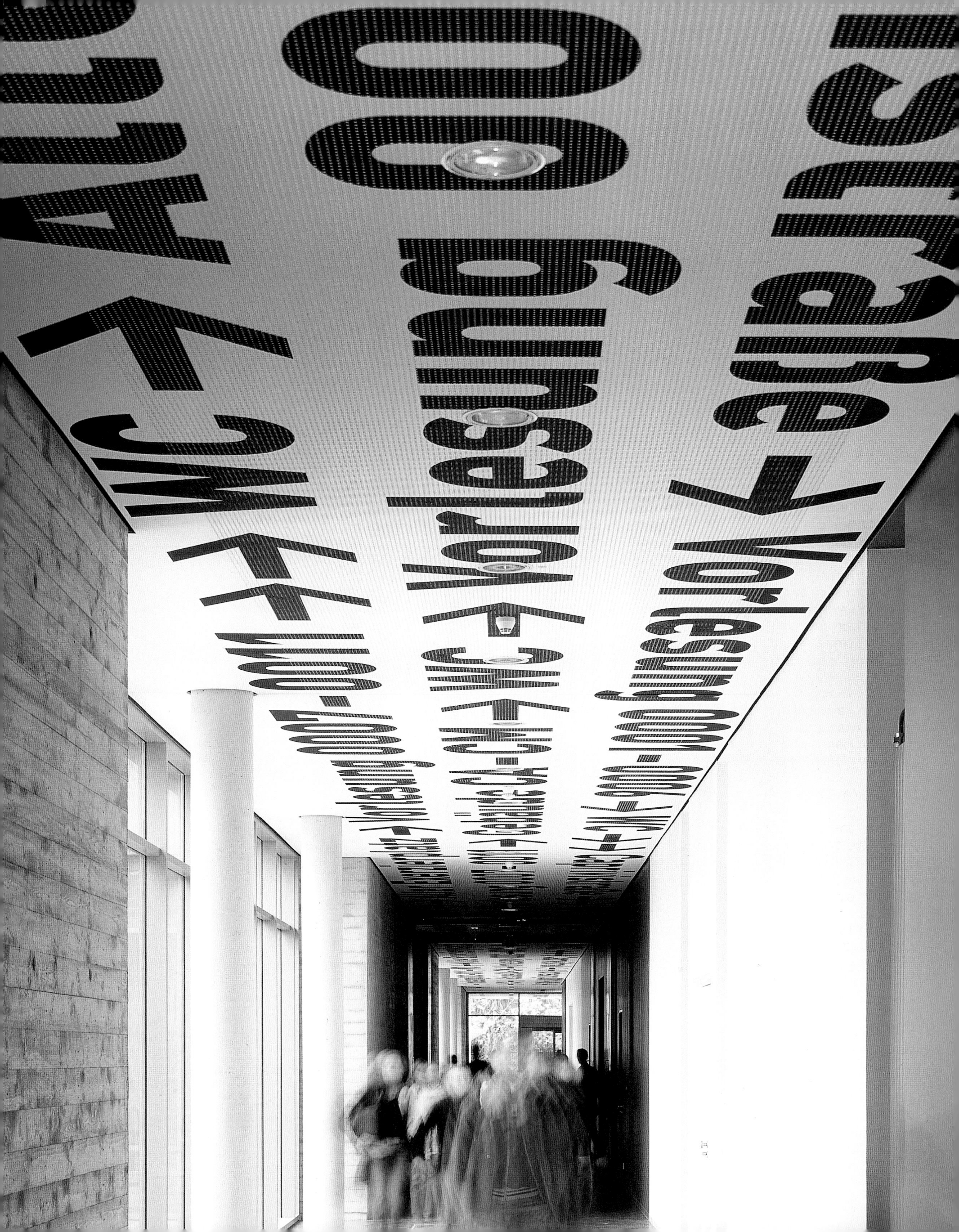

Signage – effective orientation

Jimmy Schmid

Signage combines different design specialisms with the communicative intent to give a location a distinctive visual profile. Intelligent and sustainable orientation and information systems not only guide and inform, they also allow the user to experience the space, thus creating identification and identity. The discipline thus operates at the interface between graphic design, communication design and industrial design, theatre design, architecture, interior design and landscape architecture, as well as urban and regional planning. This interrelationship demands an interdisciplinary perspective and approach to working and exploits the necessary contrast between two- and three-dimensional viewpoints.
Projects involving signage require knowledge and skills in various fields as well as networked thinking. Because of their complexity, they demand specific competencies, both regarding management and strategic and conceptual aspects as well as in the sphere of design and implementation. Aspects of the psychology of perception and information also play a role. This interdisciplinary cooperation is all the more successful and productive if the different specialisms collaborate at an early stage and are included in the planning process. In this way, in extensive outdoor projects or complex building developments, coherent spaces are created in which people feel at ease, are happy to spend time and can find their way around.

The Picket Fence

One time there was a picket fence
with space to gaze from hence to thence.

An architect who saw this sight
approached it suddenly one night,

removed the spaces from the fence,
and built of them a residence.

The picket fence stood there dumbfounded
with pickets wholly unsurrounded,

a view so loathsome and obscene,
the Senate had to intervene.

The architect, however, flew
to Afri- or Americoo.

These lines by the German poet Christian Morgenstern (1871–1914, English translation by Max Knight) suggest that the intervening space came before the actual building, that it serves here, so to speak, as a starting point, even as a possible centre for a piece of architecture. Seen from this perspective, the intervening space, or in this case the access space, is of eminent significance and importance.
However: does this reflect today's reality? Do access spaces not in fact have a purely functional significance? Should they really be nothing more than transit zones to be crossed as quickly as possible in order to arrive at the intended destination?

Access space as a space for experience

Architecture (in this sense also always understood to embrace interior design and landscape architecture) and access spaces are inseparably connected with one another. Outside the building, we are met by an entrance courtyard or forecourt with an ambience that welcomes visitors and sets the atmosphere of the building. In the interior, entrance halls, stairwells and lifts, foyers and passageways radiate qualities which have a significant influence on the atmosphere of the architecture. Each individual determines for themselves the manner in which they negotiate these spaces, and the speed at which they do so: in this way, a supposedly transitory area can become a space in which to linger or to wait, that is to say a space for communication. "All of our experiences take place in spaces. Not just in the reality of the visible and measurable, but also in the ephemeral spaces of the imagination, memory, association. Spatial perception is a process, not a freeze-frame. Its underlying principle is montage", writes the Swiss architect Elisabeth Blum [1].
The principal function that an access space assumes influences the architectural design: the arrangement of building volumes, visual axes and spatial relationships, the materials used, the choice of colour, the lighting – all these things have an effect on the quality and the atmosphere of the architecture. The resulting "resonance" of a space affects the way in which it is used. This can be deliberately planned, and can for instance ensure that an access space becomes more than simply a transit space.
An example of this can be seen in the design by the OMA/Böll/Krabel consortium for the Ruhr Museum on the site of the Zollverein colliery in Essen, which was opened in 2010. In the former coal pit shaft which defines the ambience of the exhibition, a staircase was created which links the reception foyer with the three levels of the exhibition (fig. 3, p. 37). The material qualities of the stairs contrasts with the pit atmosphere and the severe and precise lighting on the balustrades

guides the visitor to the different exhibition areas. But not only that: the atmosphere in the stairwell also forms a contrast with the exhibition, and in this way makes it possible to separate the different themes of the exhibition which are presented on the different levels.
In contrast, the Atelier Integral Ruedi Baur conceived the identity and the signage for the 125th Assembly of the Inter-Parliamentary Union (IPU) in Bern in 2011 as a grand welcoming gesture. The basic idea behind the overall concept was to focus on the theme of the voters in the outdoor areas and on that of the parliament and parliamentarians in the interior. Floor installations on the congress site showing the population sizes of all IPU countries accompanied the visitors from the bus stop to the entrance (fig. 2). The large-format photographs which were positioned on the left and right sides so as to flank the route to the entrance like an avenue showed voters from all continents. In the congress building, the walls of the rather anonymous and cool access spaces were covered with large-format information graphics (fig. 3) depicting different statistics relating to the theme of the congress, "Parliaments": comparisons of gross domestic product with IPU votes, the first women in politics, the proportion of women in the different parliaments etc. These gave the space a positive atmosphere and encouraged visitors to stop for a while and view them.
Another possible way of using signage in space is demonstrated by the temporary design for the entrance area of the Assembleia da República in Lisbon (fig. 4) by P-06 Atelier, in which the stairs served as an information carrier. Viewed from a distance, the exhibition title "Arquitecto Ventura Terra 1866–1919" applied to the steps themselves served to attract visitors; the more closely they approached the exhibition venue, the more the title dissolved into a sort of carpet of pixels leading to the museum entrance.

Distinguishing features assist orientation
Clients and architects define the functions of the individual spaces and the resulting requirements; the same applies to access spaces. The brief may require cool objectivity, stylish elegance or staged experiences. The more complex the organisation of the space, the more difficult it is for users to orient themselves and the more support they need in order to be able to read the structure and find their way around the space. Architectural aids in the form of visual axes and spatial relationships, lighting, different material properties etc. make possible tor at least support orientation.
However, complex spatial arrangements often require an additional orientation and information system – a signage that points out directions and informs. The dialogue between architecture and signage forms the basis for user-friendly orientation aids in spatial situations. Fundamentally, there is a relationship between the quality of the architecture and the need for orientation systems – the higher the spatial quality ofthe building, the fewer signage elements are necessary.
Just as architecture and its functions can be extremely diverse, so too must signage be defined in different ways. The use of signage is always dependent on the context – it can be understated and subtle or loud and dominant. Frame conditions such as building type, layouts, user requirements and route ways must be analysed first to make sure that the signage measures are appropriate.

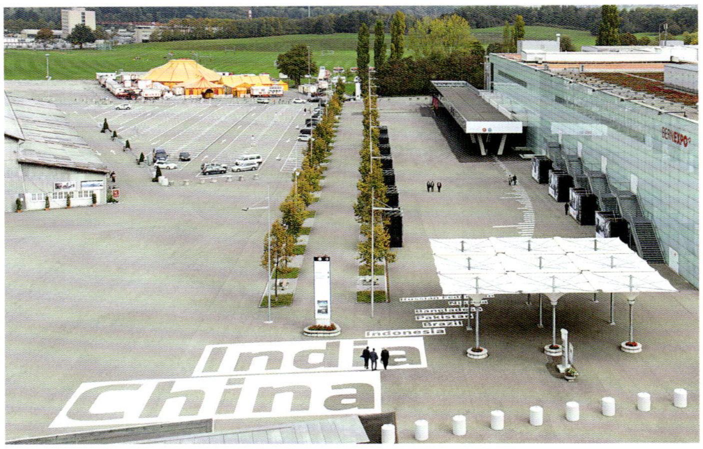

2

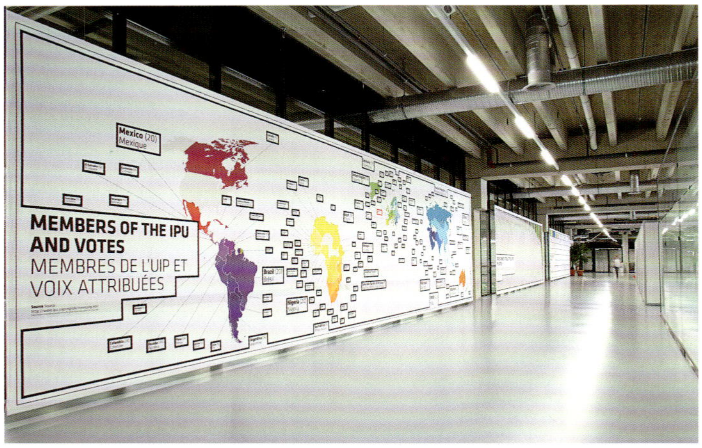

3

1 Osnabrück University of Applied Sciences (D) 2004, architecture: Jockers Architekten, signage: büro uebele visuelle kommunikation
2, 3 125th Assembly of the Inter-Parliamentary Union (IPU) at the Bernexpo Convention Centre in Bern (CH) 2011, signage and temporary appearance: Integral Ruedi Baur
4 Exhibition "Arquitecto Ventura Terra 1866–1919" in the parliament building, Lisbon (P) 2008, signage: P-06 Atelier
5, 6 Schoolhouses, Biel (CH) 2005, architecture: A + P Architektur und Planung, bauzeit architekten; signage: 2. stock süd netthoevel & gaberthüel

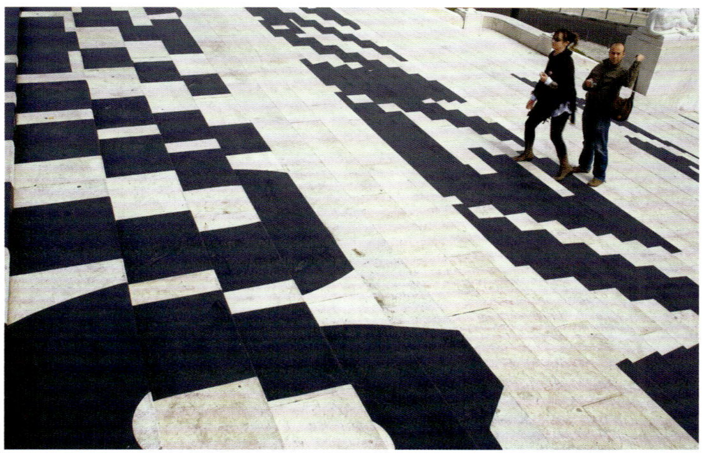
4

During the course of renovation work on the site of various school buildings in Biel, the design office 2. stock süd, netthoevel & gaberthüel also developed a signage for the Poststraße school building. In order to identify the buildings, half numbers and letters were applied directly to the facades, with clear colour differentiation assisting orientation. Each building was assigned a clear identification scheme (for example 1A yellow). The colour scheme is also continued into the interior of the different school buildings. On the projecting roof dating from the 1960s, the letters of the text frieze – an unbroken sequence of words with 26 terms containing all the letters of the alphabet – were blasted out of the weathered patina. The row of words will disappear over the course of time through the effects of weathering (fig. 5). The opposite approach was taken in creating the distinguishing features on the newly built facade. The numbers 442 and 446 sandblasted into the bare concrete state the schoolhouse's altitude above sea level. Rainwater collects on the slightly concave flat roof and then trickles down over the facade. Through the effects of weathering, the numbers will stand out more and more distinctly over the years, as moss and lichen colonise the textured surface (fig. 6).

Communication in space

Not only should the function of the architecture be defined during the planning, the relationship between the signage and the building and the nature of the dialogue between the orientation and information system and the built exterior or interior should also be defined. Two different approaches to this question can be discerned: on the one hand, the signage can be subordinated to the architecture, in which case the orientation and information system represents an integral part of a spatial situation. On the other hand, it is also possible to regard signage as making an independent contribution to the architecture and allow it to function autonomously, like a second level. Whichever approach is taken, it is essential to ensure at the planning stage that a location-specific system of signage which creates identity is developed. The visual perception of space and its signage also influences the perception of the company, institution or public facility occupying the building.

As a rule, a signage system is developed from an existing corporate design: from a two-dimensional image, three-dimensional spatial elements are developed which create a three-dimensional corporate identity. However, there are also projects (including competitions) in which the signage concept and corporate design are explicitly developed jointly.

The international competition for the town of Yverdon-les-Bains is a good example of this: LM communiquer and JJR Architecture developed a location-specific and identity-defining modular urban signage system. This is intended to help the local authorities communicate externally with tourists, hotel guests, visitors etc. and internally with residents. In the design, which has not yet been implemented, the road network and restored canal system form a coordinate network covering the entire city. All traffic nodes and bridges are transformed into multi-sensory experiences, using different materials, plants, water features, light etc. Each canal is given its own identity, helping visitors orient themselves within the city (fig. 7, p. 44).

Sender – medium – receiver

Signage concepts are based on the classic communication model which states that there is always a sender and a receiver. An orientation and information system transmits signs and signals which the receiver decodes and understands. It is therefore essential to take into account the user's ability to receive, i.e. signage is dependent on the target public.

The different elements of the signage represent the mediating medium, so to speak, between the sender, who wishes to communicate a message, and the receiver, who needs to understand this in order to be able to orient themselves within the interior or external space. The information carriers can be formulated in different ways depending on the context: using colours and shapes, pictograms, text and images on a wide variety of materials.

One example of this is the orientation system for the Novartis Campus in Basel developed and realised by Mifflin-Schmid Design: 26 letters and 26 personalities from the history of science are used to name the streets. The names of important scientists, inlaid in the kerbstones in the campus font designed by Bruno Maag, indicate the street names and commemorate earlier researchers, the inscriptions on the park benches provide a brief resumé of their scientific contributions and achievements (fig. 9, p. 44).

One typographical implementation is the AB Eiffel font developed by André Baldinger, which represents an interpretation of the Eiffel Tower. This font was the basis for the winning, but unrealised, competition entry by Intégral Ruedi Baur Paris for a signage and information system for the Eiffel tower. The font, with its different styles, emphasises how the

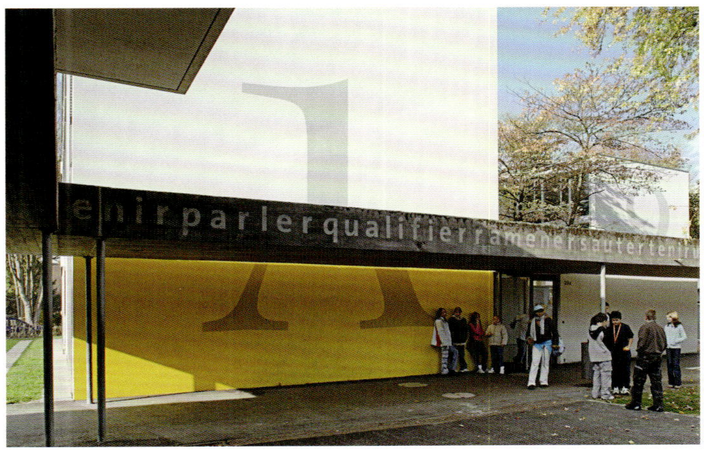

5

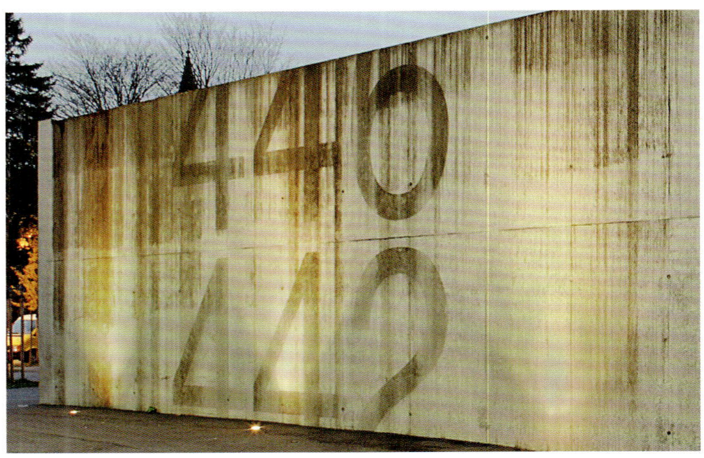

6

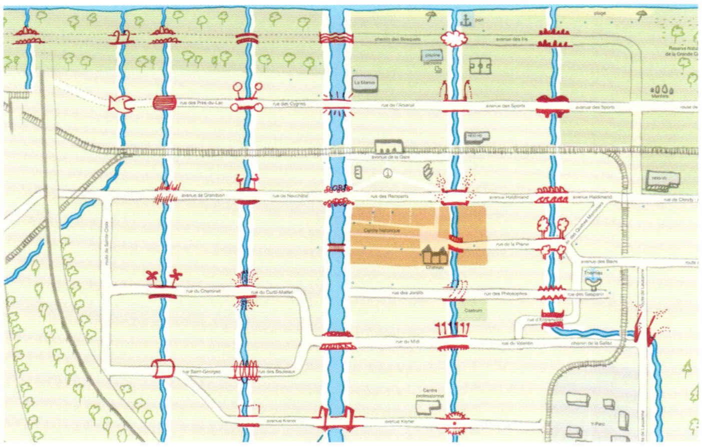

7

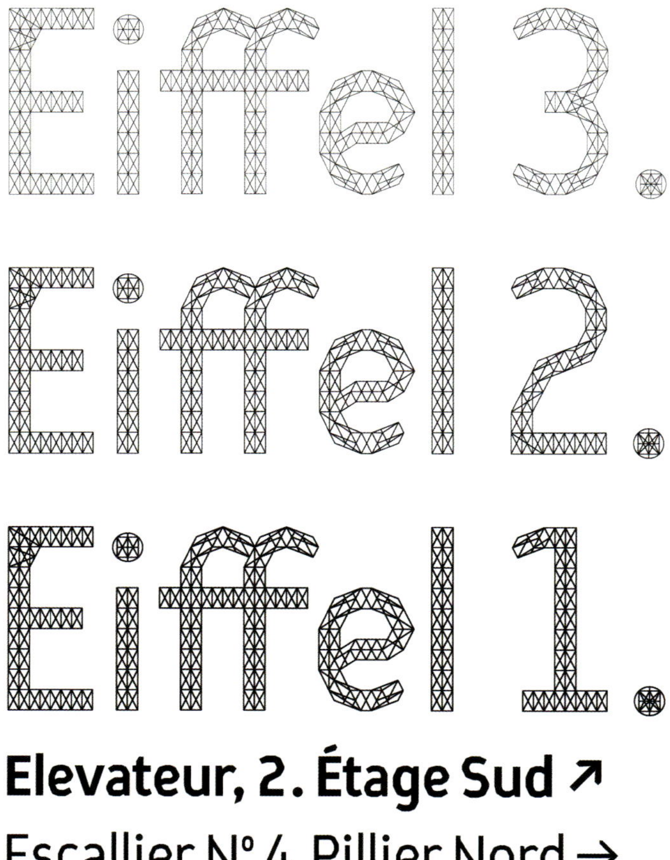

8

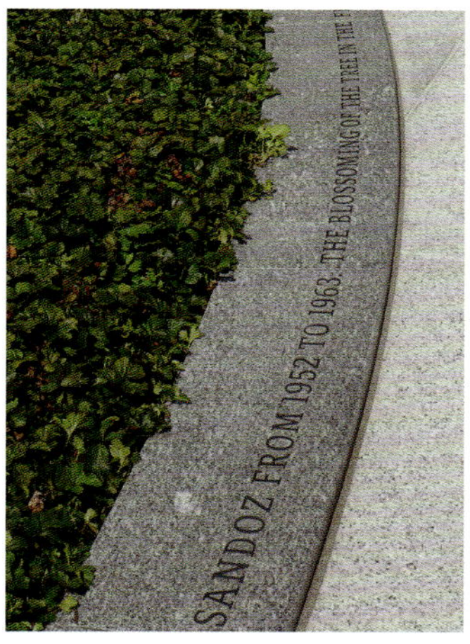

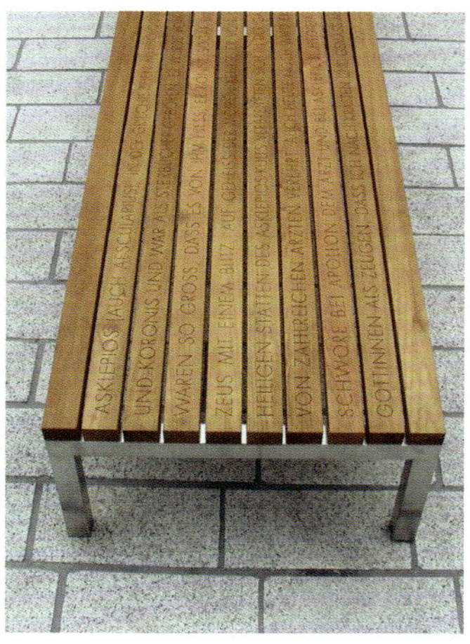

9

atmosphere on the different levels is perceived by the visitor: the higher up the tower one ascends, the more delicate the metal structure, the more the angle of view over the city opens up (fig. 8).

Information hierarchy and integrated design

The information hierarchy takes into account the fundamental principles of orientation in space: the processing of information and the processing capacity of the recipients, the use of appropriate symbols and elements, etc. Hierarchisation helps structure the information and make it understandable. For this purpose it must be organised both functionally and in terms of content. The organisational principle for signage developed by the Swiss graphic artist Theo Ballmer can serve as a basis for this. According to this principle, orientation and guidance function according to the following pattern: persons seeking information orient themselves at entry and linking points and then follow the directional instructions until they reach the destination confirmation. Following reorientation, the same process is repeated again. These three types of information – orientation, directional instruction and destination confirmation – are implemented in the network of routes which covers all destinations according to the classification and hierarchisation of the information.

This complex and also multilayered density of information rerquires an orientation system that must be planned by an interdisciplinary team from an early stage. Unfortunately, at present there are (still) too few invitations to tender and competitions in which the planning team is explicitly expected to include a signage designer. However, this would in fact open up major opportunities: a signage perspective assists the conception and planning of the space allocation plan, considerations as to which architectural and/or signage elements and distinguishing features might be conducive to orientation can be taken on board at an early stage. This would allow the nature of the materials used in the architecture, the elements of the orientation system and the fixtures and fittings to be optimally matched to one another – both indoors and outdoors. This prevents (subsequently developed) signage in the built space from being perceived as an imposed alien element. In the case of new builds, changes of use and renovations which were realised with the signage being integrated at an early stage it s noticeable that the space is perceived as a whole, as a unit, which facilitates orientation and encourages a feeling of well-being.

Polarity I: functionality versus narration

The visual design of a signage system can be absolutely functional in orientation and communicate a minimum of information in a precise and pragmatic way. One example of this is the concept for the Osnabrück University of Applied Sciences realised by büro uebele visuelle kommunikation. Here, all the information is applied to the ceilings, so that users are guided to the desired destination by a "sky" of black letters and numbers. The continual repetition of the information throughout the building helps prevent loss of orientation (fig. 1, p. 40).

In contrast, a narrative system tends to make it possible to experience orientation in a playful and narrative way. For example, L2M3 Kommunikationsdesign developed a communication concept for the newly built Adidas factory outlet in Herzogenaurach which makes use of different graphic elements. Road surfacings in green or orange clearly differ-

entiate the two parking levels from one another in terms of colour. Historic moves from different sports are reproduced on the floor: football on a green background, basketball on an orange background. The bright colour scheme helps visitors orient themselves, the illustrations are both informative and entertaining (fig. 11).
L2M3 Kommunikationsdesign are also responsible for the 22 metre high tree in the main stairwell of Kreissparkasse Tübingen. From this access space, leaves and animals spread out over the walls of the different floors – each is assigned its own fauna and flora. The concept provides basic orientation inside the building that is mainly used by staff, but above all it creates identity and gives the building a graphical face (fig. 10).
The important thing is always to define during the planning in which direction the signage which is to be developed will be positioned with regard to the functional and narrative polarities.

Polarity II: integration versus contrast
Another interesting and exciting question is always the extent to which the signage is going to be integrated into the architecture (for example because of conservation considerations) or whether it needs to appear as a contrasting element in relation to the environment. Naturally, a system can also exhibit both characteristics, as when two different levels of information are to be communicated in different ways.
Intégral Ruedi Baur Paris conceived the temporary signage for the Parc de la Villette and the design of the Place de la Fontaine aux Lions during the construction work being carried out on the Grande Halle in Paris as a deliberate contrast. The large letters consisting of boards nailed together were more effective than many sophisticated signs would have been. The poster-like character of these elements ensured that the new routes could be located without any problem (fig. 18, p. 47).
In contrast, the floor numberings in the new building for the University and the University of Teacher Eduaction in Lucerne developed by the agency Hi Visuelle Gestaltung are subtly integrated into the different wall structures. Although the numbers, ground in different textures, have room-filling dimensions, they integrate harmonically into the different spatial situations (fig. 12, p. 46).
Both concepts are combined in the name logo of the City Garden four-star business hotel in Zug. The central theme of the architecture is reflection. Bringolf Irion Vögeli realised a signage concept on the facade which integrates with the background by day but stands out in contrast at night (fig. 16). The sculptural facade has been adapted for the lettering by its zigzag projections and choice of materials, and implemented in three dimensions – the letters extrude from the facade. By daylight, the lettering elements appear as sculptural objects, whereas in the darkness the illuminated fronts of the letters are visible from far off.

Polarity III: static versus dynamic systems
The need to update information varies greatly. There are places in which up-to-date information is very important and which therefore require a dynamic system. In other situations, information retains its validity for years. Often, a static signage system is supplemented by a dynamic system in order to achieve optimal flexibility in the communication of orientation and information.

7 Competition for a signage system for the city of Yverdon-les-Bains (CH) 2009, design: LM communiquer, JRR Architecture
8 Competition for an information system for the Eiffel Tower, Paris (F) 2002, design: André Baldinger
9 Novartis Campus, Basel (CH) 2007, signage: Kaspar Schmid, Mifflin-Schmid Design
10 Kreissparkasse Tübingen (D) 2006, architecture: Auer + Weber + Assoziierte; signage: L2M3 Kommunikationsdesign
11 adidas factory outlet, Herzogenaurach (D) 2003, architecture: Wulf Architekten, signage: L2M3 Kommunikationsdesign

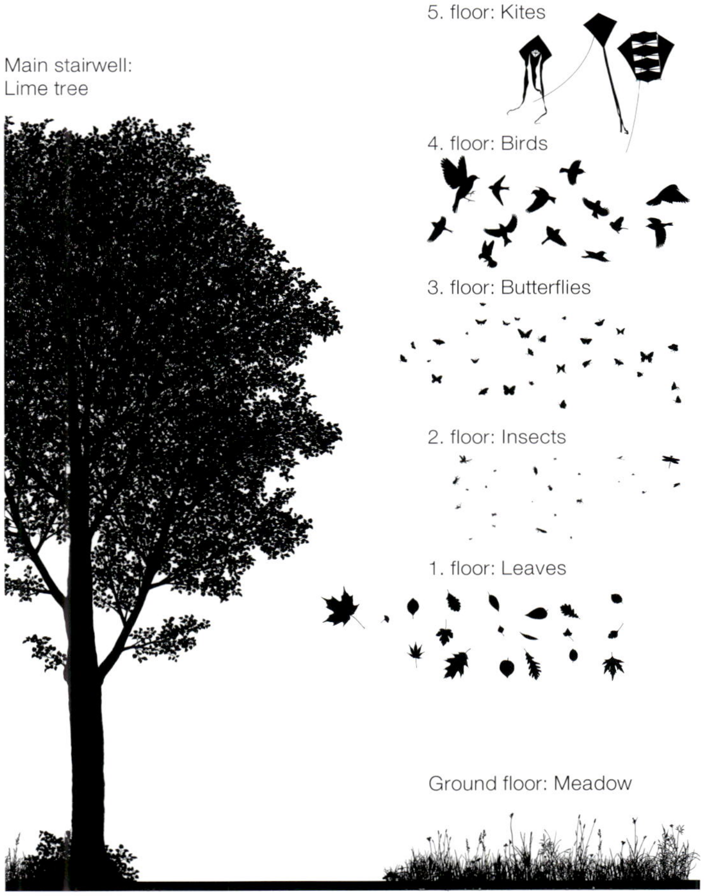

10

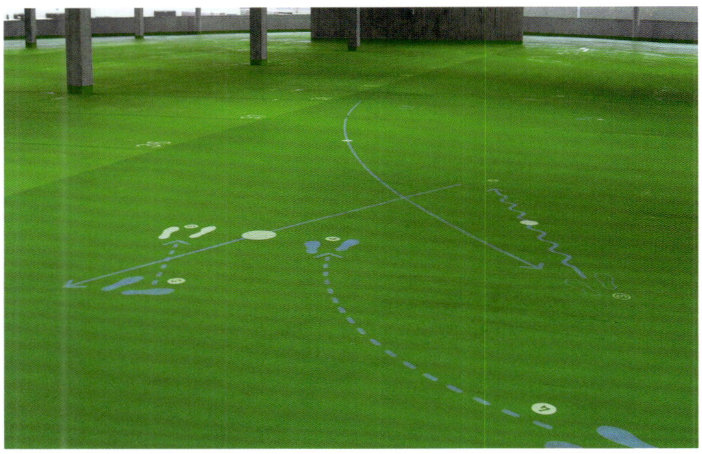
11

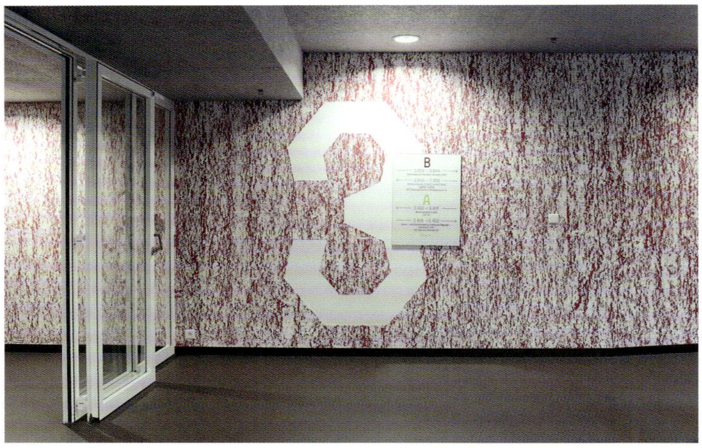

12

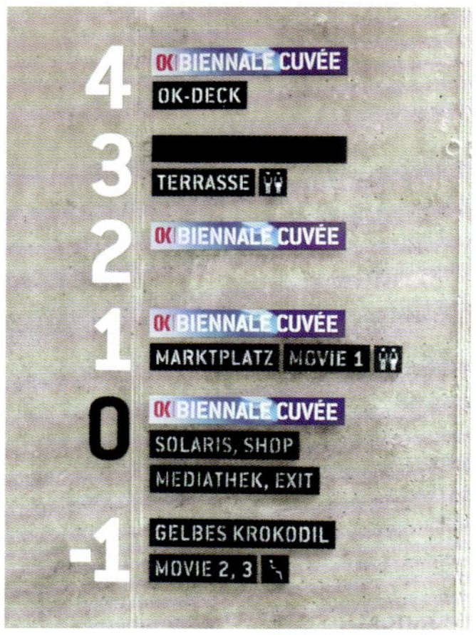

13

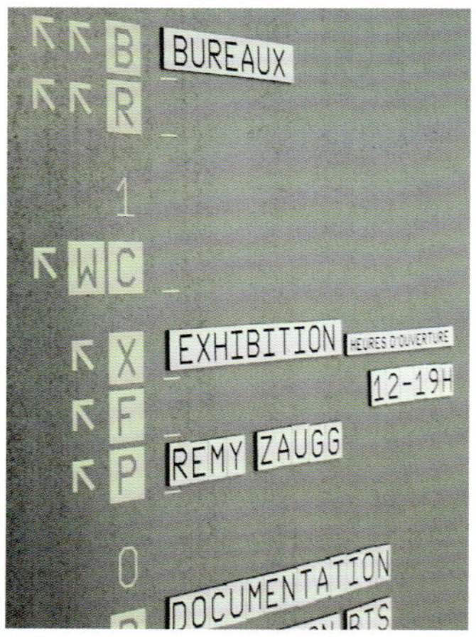

14

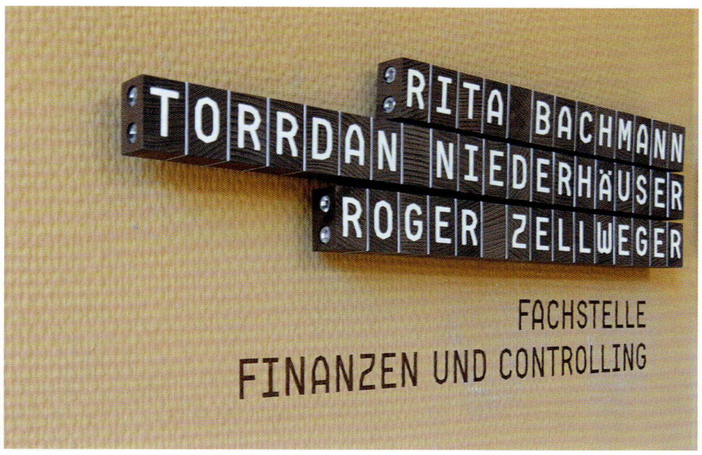

15

That the dynamic communication of information need not necessarily be electronic or digital is shown by two projects in which the contents can be quickly added to and updated using individual word stencils or flexible magnetic letters.
For the Offenes Kulturhaus Oberösterreich, a new cultural quarter containing galleries, restaurants, offices and a large square in Linz's historical city centre, bauer concept & gestaltung developed a flexible signage design for indoor and outdoor spaces (fig. 13). This allows variable information about the temporary exhibitions and events to be integrated quickly and seamlessly in the overall concept.
The agency re-p realised a concept for the FRAC (Fonds régional d'art contemporain) Lorraine in Metz which informs visitors of current events using changeable magnetic letters among other things (fig. 14).
Hi Visuelle Gestaltung developed a combination of static and dynamic systems for Werkheim Uster (fig. 15). Fixed information such as floor numberings, departments etc. is applied directly to the walls with paint or using glued-on 3D elements. In contrast, the changing information (employees, residents) is formed of individual letters. These elements can be manufactured in the in-house workshop.
In a cooperation project between the University of the Arts in Bern and the Tongji University in Shanghai (College of Design and Innovation and College of Architecture and Urban Planning), a signage workshop was held in Shanghai in the autumn of 2010. The participants worked in groups with Chinese and Swiss students to develop various bilingual – Chinese and English – orientation and information systems for the Tongji University Siping Lu campus in Shanghai. During the two-week workshop, the students jointly formulated and visualised analyses and conceptual approaches which the Swiss students then developed in more detail in Bern.
An international project team developed a campus map as a mobile application for tablets and smartphones (fig. 17). The technical possibilities offered by mobile communications were explored without neglecting the emotional qualities. In addition to the map function, the users can access lots of additional information, such as the history of the campus, information on its architecture and lecture timetables. This application was one of several measures embedded in a comprehensive overall concept.

Reference:
[1] Elisabeth Blum: Atmosphäre. Hypothesen zum Prozess der räumlichen Wahrnehmung. Baden 2010

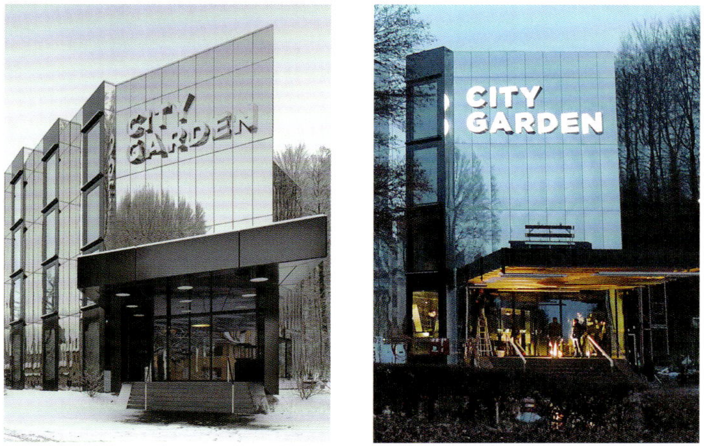

16

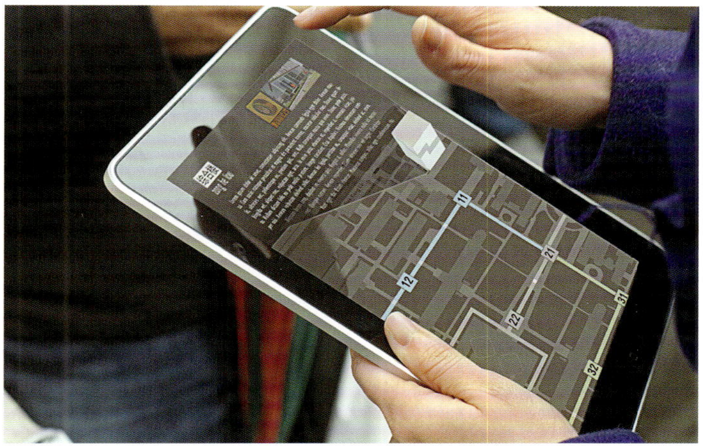

17

12 University and University of Teacher Education Lucerne (CH) 2011, architecture: Enzmann Fischer, signage: Hi Visuelle Gestaltung
13 Offenes Kulturhaus Oberösterreich, Linz (A) 1998, extension in 2007, architecture: Riepl Riepl Architekten, signage: bauer – concept & design
14 FRAC Lorraine, Metz (F) 2004, architecture: Bodin & Associés Architectes, signage: re-p.org, Maia Gusberti, Nik Thoenen
15 Werkheim Uster (CH) 2011, architecture: Hodel Architekten, signage: Hi Visuelle Gestaltung
16 Facade during day and night, City Garden Hotel, Zug (CH) 2009, architecture: EM2N, signage: Bringolf Irion Vögeli
17 Map of the campus of Tongji University as a mobile app for smartphones and tablets, Shanghai (CN) 2010, project team: Zhou Zhe Yuan, Huang Shuting, Heinz Anderhalden, Pascal Kirchhofer, Carmen Meyenhofer
18 Temporary signage for the Parc de la Villette and the Place de la Fontaine aux Lions during the building work on the Grande Halle, Paris (F) 2005, signage: Intégral Ruedi Baur Paris

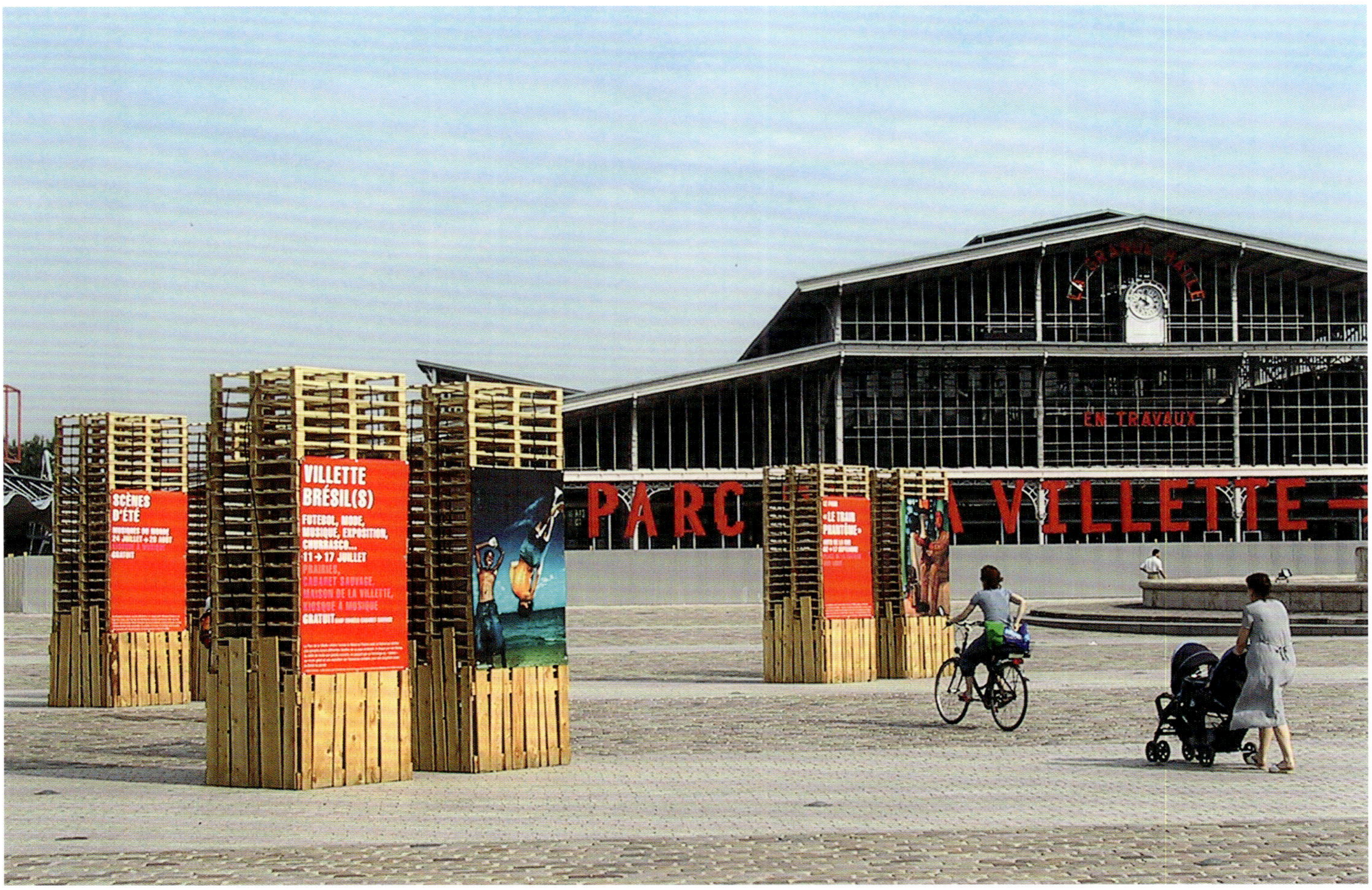

18

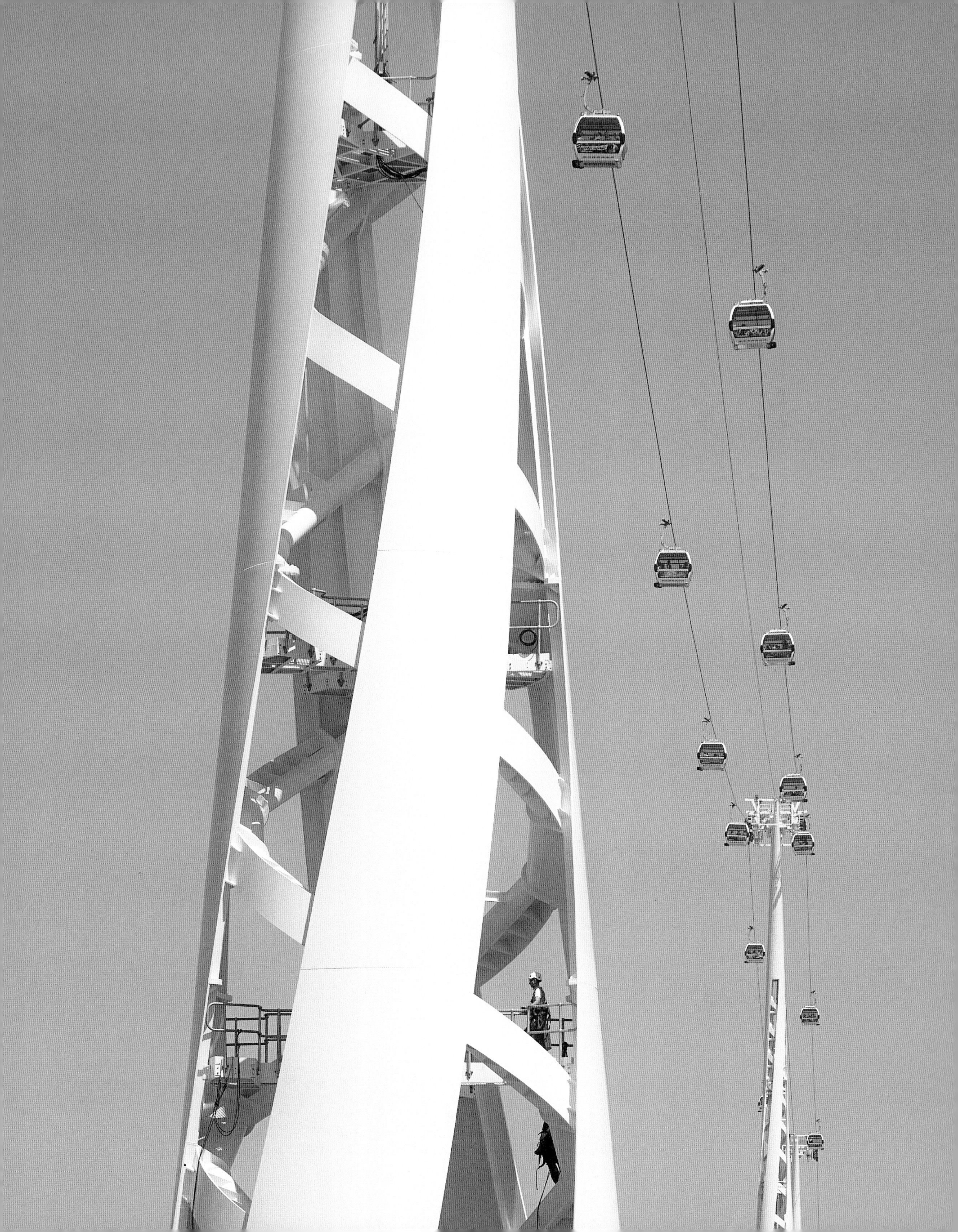

Cable cars in the city – large-scale aerial urban circulation systems

Oliver Herwig

Everybody may be talking about electric vehicles and e-mobility, but the real mobility revolution is in fact taking place a distance of some metres above the ground. Urban cable cars could change the appearance of our cities. These inner city circulation systems capable of serving extensive metropolitan areas know neither traffic jams nor bad weather. They function like a conveyor belt – slow but steady – and are therefore much faster on average than most methods of transportation available in the city. From a technical point of view, there are two types of systems relying on cable haulage. The first type includes ground-based cable cars, also called cable railways (such as the famous cable cars in San Francisco). These have an infrastructure on the ground similar to trams or streetcars, which however carry their own propulsion equipment (the electric motors at least). The second type refers to aerial cable cars, also known as areal tramways. Both combine a high degree of flexibility with very good environmental figures. This article is concerned with the aerial routes offering urban planners and architects a new type of freedom.

Caracas, London, Medellín, Singapore or Vancouver – urban cable cars not only show how topographic obstacles such as mountains and rivers can be overcome elegantly with a reasonable amount of technical effort, but also demonstrate that there may be an alternative to the conventional motor vehicle-based transport infrastructure in cities in the future. Although the typical images linking gondola lifts to Alpine regions and skiers queueing to get in are still in our minds, things have long changed in reality. Urban cable cars compete with other forms of local public transport, replacing buses and trams – and in an extremely environmentally friendly way at that. A study by the strategy consulting firm ClimatePartner Austria conducted in 2009 showed that, at a capacity utilisation of 50 % and above, the continuous conveyor system causes the lowest emission of CO_2 out of all the means of travel considered in the study [1]. Modern cable car cabins transport up to 8,000 passengers per hour and demonstrate the advantages of "air traffic": little infrastructure (pylons, stations) and big flexibility. Urban cable cars are being increasingly integrated in the infrastructure of cities. What was once an isolated solution is turning into a normal method of transportation. Cable car gondolas are conquering cities and spreading from the so-called Third World – where they are already solving infrastructural problems in several high population density areas – to our global metropolises.

Garden shows and major sporting events as catalysts

Big events often act as catalysts of change, as pacesetters for new developments. They are subject to their own special regulations and these ensure that new connections and means of transportation can practically appear overnight. Instead of waiting 10–15 years for a tram line or an interurban railway track, fast planning is of the essence during such events. And presto, suddenly even more unconventional transport projects have a chance, as illustrated by the cable car systems at the World Exhibitions in Lisbon (1998) and Hanover (2000), the Federal Garden Shows in Munich (2005) and Koblenz (2011), or the Olympic Games in London (2012).

The Cologne Rhine cableway has been crossing high above the Rhine since April 1957: 16 million passengers have been transported between Zoo and Rheinpark (Rhine Park) since then without an accident in Cologne's safest means of transportation. The cable car wasn't originally intended to be a regular mode of transport, but rather an attraction for the Federal Garden Show held on the right side of the Rhine, in today's Rhine Park. The service created for about 1.5 million Deutschmark was so popular that the people of Cologne didn't want to miss it after the Garden Show. The cableway was however initially closed down and dismantled in 1963, because its support on the right side of the Rhine stood on ground needed for the new Zoo Bridge. After it was ensured that the hovering cable cars didn't irritate car drivers too much, the 50.3-metre-high pylon on the right side of the Rhine was moved southwards and the service was also extended to include another station further inside the Rhine Park, requiring the addition of another support, this time only 34 m in height. The new Rhine cableway re-opened in August 1966. Its 41 four-seat cabins can transport up to 1,600 passengers per hour. The distance of 935 m is covered within a

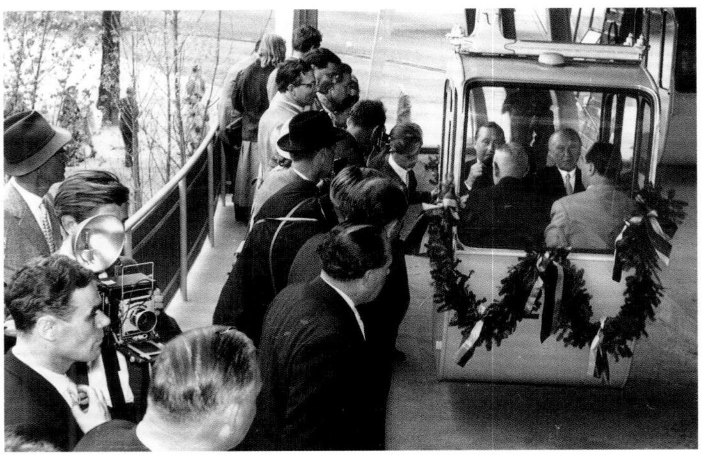

2

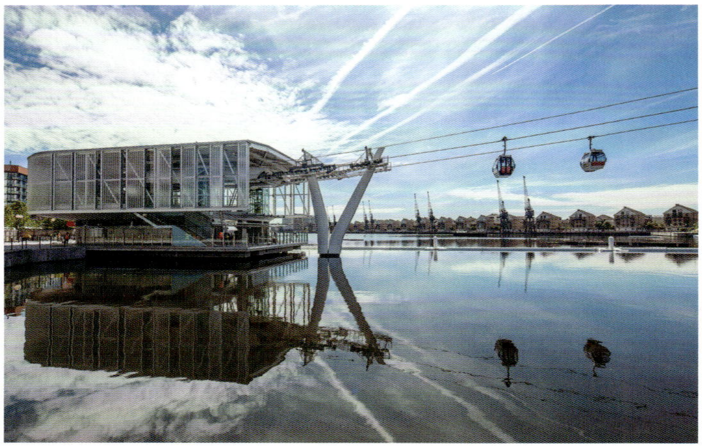
3

1 Thames cable car, London (GB) 2012
2 Inauguration of the Rhine cableway, Cologne (D) 1957
3 Royal Docks terminal of Thames cable car, London (GB) 2012, Wilkinson Eyre Architects
4–6 Metrocable, Caracas (YV) 2010, Urban-Think Tank

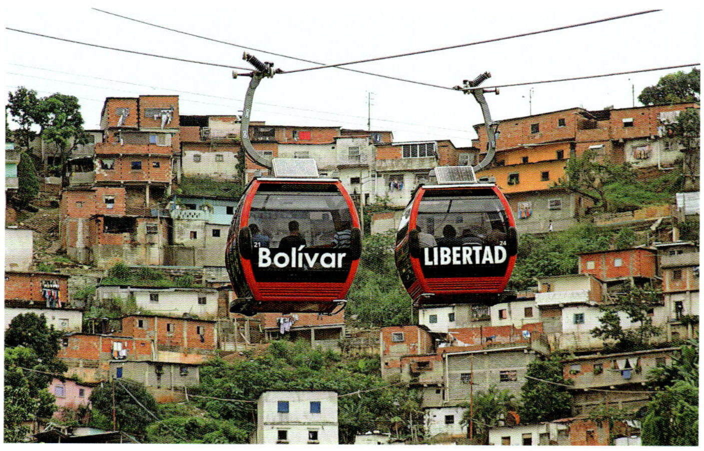

4

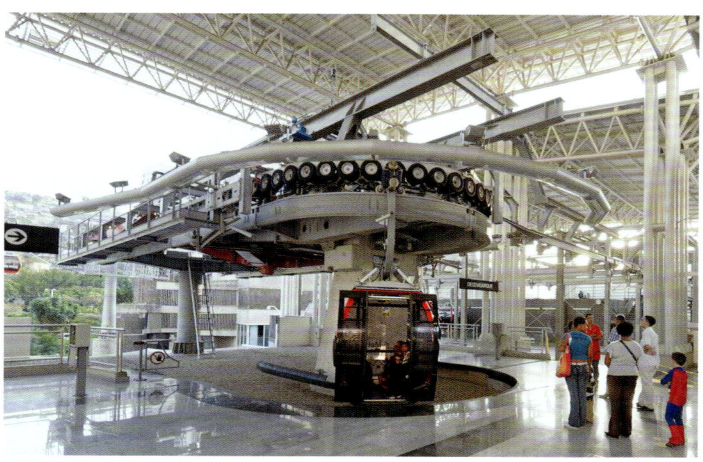
5

travelling time of six minutes, which corresponds to an average speed of about 10 km/h.
When Federal President Theodor Heuss and Federal Chancellor Konrad Adenauer (former Lord Mayor of Cologne) squeezed themselves in one of the tiny gondolas (fig. 2, p. 49), the Rhine cableway service was still the only aerial cableway over a river in the whole of Europe. Since 2011, the Rhine can also be crossed via cable car in Koblenz, in a 12-million-euro construction completed within just 14 months for the Federal Garden Show.
London's most modern means of public transport, the Thames cable car, was ready in time for the Olympic Games in 2012. During the Games, the cable car service, named Emirates Air Line after the main sponsor, linked venues like the O_2 Arena in Greenwich and the ExCel Exhibition Centre in the Royal Docks in East London. Neither bridges nor tunnels had to be constructed, and towers and stations only had to be built in Greenwich and the Docklands – yet the system cost approximately 60 million pounds. The 34 cable car cabins designed to hold ten persons and two bicycles each, can convey up to 2,500 passengers per hour, replacing about 30 buses (fig. 1 and 3). The sober efficiency signalised by the industrial glass cladding of the gondolas is an outward demonstration of the technology contained inside. Even though the final cost was more than double the original budget of 25 million pounds, the cable car service was a welcome relief for London's chronically overloaded traffic system, especially in anticipation of the millions of additional visitors heading for the Olympic Games in buses and tubes. On 28 June 2012, the Mayor of London, Boris Johnson, went on the first urban cable car ride (also referred to as "flight") in the United Kingdom, hovering 1,103 m across the Thames. Although over half a century lies between the cable car services established in Cologne and London, the task they had to fulfil is similar: how could a river be crossed as quickly and as simply as possible, and how could it be ensured that people got to events staged on either side? Rhine and Thames are enclosed by quay walls and quayside express roads – ruling out both pedestrian bridges and lots of ferries going back and forth as traffic handling options. The cable car seemed like a suitable solution – and people love it. It goes without saying that the gondolas have become more comfortable in the course of time, and larger, but the principle is still the same: a safe, barrier-free means of transport, that can be planned and realised quickly. Odd then that despite the great popularity of the cable car service in Cologne, it only started to operate in the black in 2004.

Between favelas and modern urban development

Bright red, with roof-top solar cells and single word imperatives adorning the metal exterior: Amor (Love), Participación (Participation), Libertad (Liberty). The cabins of the Caracas Metrocable are more than just a part of an infrastructure project – they are a symbol of change in the Venezuelan capital (fig. 4). Connecting the informal settlements located on the terraced slopes and the infrastructure of the officially planned city, the cable cars float down towards the valley via five stations suspended on pylons of steel – or, looking the other way round, back up to the poorer districts (fig. 6). And that's exactly what it's all about: a paradigm shift in urban planning and transport infrastructure. Venezuela, a country big in oil and motor vehicles, goes for a mode of transport that is not driven by a combustion engine.

Caracas, an oil industry city that experienced a veritable building boom in the 1950s, has two faces: skyscrapers and representative constructions alternate with settlements in the hills – high density housing of former migrant workers who stayed, without central heating, water supply or sewage system. No trace of some of these so-called barrios can be found in maps of the city, with many having come to exist without paying heed to any building regulations. The few hundred metres that separated these areas from the electricity and sewage system as well as from street network and public transport facilities, not only meant a time-consuming climb up and down – those few hundred metres also symbolised the marginalisation of entire strata of the population. The people in these settlements had no voice, no representation for a long time. All this changed with a project by the Austrian-Venezuelan design practice Urban-Think Tank (U-TT).
Construction of the first cable car line commenced in April 2007 and the official opening took place in January 2010. Starting from the district of San Agustín, the service takes passengers to Parque Central station, which is 1.8 km away and linked to the city's subway network. Metrocable was built by order of the Ministry of Infrastructure according to plans by the U-TT architects. A small structural as well as design revolution lies behind this.
The founders of U-TT, Alfredo Brillembourg and Hubert Klumpner, were fascinated by structures that function completely autonomously with a minimum of energy, such as the barrios in their adopted South American home city Caracas. In order to realise their objectives – the transformation of cities and connection of informal to planned city – U-TT struck a new path. Instead of waiting for architectural competitions, they went to the barrios themselves and talked to the local representatives about the actual problems. After that, the architect social workers presented their proposed solutions to the urban administration and propagated a kind of mobilisation of the slum dwellers by means of urban cable cars. The unconventional approach turned out to be successful. At a cost of about 300 million dollars, Caracas Metrocable entered the test phase in December 2009 and is fully integrated in the metro network today. The system transports an average of 1,200 persons per hour since it opened in 2010. There is space for eight passengers to sit and two to stand in every gondola. Caracas Metrocable was designed for 3,000 passengers per hour and direction. It is the first public metropolitan cable-based transport system to use deflection bullwheels to implement up to 90° turns. Two separate lines were created in this way. Gondolas can switch from one line to the other at the middle station, i.e. either continue or go back as required. This also increases the safety of the system, because one line can carry on operating in case of an accident.
The stations are not simply cable car entry and exit points, but more like centres of city districts, thanks to the designs by Hubert Klumpner and Alfredo Brillembourg, who hold a Chair of Architecture and Urban Design at the ETH Zurich. The architecture draws on international high-tech design and calls to mind the ellipsoidal steel roofs familiar from airports (fig. 5). An appropriate association, since these cable car stations are after all places of arrival and departure for air travellers. The municipality of Caracas has in the meantime continued along this route: a second, much more extensive Metrocable facility is to be put into operation in 2013.

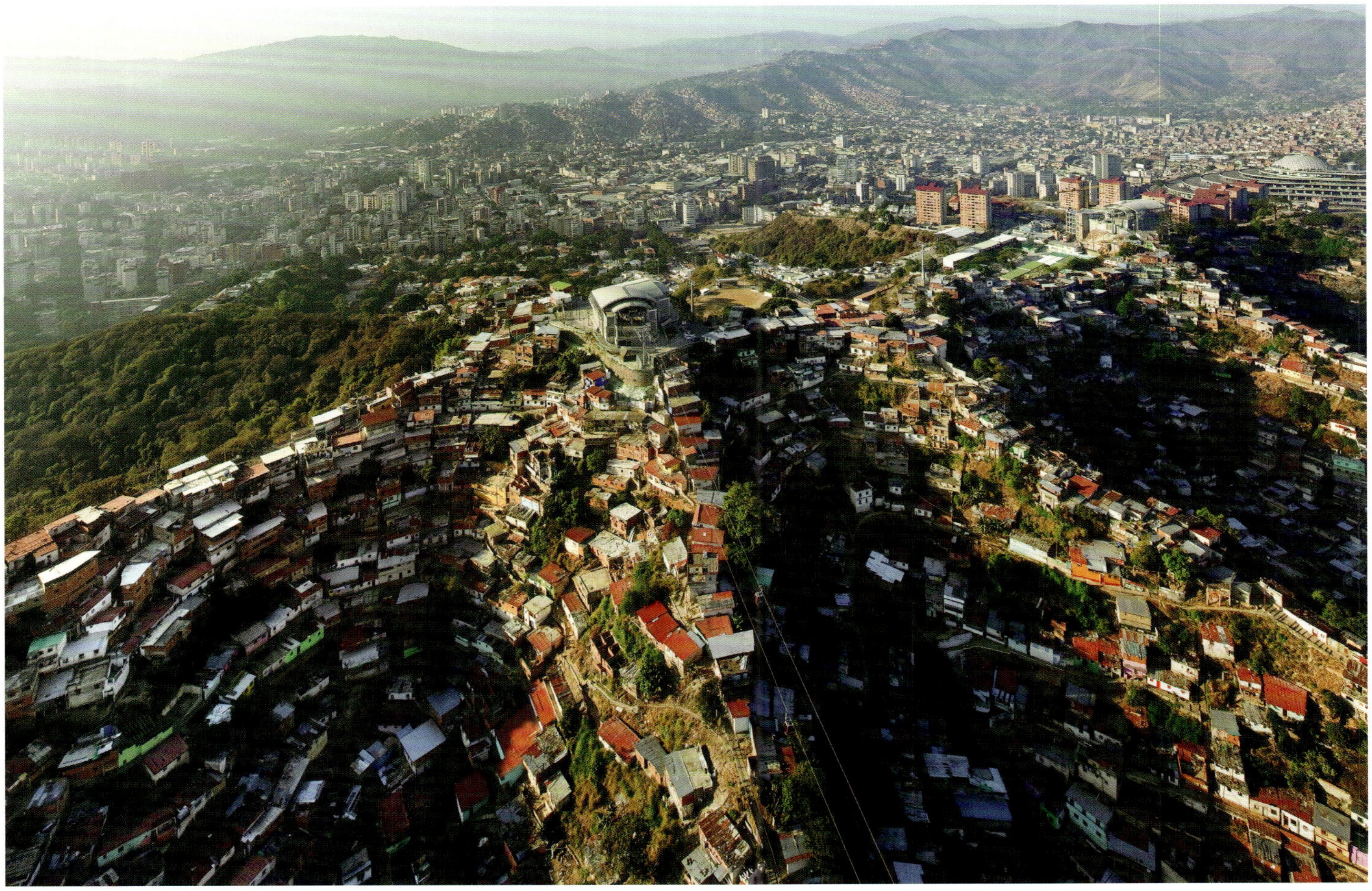
6

Cable cars as part of a South American success story? In a certain sense they are: an economical alternative for combating the segregation of city districts, for giving the excluded classes access to general society, and hence upgrading the barrios. Back in 2004, a system overcoming an altitude of almost 400 m was put into service in the Colombian city of Medellín, connecting the districts Santo Domingo and San Javier with the city in the Aburrá Valley, as well as with the urban train network (Medellín Metro), which carries about 100 million passengers per year on a total line length of 42 km (fig. 7). Before the construction of the Metrocable, the people living in the Santo Domingo barrio had to spend two-and-a-half hours to get to work. Now they can travel to the city comfortably and very safely at a speed of 16 km/h. The figures speak for a rapid extension of urban cable car systems. If a serpentine road had been built in Medellín, a third of the illegally built houses would have had to have been pulled down, reckons Günther Ecker in an article in a blog concerned with the future of mobility, called "Zukunft Mobilität" [2]. Cable cars as part of a rapid infrastructural development while keeping the existing buildings extensively intact – great food for thought for politicians and transport experts in western industrialised countries with their incredible planning lead times and (justified) objections by residents and affected individuals.
In the meantime, Medellín – with 2.2 million inhabitants the second largest metropolitan area in Colombia – finances the operation and extension of public local passenger traffic in accordance with the UN concept for climate protection by emission trading: the city saves around 20,000 tons of CO_2 per year and sells the corresponding emission certificates. An (environmental) profit for all. Since 2010, the Metrocable system in Medellín has three lines, and the cutback in exhaust fumes has gone hand in hand with a drop in criminality.
Two metropolitan areas, two success stories that show how a transportation method can be implemented and/or introduced to the outskirts of cities without destroying established settlements, as roads would have done. And that is the key issue: cable cars show how new elements can be integrated in the existing infrastructure, supplementing and extending it. The popularity is growing: after Medellín followed Manizales (2009), Caracas (2010) as well as Rio de Janeiro (2011), where considerations for further extension are in full swing – apparently there are already plans for three more lines. Obviously a concept with a future.

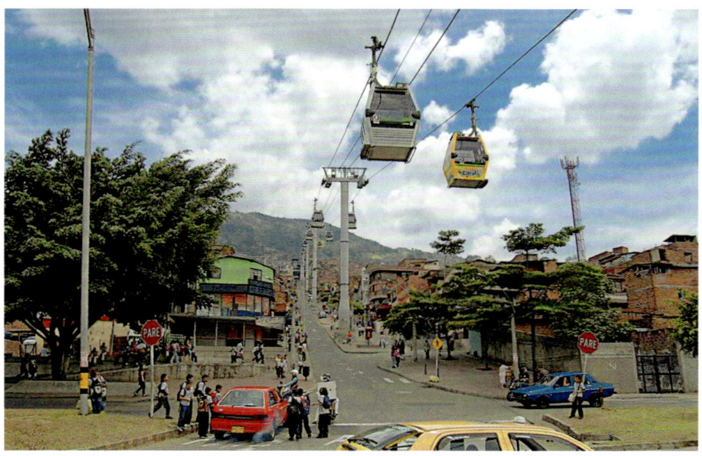

7

A perfectly normal mode of transport

When does something special become normal? Would it suffice to note that research on urban cable cars is also being conducted in the so-called First World, that experimental infrastructures and plans (like in the threshold countries) would also appear to function in these parts? Or is proof required that work on such systems is already in progress in Vancouver and Montreal? And that architecture suddenly plays an important role, or even sets a sign that can be marketed, as happened in the USA?
Portland has had such a system since January 2007. It has transported over three million passengers so far, even at wind speeds of up to 80 km/h. The Portland Aerial Tram – the second aerial tram for public local passenger transport in the USA after the Roosevelt Island Tramway built in New York City in 1976 – was designed by Sarah Graham and Marc Angélil of agps.architecture in 2003 and planned by Arup. It is one of the most elegant creations in recent cable car history. The much acclaimed means of transport connects the Oregon Health & Science University (OHSU) with the South Waterfront. An altitude difference of 145 m is bridged by 1,033 m of cable. The steel edges and angles of the expressive upper terminal (fig. 8) offer a maximum contrast to the egg-shaped twin gondolas (calling to mind Richard Buckminster Fuller's Dymaxion Car) accommodating 78 passengers. The design scope associated with the construction of cable cars is clearly demonstrated by the 60-m-high support, a buckled steel structure projecting outwards at the top with a concrete core resting on up to 16-m-deep piles (fig. 8). Architecture critics believe that the Portland Aerial Tram has the potential to become a new landmark of the city – together with Michael Graves' Portland Building.
In New York City, the Roosevelt Island Tramway mentioned earlier is being renovated for approximately 17 million dollars (fig. 10), in the Czech Republic a new aerial tramway in Ústí nad Labem connects the shopping centre Forum with the Větruše Hill, and in Bolivia, a 10-km-long cable car connection between La Paz and El Alto is under consideration. What can be concluded from that? Well, old certainties – the motor vehicle as the sole mode of transportation – are gradually becoming obsolete. The future of urban mobility will be a diverse mixture with coordinated options that complement each other. The urban cable car is one of them.

Outlook: Go green in a gondola?

What role will cable cars play in the city in future? And what is the possible benefit of this still rather unusual (at least from a European point of view) form of urban transportation? A comparison may be helpful here. An attempt to explain the air traffic boom that occurred over the past decades could refer to subsidized kerosene and the still progressive image of flying, but maybe it's simply the fact that no one had to build motorways between the continents and that air traffic hubs are limited to just a few per country. Air traffic infrastructure is reasonably manageable. The same goes for investment in cable cars.
Fans advocate the advantages of an environmentally friendly and safe (apparently: the safest) traffic system. Heiner Monheim, Professor emeritus of Spatial Development and Land-Use Planning at Trier University, is one of these. He predicts a great future for urban cable car systems and believes that they have outstanding potential. Typical classical traffic problems, such as a lacking connection, can be resolved very

quickly with simple planning systems in a minimally invasive manner, as well as requiring relatively little building effort. [3] A cable car, according to Monheim, neither demands many prerequisites nor a lot of space, simply floating above the rest of the action. It's also a winner in terms of transport capacity. The cable car service in Koblenz transports up to 7,600 passengers per hour in its 35-person cabins – a world record (fig. 9). An urban cable car may be a little slow (around 20 km/h), but the high gondola frequency and barrier-free access beat bus and rail services, on top of the fact that the route through the air doesn't have to deal with the traffic situation below. A metropolitan cable car moreover offers something that few modes of transport can claim: spectacular views.

Experts laud the comparatively low investment in lines and stations that is required, the fast construction and economical operation, as well as the favourable energy balance of cable cars. In a study comparing car, bus, train and cable car, the cable car was shown to give rise to the lowest emission of CO_2 at a utilisation of at least 50 % (248 g CO_2 on average per person and kilometre (g/pkm) for a petrol-powered motor vehicle, 38.5 g/pkm for a diesel-powered bus, 30 g/pkm for a train with electric locomotive, 27 g/pkm for a cable car) [4]. Propulsion with power from renewable energies naturally allows CO_2-neutral transport (this of course also applies to other modes of transportation).

The problems posed by increasing urbanisation can no longer be solved with European urban planning methods. If the proportion of persons living in cities increases to about 70 % within less than one generation, the express roads will be pushed to their limits. Something completely new is needed, something fast, flexible and efficient – like urban cable cars. They are the transport planning answer to the change taking place in urban centres. Cable cars connecting city centres to deprived areas result in improved social cohesion. And suddenly architecture is given a new key task – to design the visible signs of this change for everyone to see.

In 2010, there was an exhibition in the Museum of Modern Art (MoMA) in New York entitled "Small Scale, Big Change". Its subtitle "New Architectures of Social Engagement" focused on an aspect of building that has increasingly slipped into the background since the late 1970s: social housing, communal responsibility, commitement to social causes. Why should what happened in such an exemplary manner in Caracas and Medellín not be a model for Central Europe, the USA or China? The cited examples illustrate that cable cars are neither tied to mountains nor to snow. They are in fact often the answer to traffic volume problems or to deficits in the existing circulation structure. This is confirmed by the increasing appearance of cable car systems and their successful utilisation all over the world. We may be standing at a turning point of urban transport planning.

References:
[1] Climate Partner Austria: Emissionsermittlung für Seilbahnen. Study commissioned by Doppelmayr Seilbahnen GmbH. Vienna 2009
[2] Ecker, Günther: Luftseilbahnen als innerstädtische Massenverkehrsmittel. Guest article in the Blog "Zukunft Mobilität", 2011. www.zukunft-mobilitaet.net/5332/umwelt/luftseilbahnen-als-innerstaedtische-massenverkehrsmittel
[3] Monheim, Heiner: Mobilität in den Städten. Urbane Seilbahnen für alle. Lecture in Koblenz, 2012. Transcript by Martin Randelhoff. www.zukunft-mobilitaet.net/11998/konzepte/vortrag-heiner-monheim-zu-urbanen-seilbahnsystemen
[4] Seilbahn hat die Nase vorn. In: Mountain Manager 07/2009, p. 18f. www.eubuco.de/mm/archiv/pdf/mm_0709.pdf

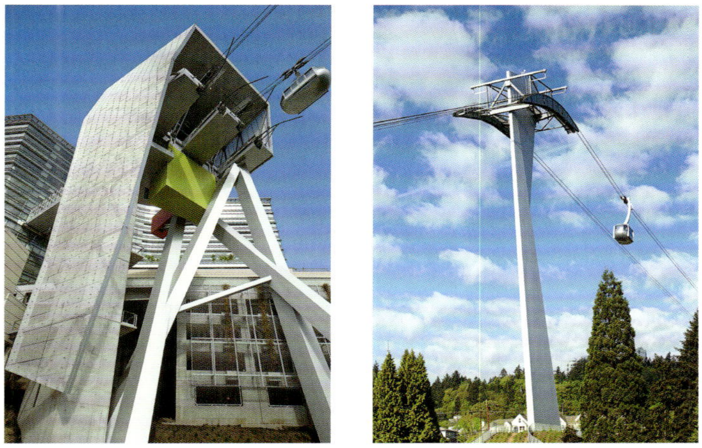
8

8 Metrocable, Medellín (CO), since 2004
9 Upper terminal and pylon, Portland Aerial Tram, Portland (USA) 2007, agps.architecture
10 Gondola of the cable car service across the Rhine, Koblenz (D) 2011
11 Roosevelt Aerial Tramway, New York (USA) 1976

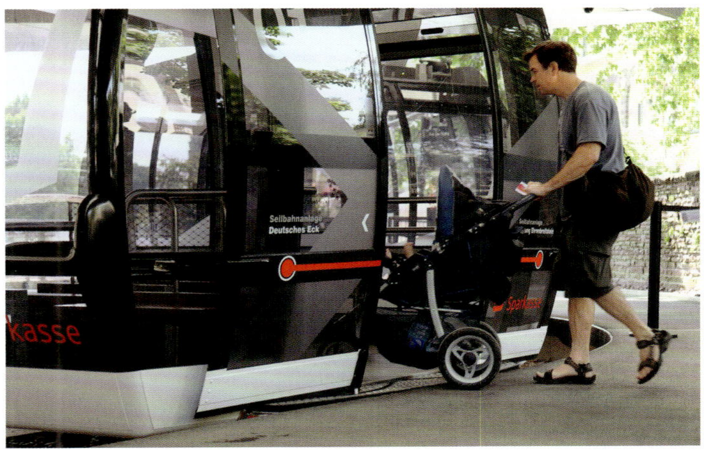
9

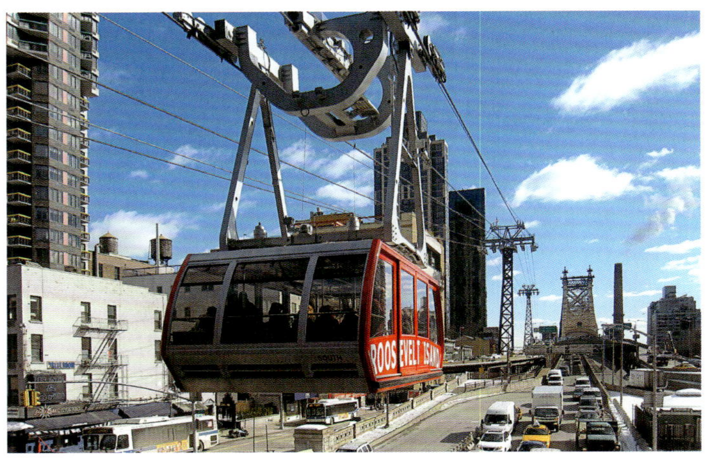
10

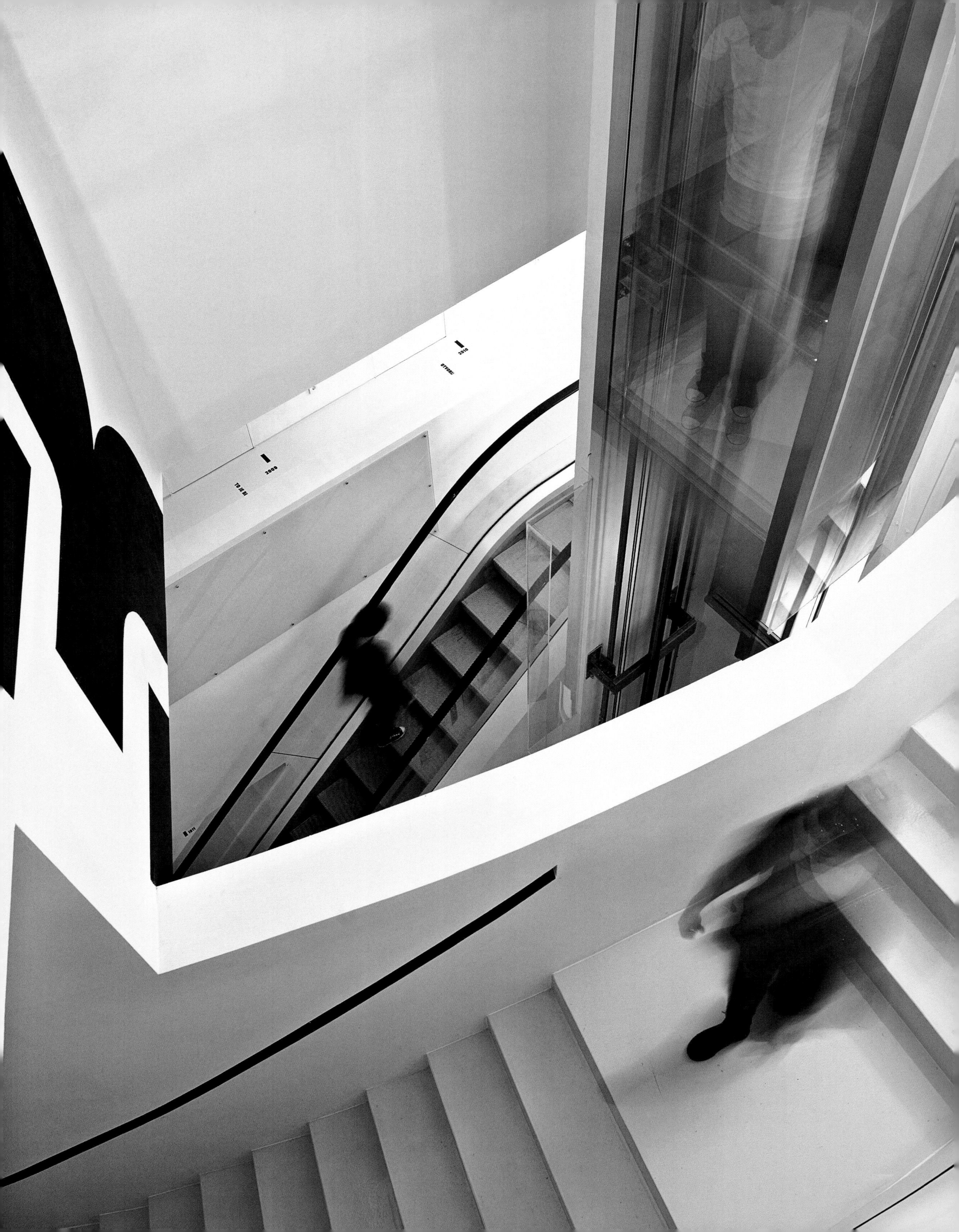

Summary of Projects

Page	Project	Architects	Use of the building	Access
56	Apartment building in Zurich	Graber Pulver Architekten, Zurich/Bern	Housing development	Central staircase, elevator
60	Residential development in Berlin	zanderroth architekten, Berlin	Housing development	Lobbies, lifts, stairs, courtyard
66	Haus am Weinberg in Stuttgart	UNStudio, Amsterdam	Single family house	Stairs, lifts
70	»House before House« in Utsunomiya	Sou Fujimoto Architects, Tokyo	Single family house	Stairs, ladders
74	Centre for senior citizens in Maienfeld	Arbeitsgemeinschaft Isler Gysel/ bhend.klammer architekten, Zurich	Home for senior citizens and care home	Central hall, stairwells, lifts, peripheral corridors with common areas
77	Apartment mound in Copenhagen	BIG – Bjarke Ingels Group, Copenhagen; JDS Architects, Copenhagen	Housing development, traffic	Sloping elevator, stairwells with lifts, connecting walkways
81	Hostel in Split	STUDIO UP, Zagreb	Hotel	Panoramic lift, escalator, stairs
86	Hotel am Domplatz in Linz	hohensinn architektur, Graz	Hotel	Foyer/atrium, access balconies
89	Children's hospital in Basel	Stump & Schibli Architekten, Basel	Hospital	Stairwells, lifts, corridors used as common areas
93	Cooper Union in New York	Morphosis Architects, Culver City	Education	Stairs in atrium, elevators
98	High school in Kopenhagen	3XN, Copenhagen	Education	Central spiral staircase, service cores, lifts
104	Municipal library in Stuttgart	Eun Young Yi, Cologne	Library	Stairs, lifts
108	Advertising agency in Tokyo	Klein Dytham architecture, Tokyo	Office	Staircase, "walkways" highlighted in white
110	Office building in Bolzano	Markus Scherer, Meran	Office	Stairs, lift
114	AachenMünchener headquarters in Aachen	kadawittfeldarchitektur, Aachen	Public space, office	Public flight of steps, foyer with sculptural staircase, internal connection via "Boulevard"
120	Restaurant and bar in Zurich	Burkhalter Sumi Architekten, Zurich	Gastronomy	Windbreak, spiral staircase
124	Pharmacy in Athens	KLab architecture, Athens/London	Retail	Ramp
126	Town hall in Bad Aibling	Behnisch Architekten, Munich	Office, library	Stairs, atrium, lifts
129	Café and exhibition spaces in Bragança	Giulia De Appolonia, Brescia	Gastronomy, culture	Ramps, stairs
132	Beachfront promenade in Benidorm	OAB – Office of Architecture in Barcelona	Public space	Ramps, stairs
136	Festival arena in Roman quarry in St. Margarethen	AllesWirdGut Architektur, Vienna	Culture	Ramps, stairs
140	Opera house in Oslo	Snøhetta, Oslo	Culture	Accessible roof, foyer, ramps, stairs
146	Porsche Museum in Stuttgart	Delugan Meissl Associated Architects, Vienna	Culture	Escalators, central cascading stairs, ramps and stairs in the exhibition areas
150	Armani Fifth Avenue in New York	Doriana und Massimiliano Fuksas, Rome	Retail	Sculptural staircase, lifts
154	New York Times Building in New York	Renzo Piano Building Workshop, Paris; FXFowle Architects, New York	Office	Lobby, lift shafts
158	Office building in Sydney	ingenhoven architects, Düsseldorf; Architectus, Sydney	Office	Lifts in atrium
164	Stadium in Kiev	gmp • Architekten von Gerkan, Marg und Partner, Hamburg	Sport	Security zones, turnstile systems

Apartment building in Zurich

Architects: Graber Pulver Architekten, Zurich/Bern

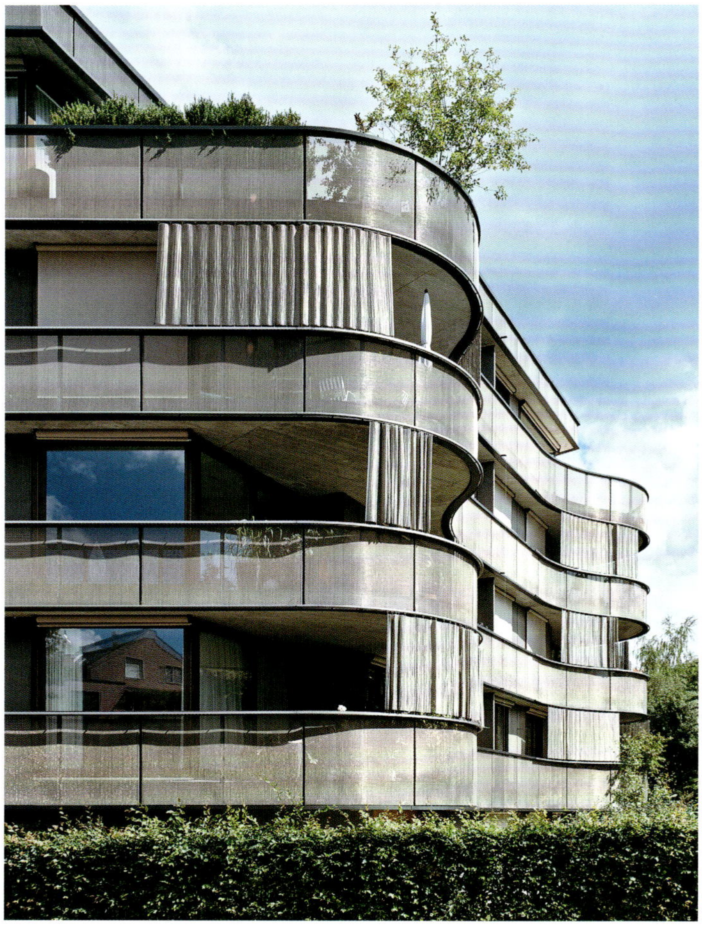

The centrepiece of the apartment building is the sculptural staircase which winds up through five stories towards the skylights.

The irregular pentagonal outline of the "Rondo" apartment building follows the footprint of the site in the middle of a heterogeneous development structure in Zurich's Oerlikon district. Peripheral balconies with rounded forms trace the pentagonal lines of the floor plan, projecting further at the corners to create deeper, comfortably spacious balcony areas. The balustrades consist of a mesh woven out of bright rolled chrome nickel steel. Elements made of the same material, which has an almost fabric-like appearance, can be moved into different positions depending on the individual's need for privacy or shade. The core of the building is formed by the central hall, the floor plan geometry of which once again echoes the external shape of the building in the interior. From here, an extremely striking stair sculpture, particularly unusual for an apartment building, ascends the entire height of the building, with natural lighting provided by three skylights. The inspiration for this feature came from different baroque staircases, but also from roads winding up to mountain passes. The multiple-flight zigzagging staircase connects the galleries which are slightly offset from floor to floor. Thus, the pentagonal airspace is also dissimilar through all the floors, which creates different specific spatial situations. This means that the route to each apartment possesses different spatial qualities, which enhances the prestigious feel of the building and is intended to encourage a feeling of identification on the part of the residents. The doors to the apartments are located in the corners of the galleries. All 22 owner-occupied apartments are arranged on the corners of the building so as to provide views in individual directions. They contain several bedrooms and an open living space.

Geometry developed using models

The flights of stairs and the surrounding galleries were interpreted as a spatial form rather than being developed in an additive way. The architects initially approximated the geometry of the stairwell on the computer using a 3-D model, but then chose to evaluate the effect of the stairwell space as well as possible visual relationships using approximately 30 working models on a scale of 1:50. The solid balustrades consist of plastered prefabricated reinforced concrete elements with an oak handrail covering; the dark step surfaces are finished in polished granolithic concrete, whereas the floor slabs are of in situ concrete.

Project data:

Use:	Residential
Access :	Central staircase, elevator
Number of floors:	5 + 1 basement level
Clear room height:	2.50 m
Gross volume:	19,861 m³
Gross floor area:	5,935 m²
Year of completion:	2007
Construction period:	18 months

Floor plans
Scale 1:750

1. Entrance atrium
2. Open-plan living space with dining area and kitchen
3. Room
4. Bathroom
5. Studio/guestroom
6. Bicycle room
7. Peripheral balcony

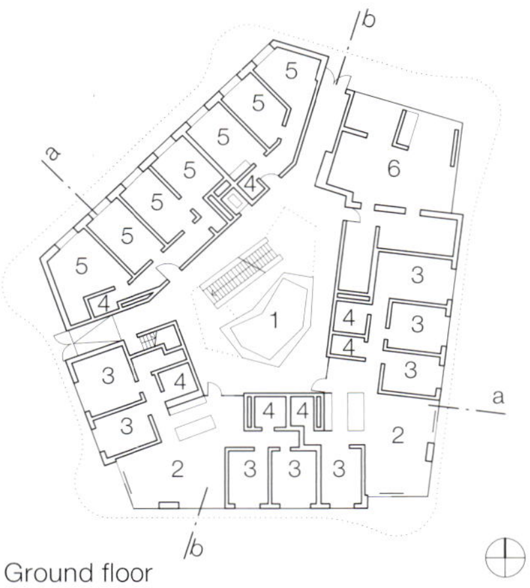

Ground floor

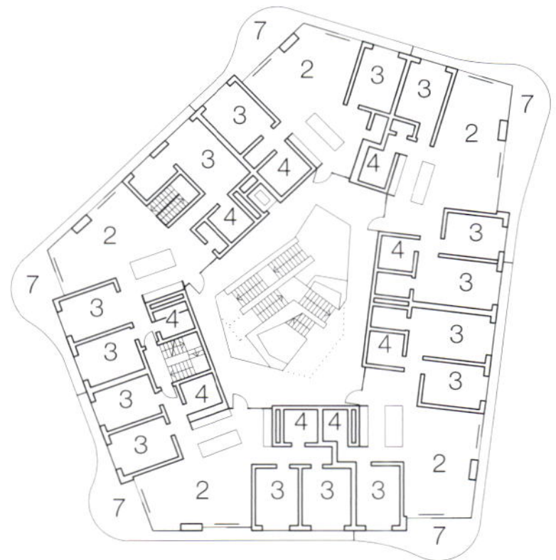

2nd floor

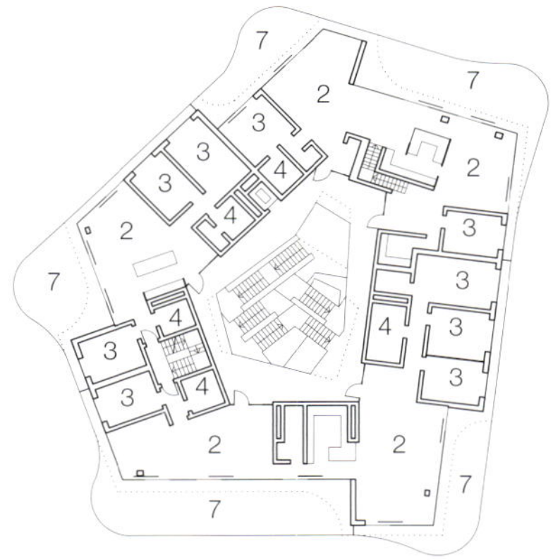

4th floor

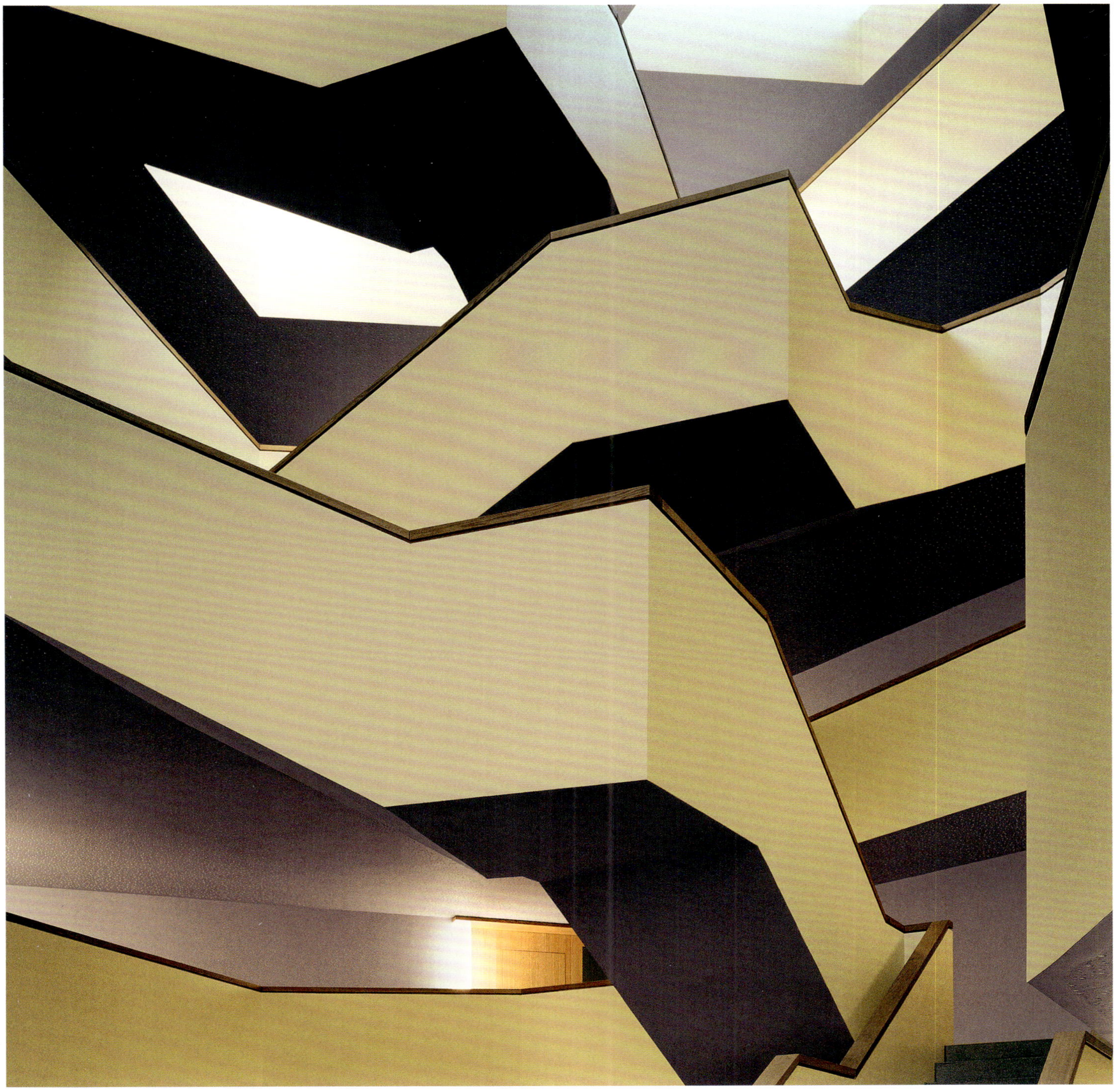

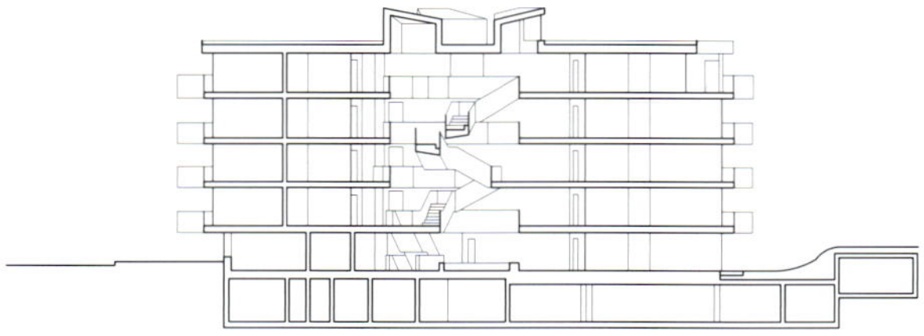

aa

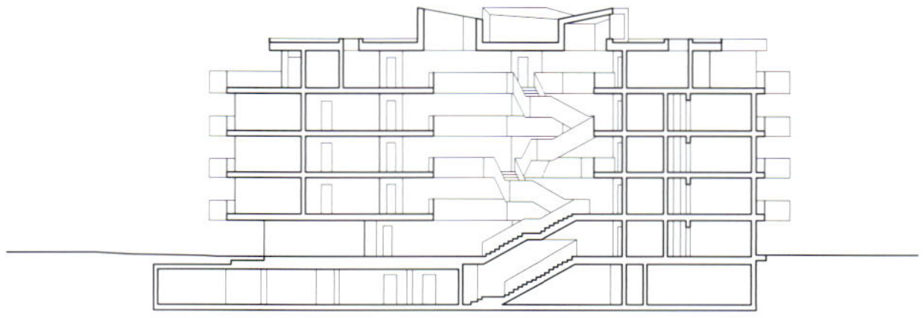

bb

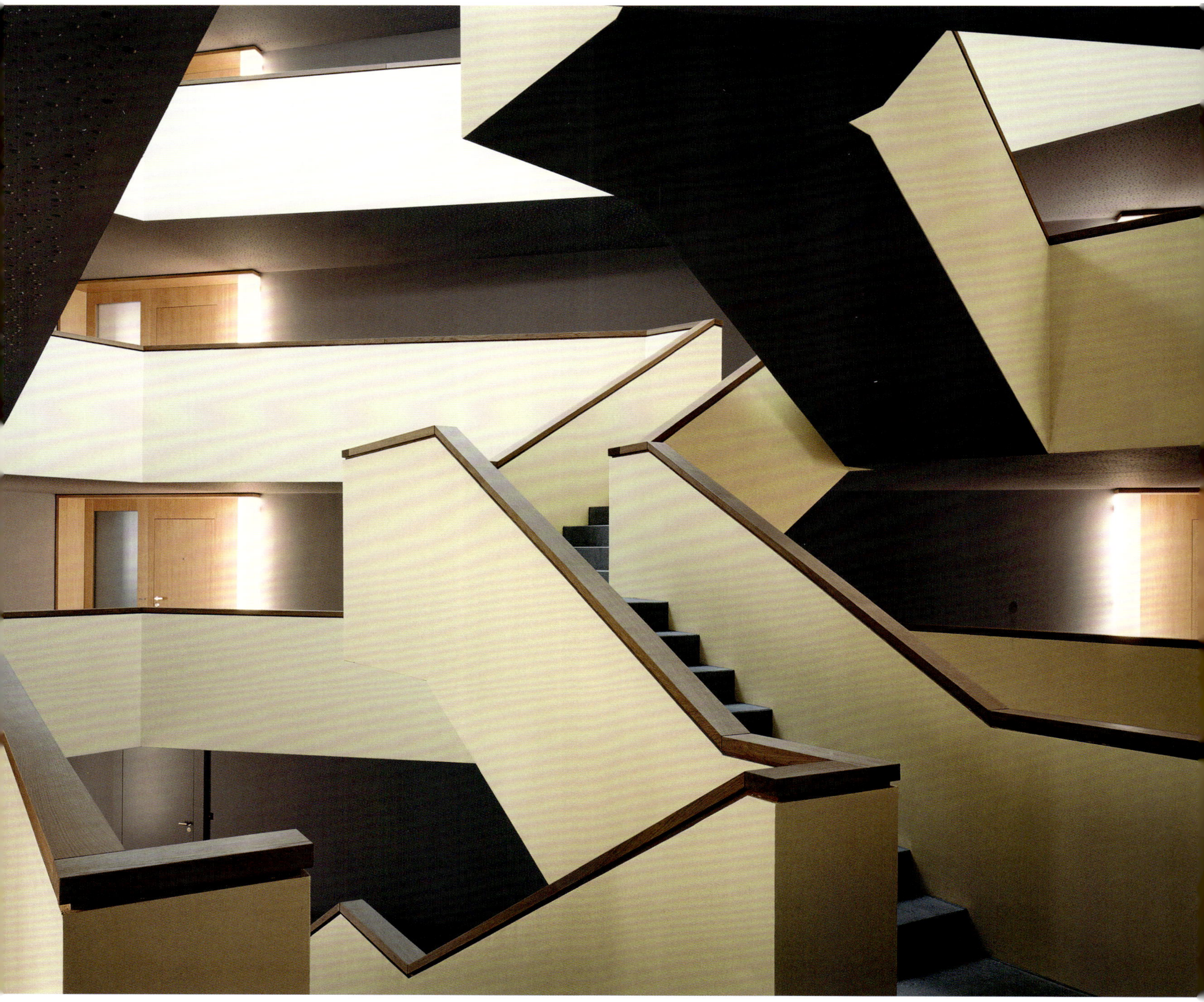

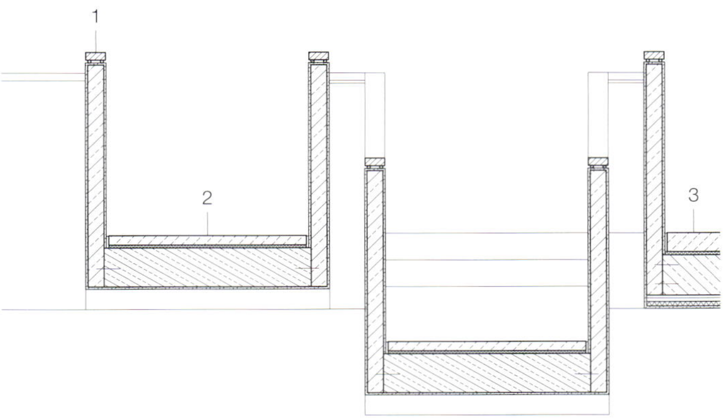

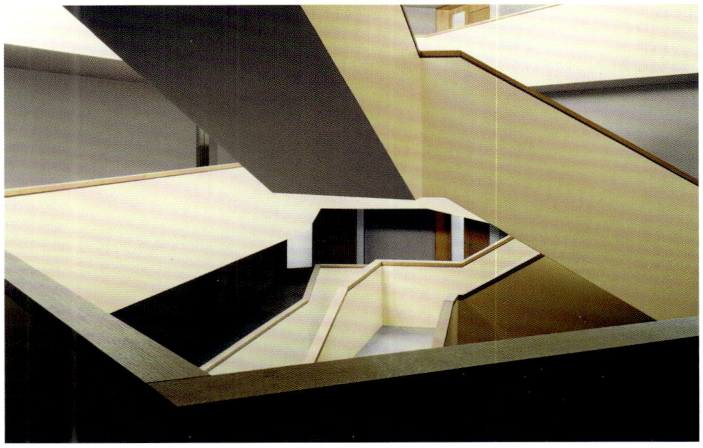

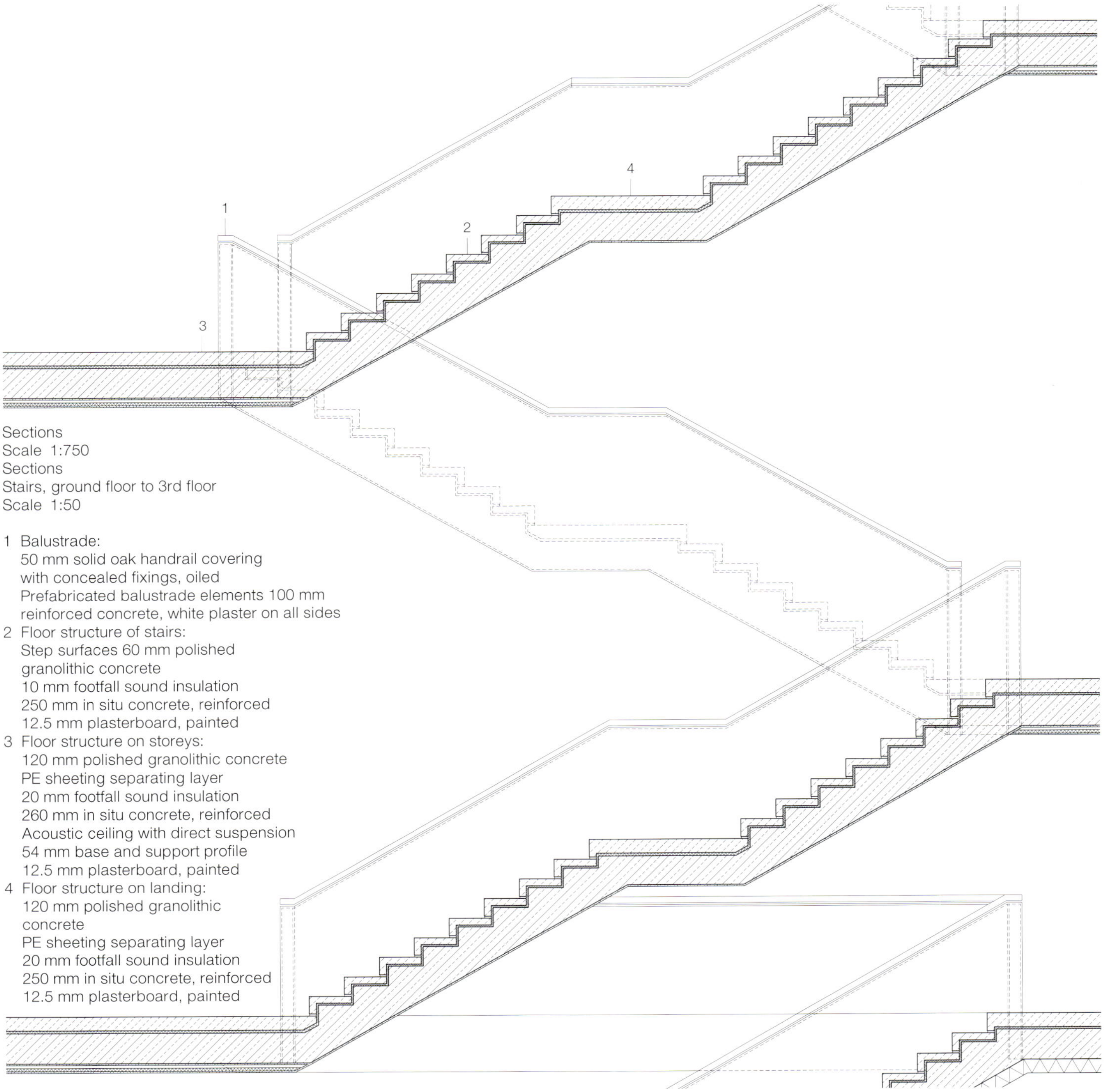

Sections
Scale 1:750
Sections
Stairs, ground floor to 3rd floor
Scale 1:50

1 Balustrade:
50 mm solid oak handrail covering with concealed fixings, oiled
Prefabricated balustrade elements 100 mm reinforced concrete, white plaster on all sides
2 Floor structure of stairs:
Step surfaces 60 mm polished granolithic concrete
10 mm footfall sound insulation
250 mm in situ concrete, reinforced
12.5 mm plasterboard, painted
3 Floor structure on storeys:
120 mm polished granolithic concrete
PE sheeting separating layer
20 mm footfall sound insulation
260 mm in situ concrete, reinforced
Acoustic ceiling with direct suspension
54 mm base and support profile
12.5 mm plasterboard, painted
4 Floor structure on landing:
120 mm polished granolithic concrete
PE sheeting separating layer
20 mm footfall sound insulation
250 mm in situ concrete, reinforced
12.5 mm plasterboard, painted

Residential development in Berlin

Architects: zanderroth architekten, Berlin

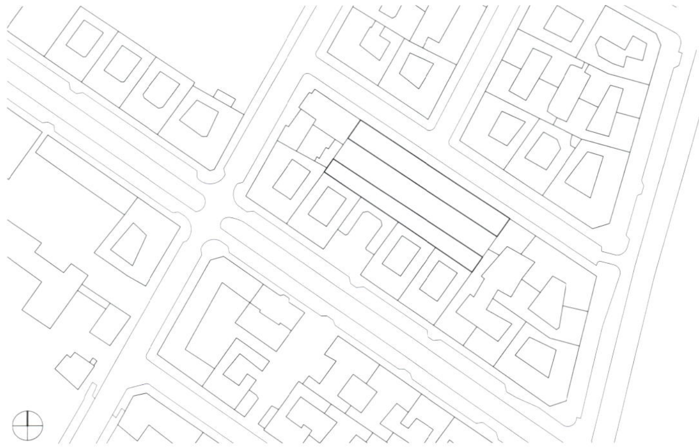

A complex access structure allows individual access to the different types of housing units.

Building cooperatives are an increasingly popular model for creating affordable and individual urban living space. They also enable cost savings of up to 25 % to be achieved in comparison with typical residential real estate developments. Moreover, in this case the site, a typical block/inner courtyard structure located in Berlin's Prenzlauer Berg, was also affordable. However, with a length of 100 m, a width of 34 m and a 22-metre-high firewall along its southwestern perimeter, it represented a challenge in terms of planning. The architects solved this difficult initial situation by creating two linear structures aligned parallel to the street and separated by an approximately 13-metre-wide interior courtyard running the entire length of the plot.

Different housing types

Three different types were developed for the individual homes, depending on their location within the development. The 23 "townhouses" in the front building are directly accessed from the street while offering spaces for commercial use along the entrance level. In addition to having access to the garden on the first floor level, they also have a direct connection to the underground parking area. The four floors of these houses are designed as open-plan, split-level spaces with a small private rooftop garden above. The facade facing the street has a rather austere, urban character and consists of prefabricated concrete elements, large glazed surfaces and openings made of larch wood. Somewhat more heterogenous in appearance are the rendered facades clad with aluminium shingles facing the courtyard which covers the underground garage. The courtyard is elevated one storey above street level in order to admit more sunlight and prevent the area between the two building blocks taking on a gorge-like atmosphere. It connects the two parts of the building while at the same time ensuring the necessary distance between them. The area is laid out so as to provide differentiated, semi-public, communal grassed areas and play areas which form niches and spaces, while deliberately not marking out any private gardens.
From the street, direct access to the courtyard level is provided via the access core located in the middle of the street-side block with a lift and stairs leading up, or via two further access points at the ends of the building. The entrances to the apartments are also located in the courtyard, on the lower floors of the courtyard buildings. These ten "garden apartments", each on three floors, are also conceived as a split-level type, with high ceilings to compensate for the unfavourable lighting, which only comes from one side. The 12 three-storey penthouses above are accessed by means of an interior corridor on the fifth floor which the residents reach via two access stairs accessible from the courtyard, or directly from the underground parking garage. In front of the private rooms on the fourth floor, a corridor open to the courtyard serves as an emergency route leading to two decentrally positioned emergency stairwells. Small courtyards along the firewall and a roof terrace offer private open spaces for the occupants of the penthouse apartments.
In this project, density isn't only a characteristic of urban life, but also the result of planning that is focused on efficiency in economic as well as architectural terms.

Project data:

Use:	Residential
Access:	Lobbies, lifts, stairs, courtyard
Number of floors:	4 (street building) and 6 (courtyard building) + 1 basement level
Clear room height:	2.85/4.20 m
Gross volume:	40,522 m³
Gross floor area:	9,210 m²
Year of completion:	2010
Construction period:	19 months

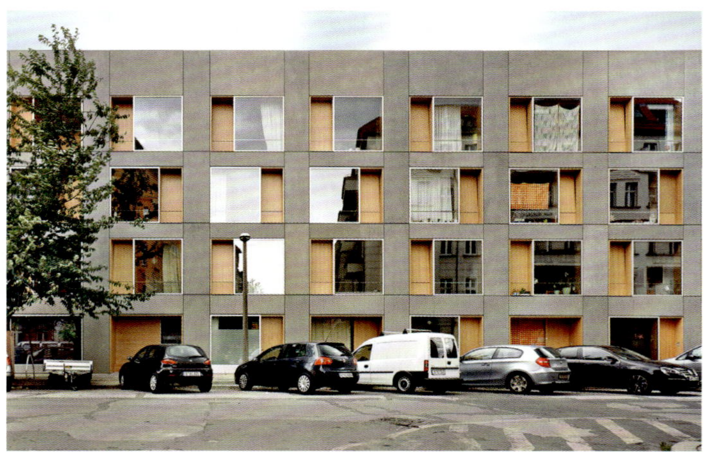

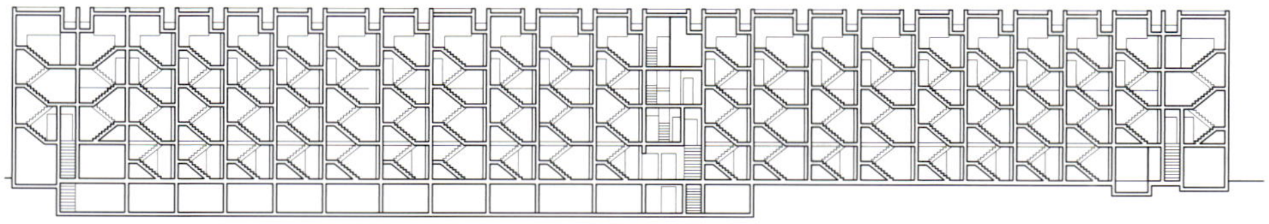

aa

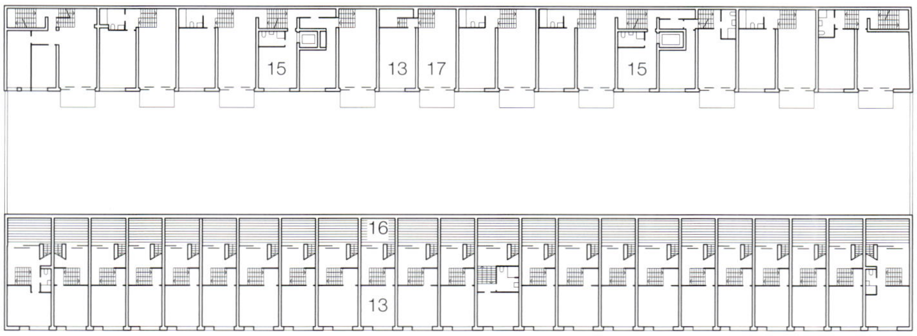

3rd floor

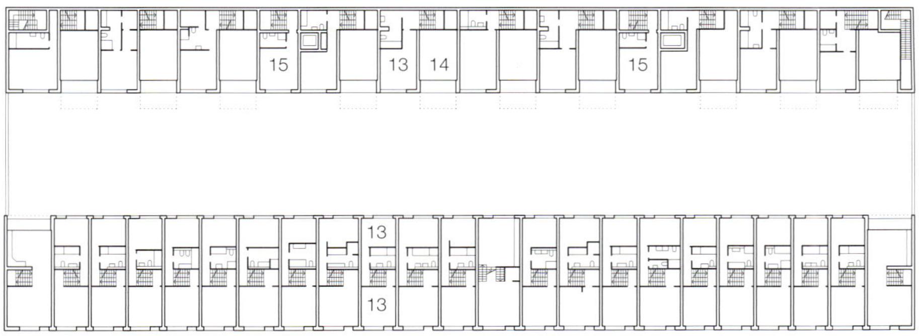

2nd floor

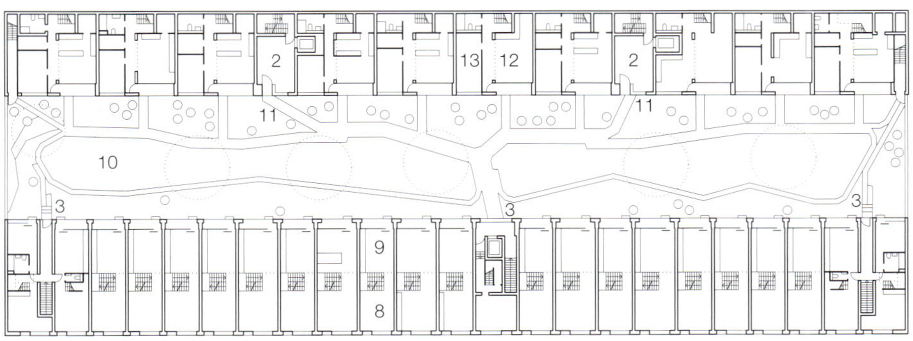

1st floor (courtyard level)

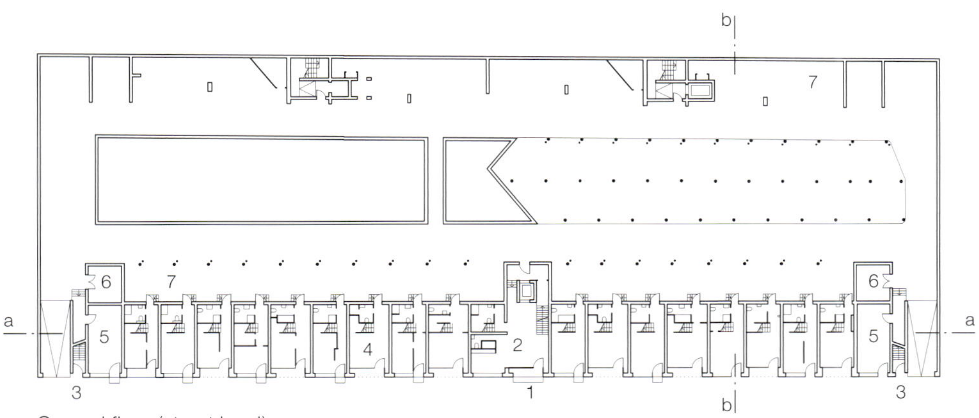

Ground floor (street level)

Section: street building
Floor plans: "townhouse homes" (street building) and "garden apartments" (courtyard building)
Section: complete development
Scale 1:800

1 Entrance to street building
2 Access core
3 Courtyard access
4 Space for commercial use/house entrance
5 Bicycle room
6 Waste room
7 Garage
8 Living room
9 Kitchen/dining area with terrace
10 Communal courtyard
11 Entrance to "penthouses"
12 Dining/kitchen
13 Room
14 Void
15 Guest apartment
16 Patio with stairs up to terrace
17 Living room

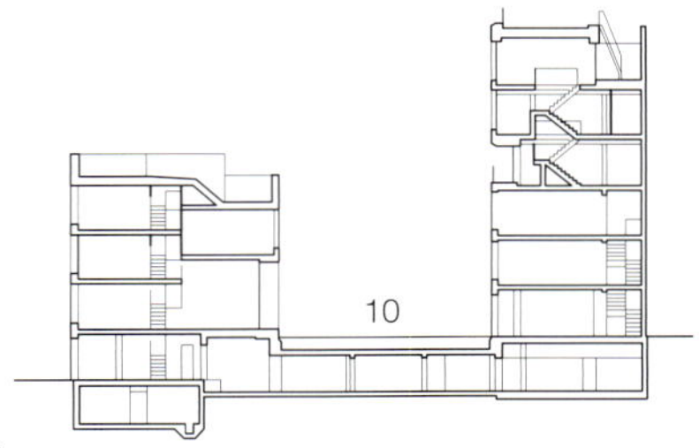

bb

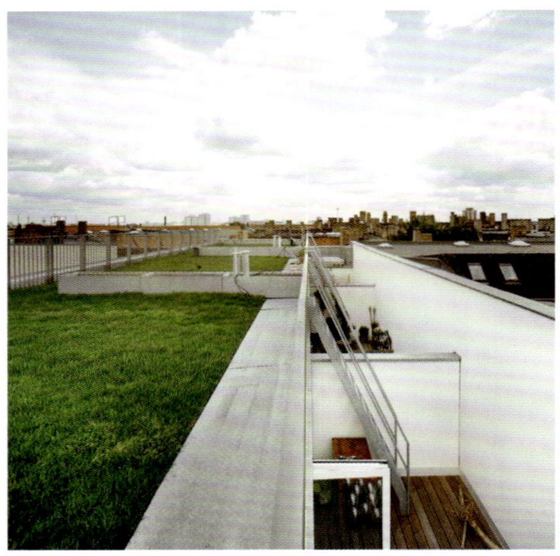

Courtyard building: "penthouses"
Section • Floor plans
Scale 1:800

1 Room
2 Corridor (emergency escape route)
3 Emergency stairwell
4 Access corridor
5 Living/dining area
6 Patio with stairs up to terrace

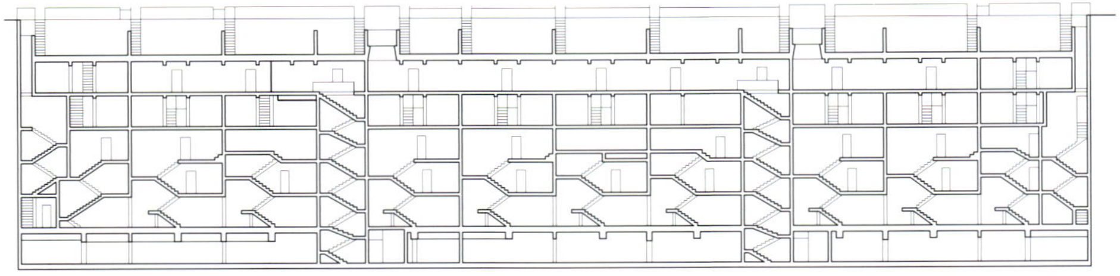

cc

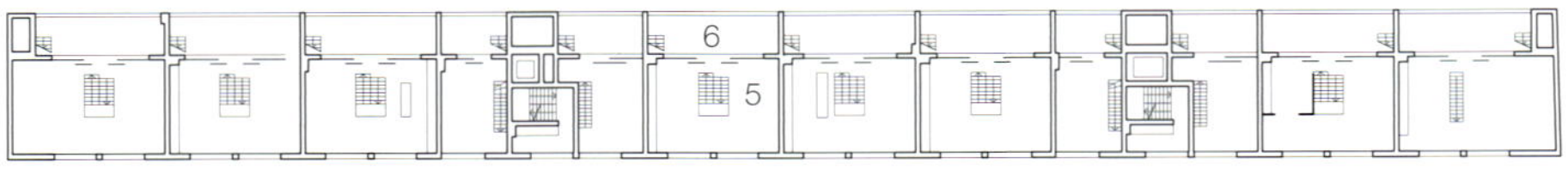

6th floor

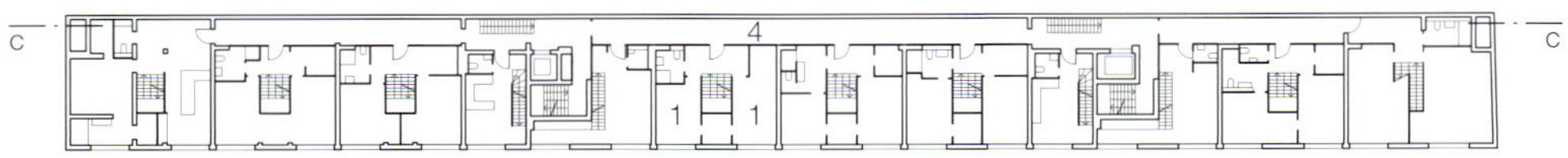

5th floor

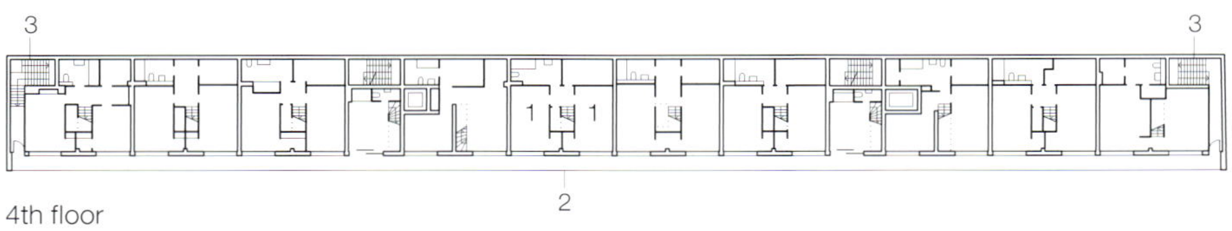

4th floor

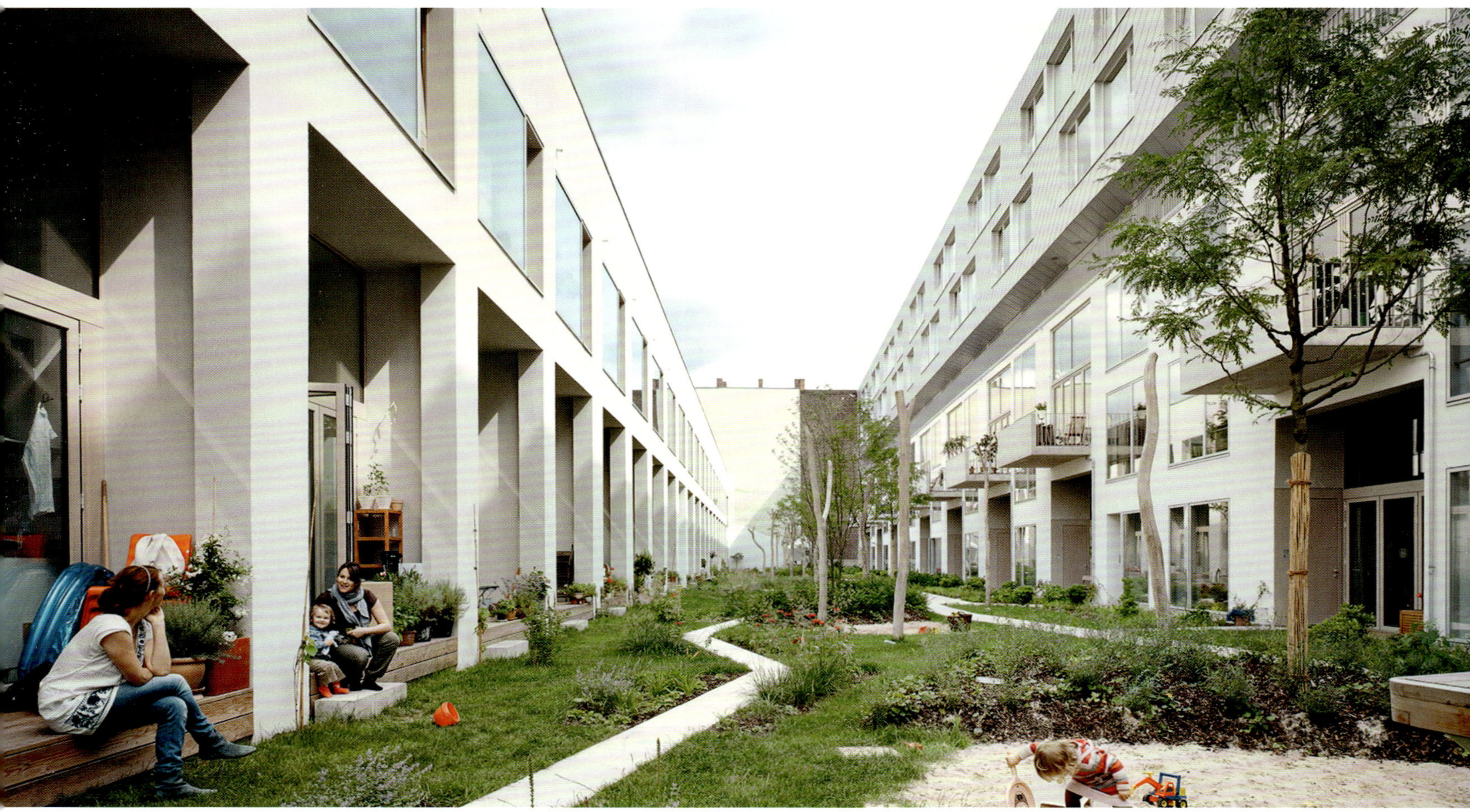

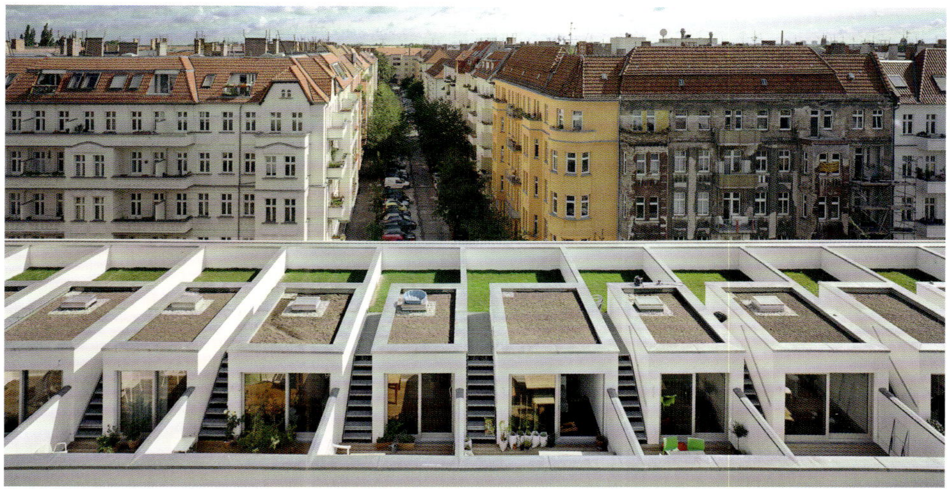

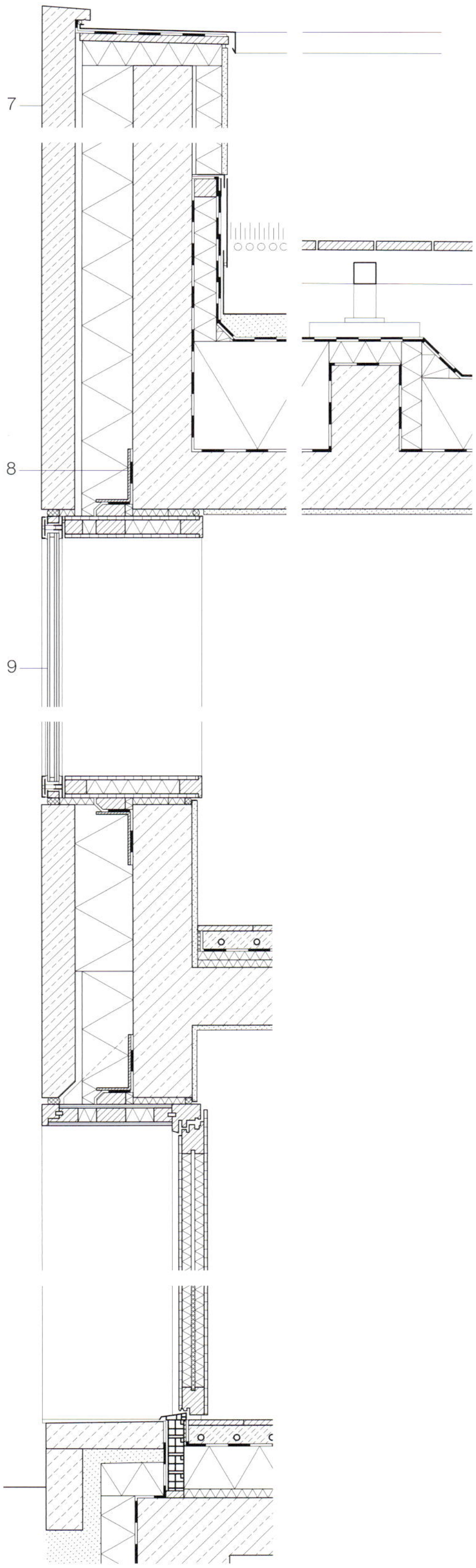

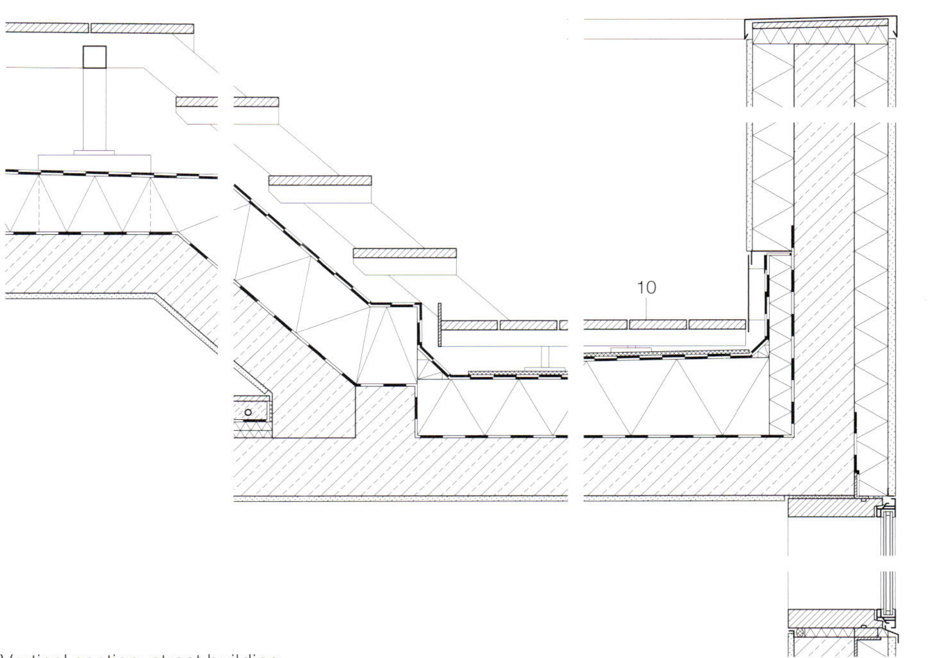

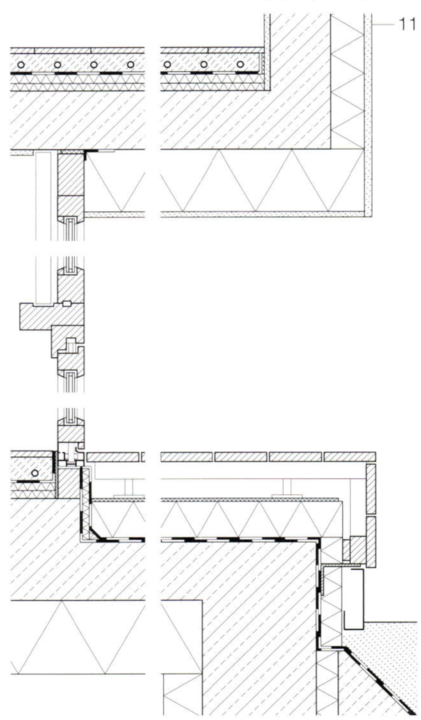

Vertical section: street building
Scale 1:20

7 90 mm precast concrete element
20 mm void
140 mm thermal insulation
160 mm reinforced concrete
10 mm levelling layer
80 mm thermal insulation
15 mm render (thermal insulation composite system)
8 160/100 mm steel angle
9 Insulation glazing in aluminium frame
10 25 mm wooden floor boards
45 mm raised wooden subfloor construction, height-adjustable
8 mm protection mat
5 mm bituminous sealant, root-resistant
5 mm bituminous sealant
2 mm vapour pressure equalization layer
215–155 mm PUR rigid foam thermal insulation, on slope
Vapour barrier
160 mm reinforced concrete
15 mm render
11 20 mm render
100 mm thermal insulation (thermal insulation composite system)
160 mm reinforced concrete
15 mm render

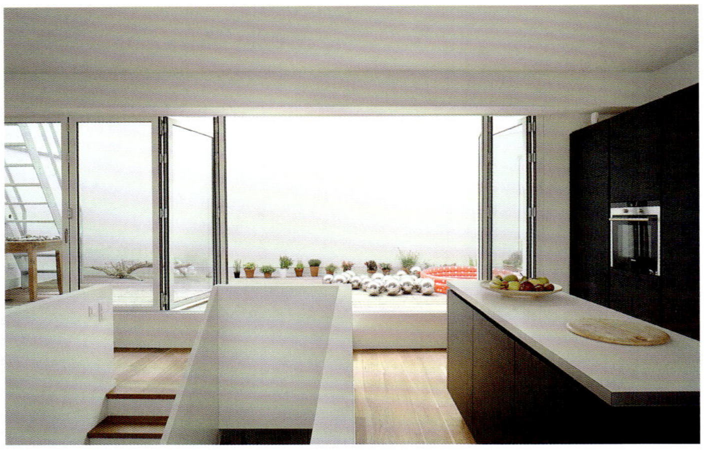

Sections through stairwell in street building and "garden apartments" in courtyard building
Scale 1:50

1 16 mm solid oak parquet
59 mm cement screed with underfloor heating
20 mm footfall insulation
20 mm thermal insulation
2 Stairs: precast concrete element with cast landings, wall connection with footfall insulation system
3 Stairs connection with footfall insulation system
4 16 mm solid oak parquet
59 mm cement screed with underfloor heating
145 mm thermal insulation
180 mm reinforced concrete
5 ISO footing bricks
6 Stairs: precast concrete element with 16 mm oak surfacing on supporting plate
7 Balustrade: 38 mm wood composite panel, fixed with screws and concrete anchors in stringer, covered with painted wooden plugs

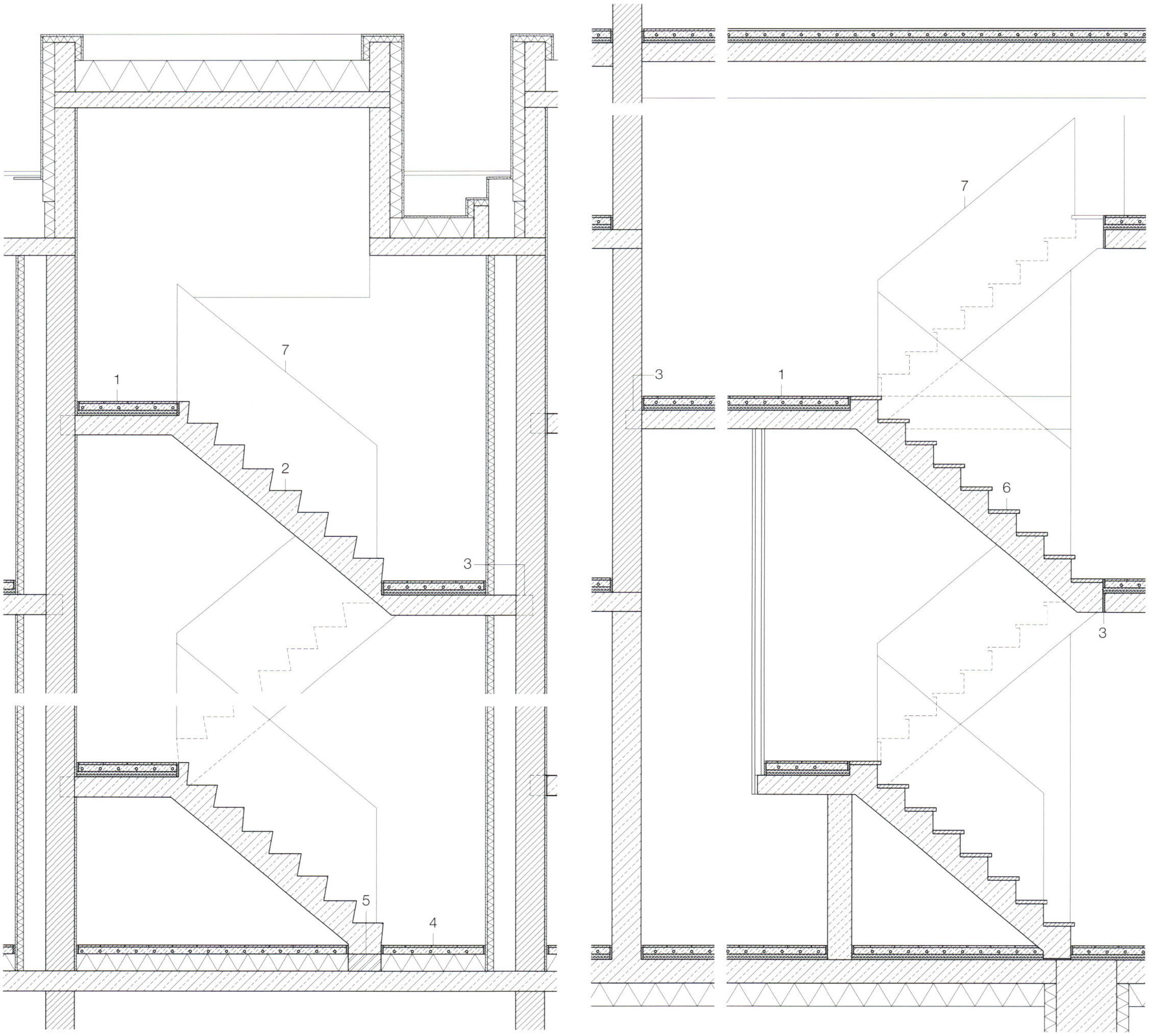

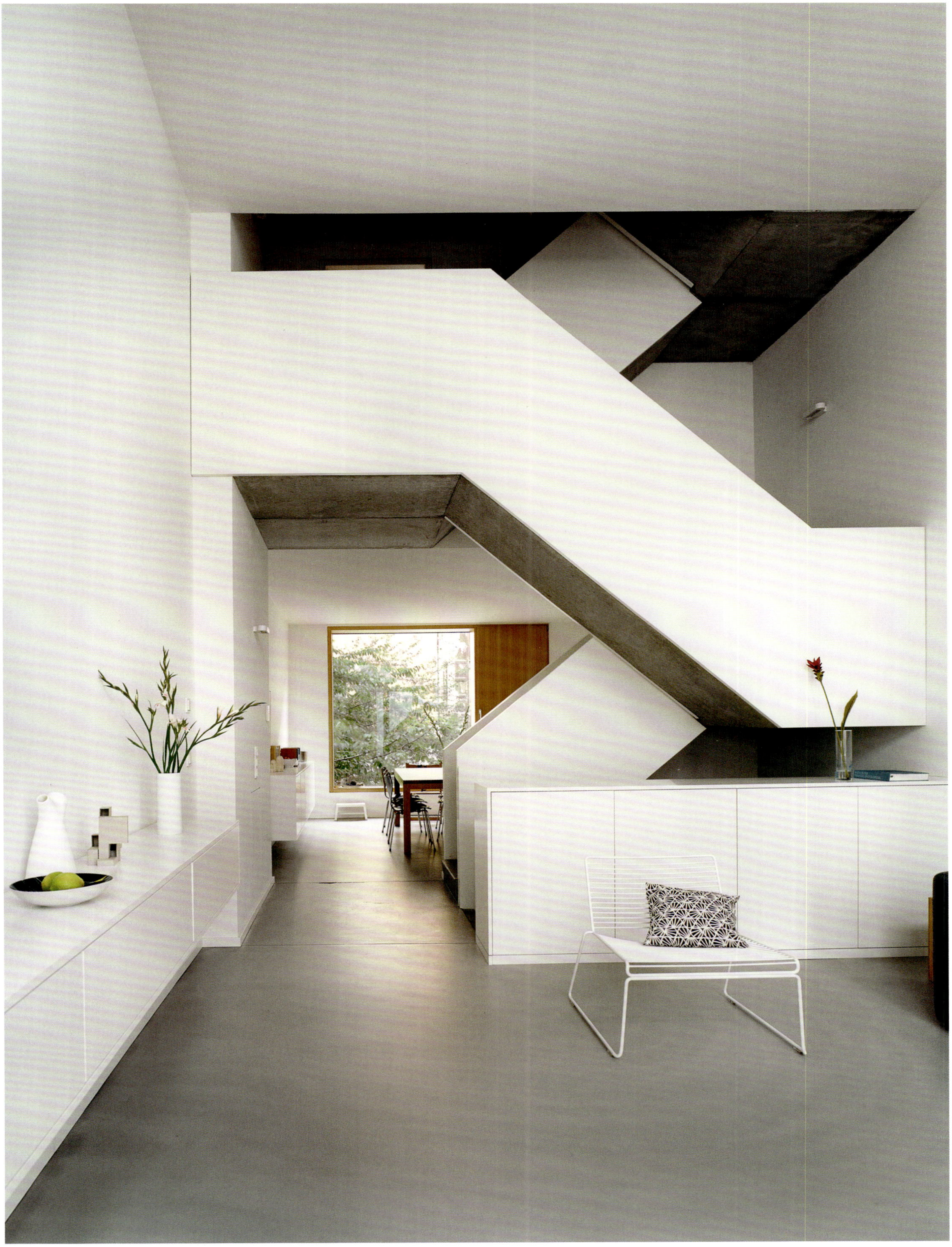

Haus am Weinberg in Stuttgart

Architects: UNStudio, Amsterdam

A curving sculptural staircase, the key spatial element, forms the centrepiece of this detached house, which is flooded with natural light.

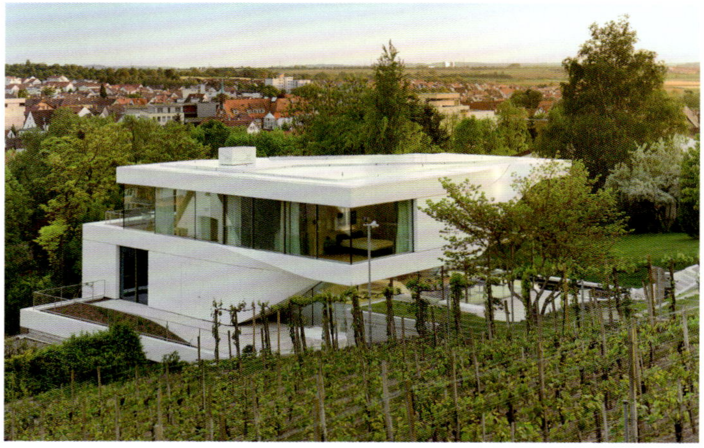

This detached house is situated on the edge of Stuttgart and surrounded by terraces of old vineyards, with a view to the south extending far beyond the city. Because the supporting concrete structure of the villa has been reduced to a minimum, natural light can pass deep into the building, and the column-free glazed corners open the rooms up to the surroundings. The bright and airy atmosphere of the villa is enhanced by the choice of materials: Jura marble, light oak flooring and white plastered walls.
The starting point for the design was the client's wish to orient themselves around the 24-hour cycle of the sun through the arrangement of the rooms. The linking of structure, space, time and movement is therefore the central theme of the house. One enters the building from the south via the entrance hall, ascending a staircase which travels diagonally to the next level, where the room opens out into a spacious dining area. The sculptural staircase, made of MDF, winds into the living area, and from here the curving stairs continue as the central access element up to the gallery on the second floor which contains the library.
The internal circulation layout, the orientation of the views and the organisation of the space allocation plan are determined by the generously dimensioned staircase, which functions as a key design element. It runs through the entire house, creating a flowing connection between all the different spaces and levels. Variations in direction introduce a dynamic element and create shifting perspectives which include the changing surroundings.
The landscape architecture of the garden continues the internal organisation of the house in that the diagonals of the floor plan are reflected in the layout of the flower beds.

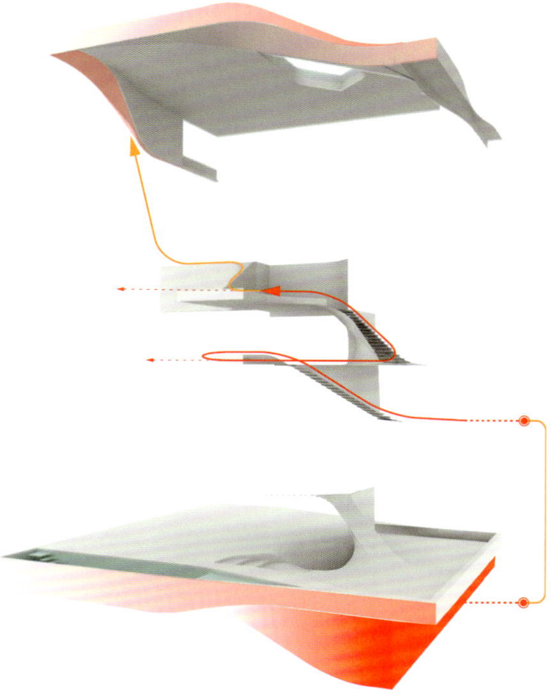

Project data:

Use:	Single family house
Access:	Stairs, lifts
Floors:	2 + 1 basement level
Gross volume:	2,839 m^3
Gross floor area:	920 m^2
Year of completion:	2011
Construction period:	20 months

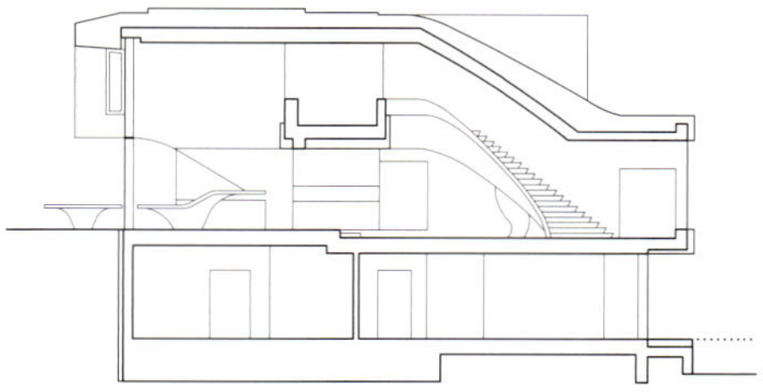

aa

Section • Floor plans
Scale 1:400

1 Entrance hall
2 Guest room
3 Guest apartment
4 Wine tasting
5 Wine cellar
6 Storage room
7 Technical services
8 Garage
9 Dining room
10 Kitchen
11 Living room
12 Multi-purpose room
13 Swimming pool
14 Gallery/library
15 Void
16 Bedroom
17 Bathroom/walk-in closet
18 Wellness area

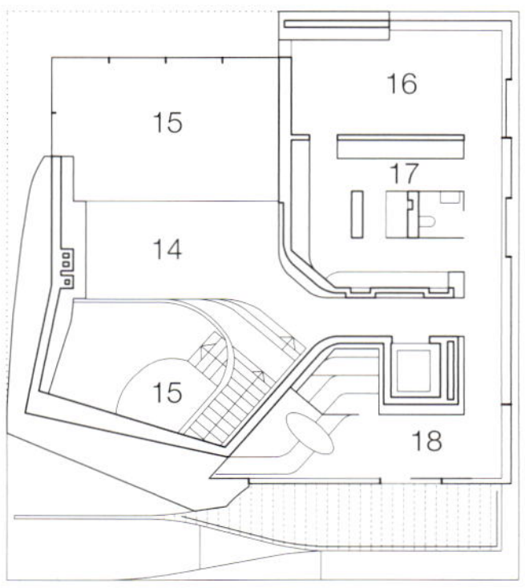

2nd floor

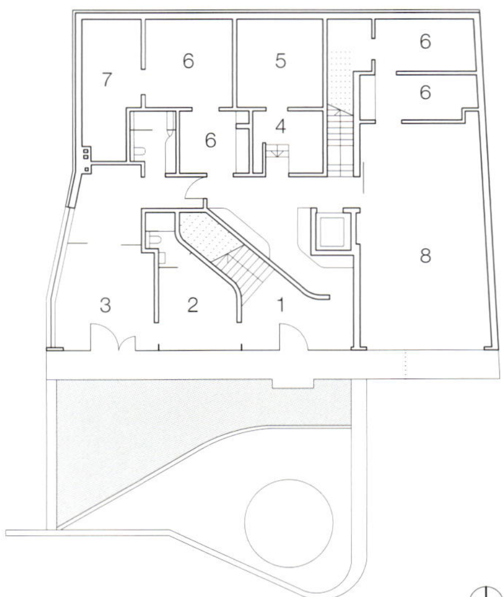

Ground floor

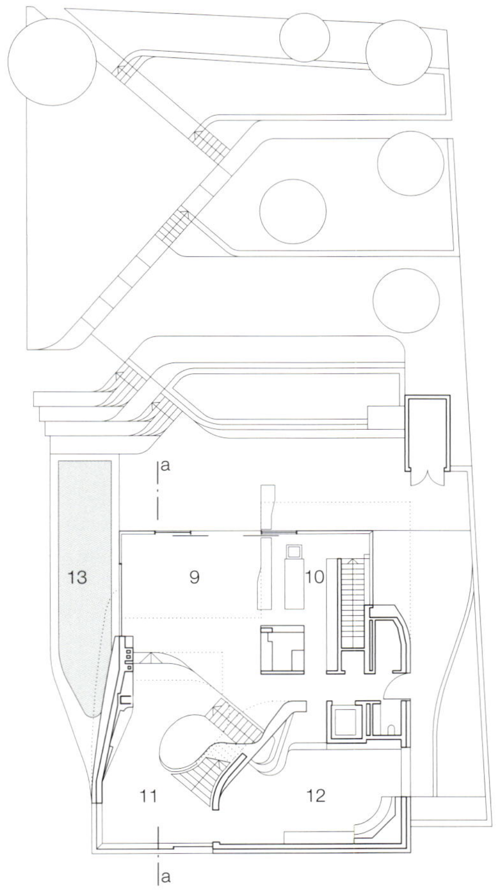

1st floor

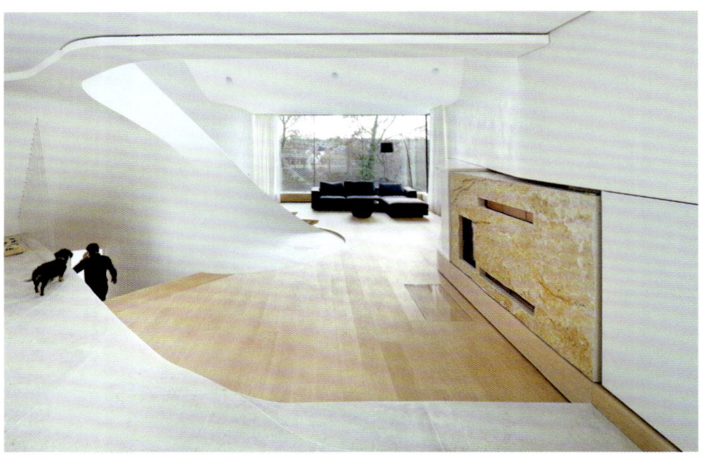

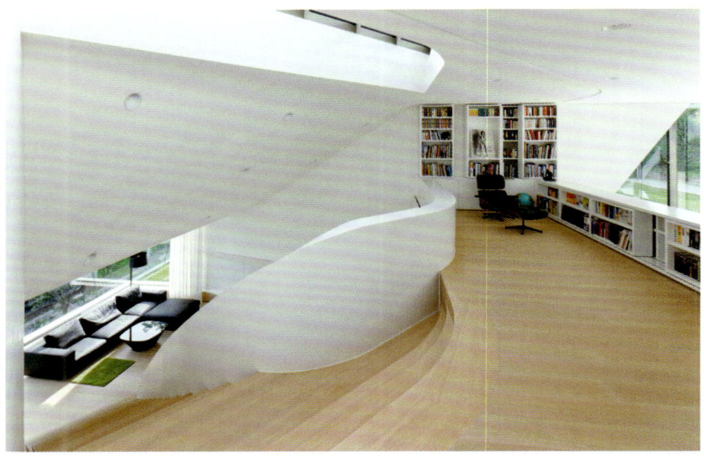

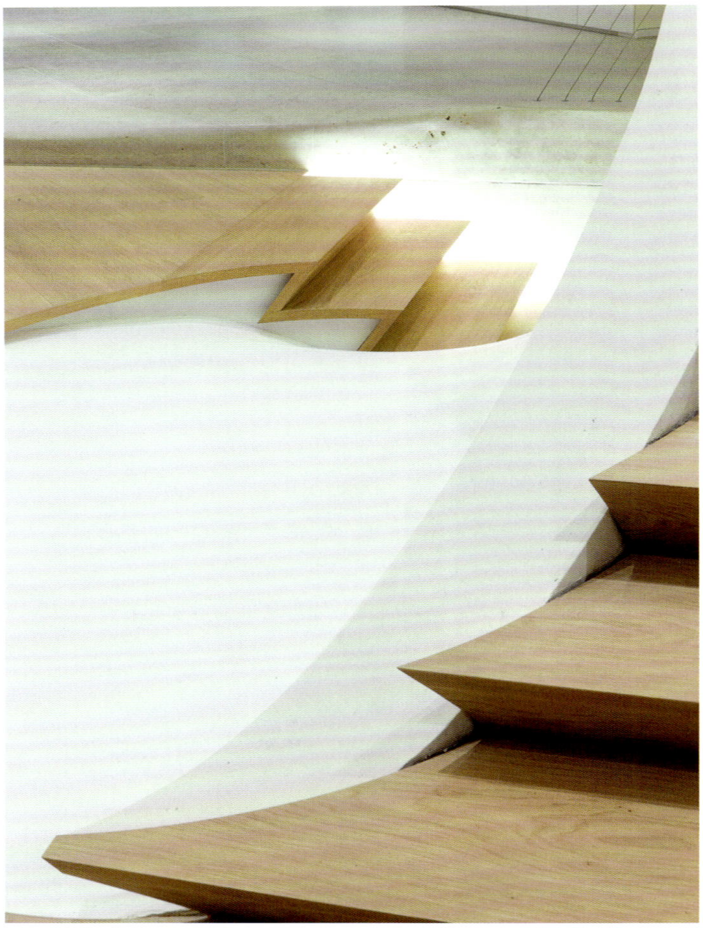

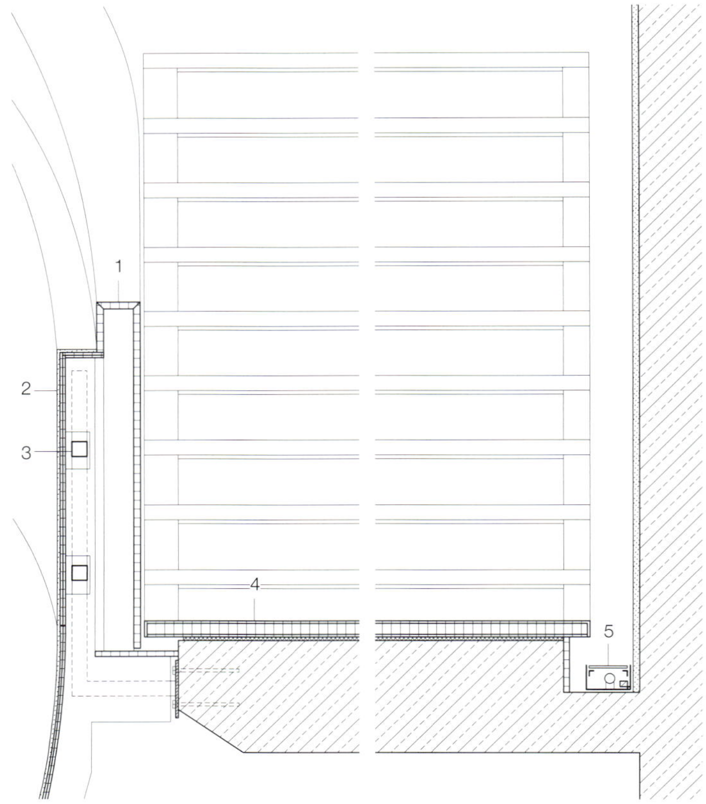
1
2
3
4
5

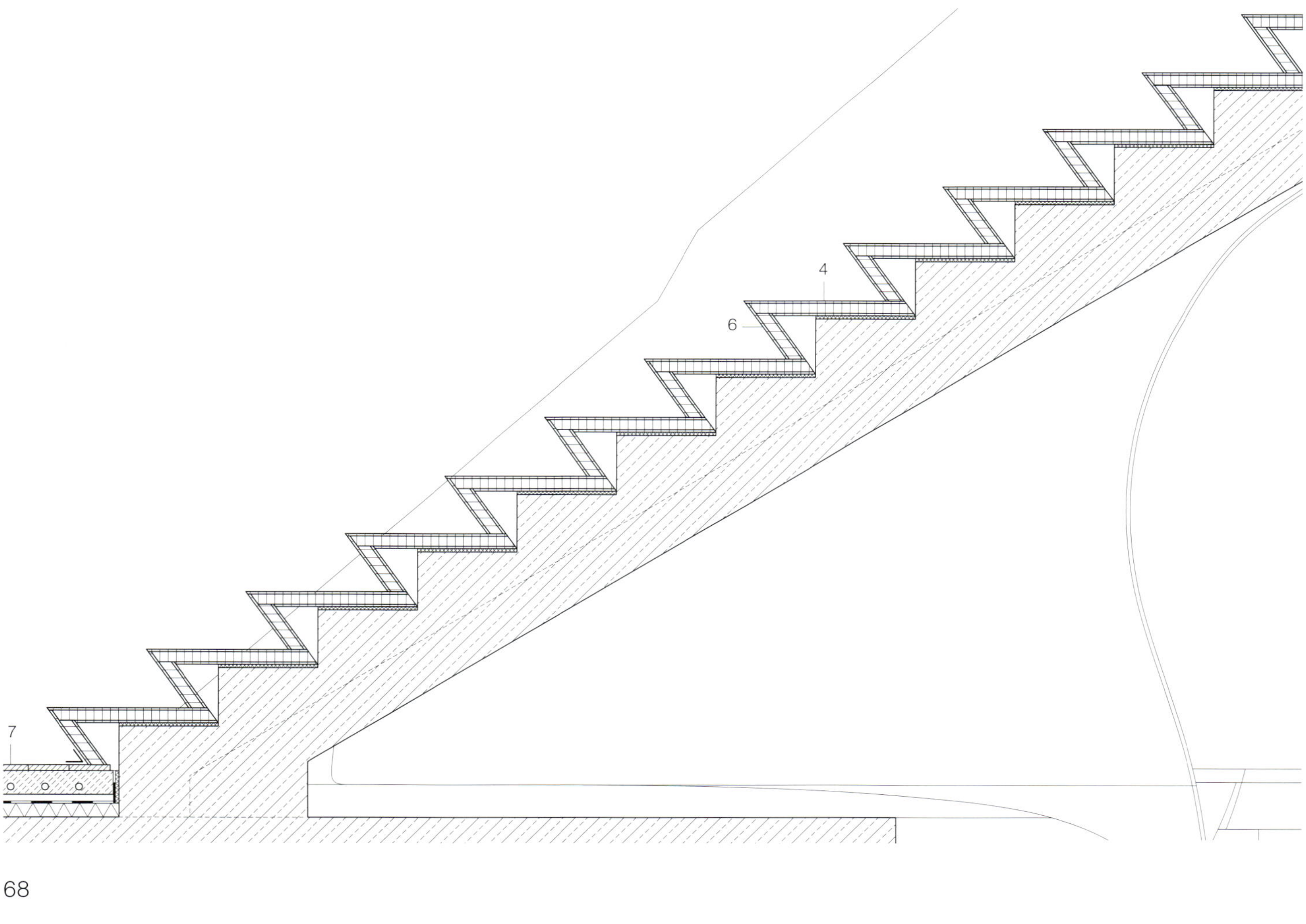
4
6
7

Sectional details
Balustrade • Stairs
Scale 1:20

1 18 mm MDF balustrade, painted
2 6 mm loam rendering
  2× 6 mm MDF
  18 mm plywood
3 40/40 mm tubular steel supporting structure, screwed together with stairs
4 Tread screwed to bare concrete step, veneered face edge
  5 mm oak sawn veneer, natural white, oiled
  30 mm birch multiplex
  5 mm oak sawn veneer, natural
  Soft rubber underlay for acoustic insulation
  Reinforced concrete
5 Fluorescent lighting, cover laminated safety glass with opaque film
6 Riser connected with tread via springs:
  5 mm oak sawn veneer, natural white, oiled
  30 mm birch multiplex
  5 mm oak sawn veneer, natural
7 Multilayered floorboard
  15 mm oak sawn veneer, natural white, oiled
  70 mm anhydrite screed with underfloor heating
  20 mm fixing for underfloor heating
  Vapour barrier
  40 mm thermal insulation
  350 mm reinforced concrete

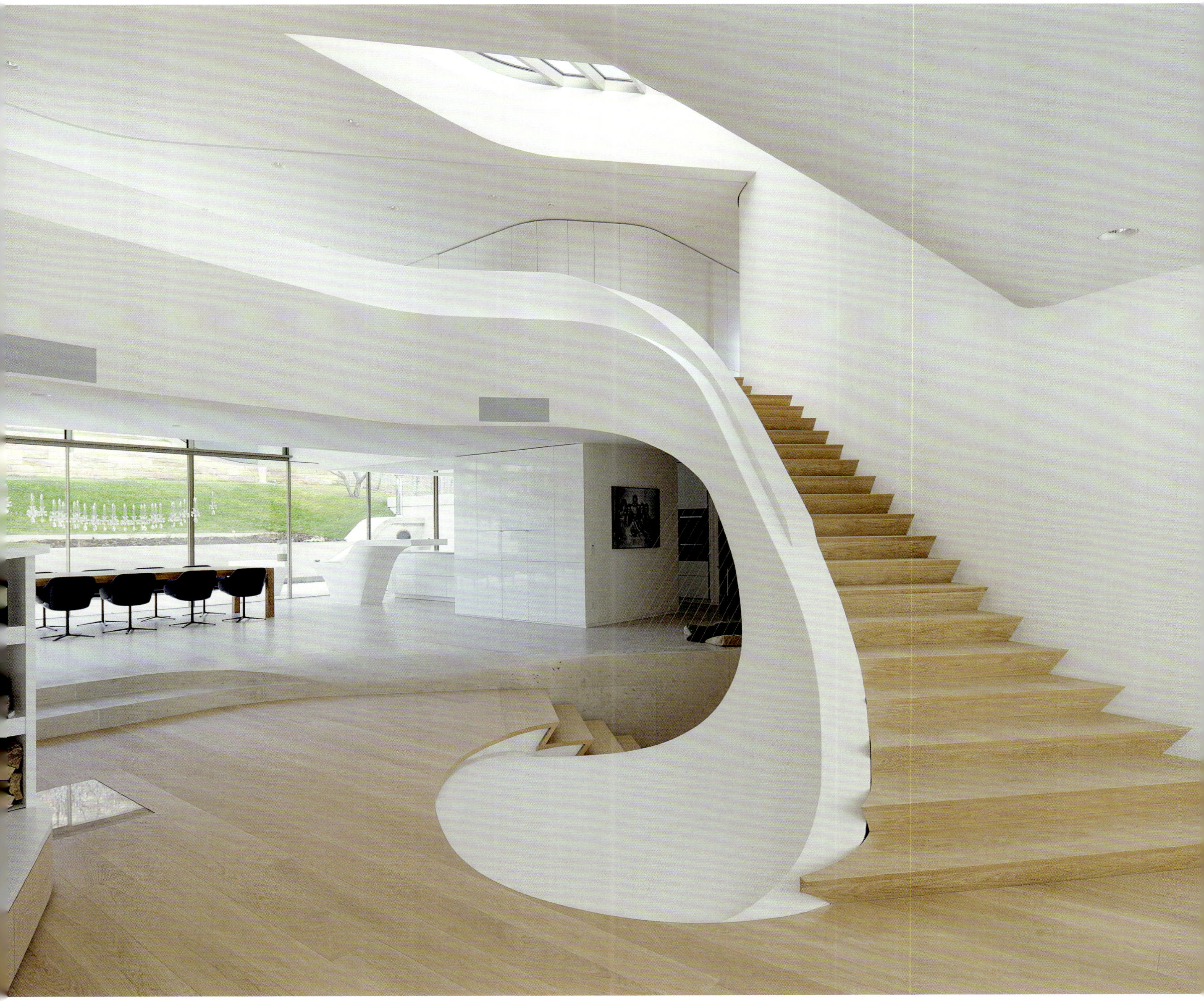

“House before House” in Utsunomiya

Architects: Sou Fujimoto Architects, Tokyo

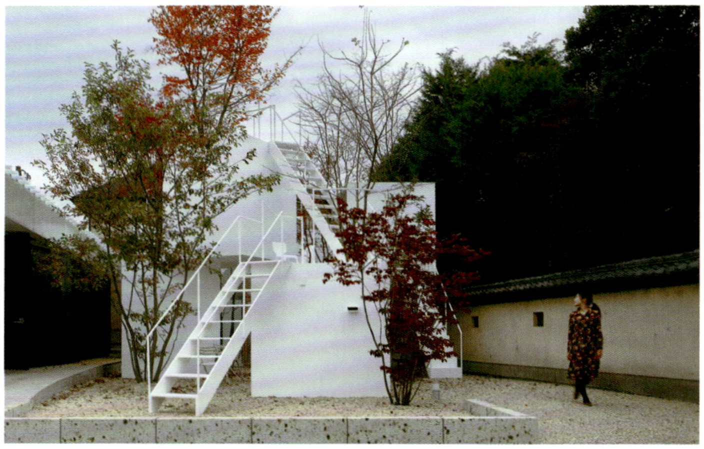

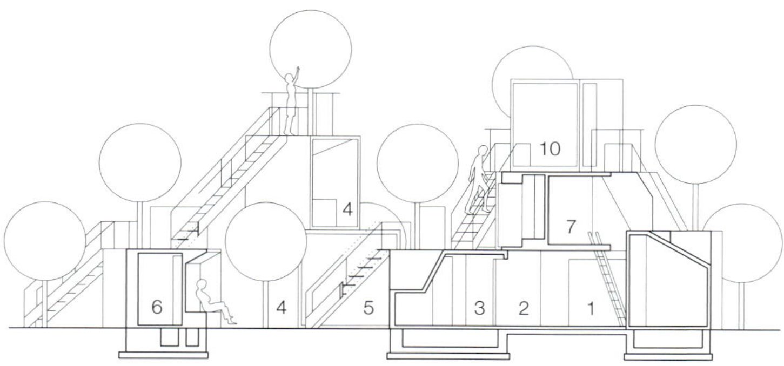

aa

Project data:

Use:	Residential
Access:	Steel stairs, ladders
Number of floors:	3
Clear room height:	1.92–3.92 m
Gross floor area:	61.31 m²
Year of completion:	2008
Construction period:	6 months

Steel stairs and ladders provide access to and connect the apparently randomly stacked white cubes to create a dwelling sculpture.

Located in a residential district in Utsunomiya, 100 km north of Tokyo, the “House before House” follows the design concept that living doesn’t just have to take place indoors, it can also extend to the outdoor environment. Rather than seeing the home as a confined space, the architect has developed it into a village-like structure with individual buildings, open spaces and stairways. Interior and exterior – house and garden – form a spatial continuum. This timeless yet futuristic interpretation reflects the traditional Japanese ideal of living, which strives for harmony between nature and man-made objects. The building is designed to accommodate two to four persons. Ten cubes of equal height, stacked seemingly at random, are scattered around a site measuring only 163 m² in area, creating both open and sheltered spaces. Simple external stairs and interior ladders provide access to and connect the individual cubes. Where they are most densely grouped, the cubes form an intersecting complex which contains the living area, surrounded by the kitchen, bathroom and bedroom. Separate access is provided to other ancillary spaces. The impression of a naturally grown environment is reinforced by trees growing from planting troughs incorporated in the roofs, by the terraced areas that extend in a series of cascading steps and by the cave-like external spaces formed between the boxes and stairs.

Simple construction

The clarity of the building structures also reflects the simplicity of the prefabricated construction method. The outer skin consists of steel sheeting welded onto a steel frame. On the building site, the boxes are lowered into place with a crane, the steel walls are sprayed on the inside with insulating material and lined with plasterboard panels and the outer shell and steel stairs are given a polyurethane coating as protection against the elements. With their extensive areas of glazing, the cubes are open to the exterior. With the exception of the staircase which leads to the children’s room, no risers or side panels on the handrails were used. The light, finely dimensioned design of the steel staircases rising between the trees allows the occupants to continually rediscover the outdoor spaces and their diverse possibilities for use.

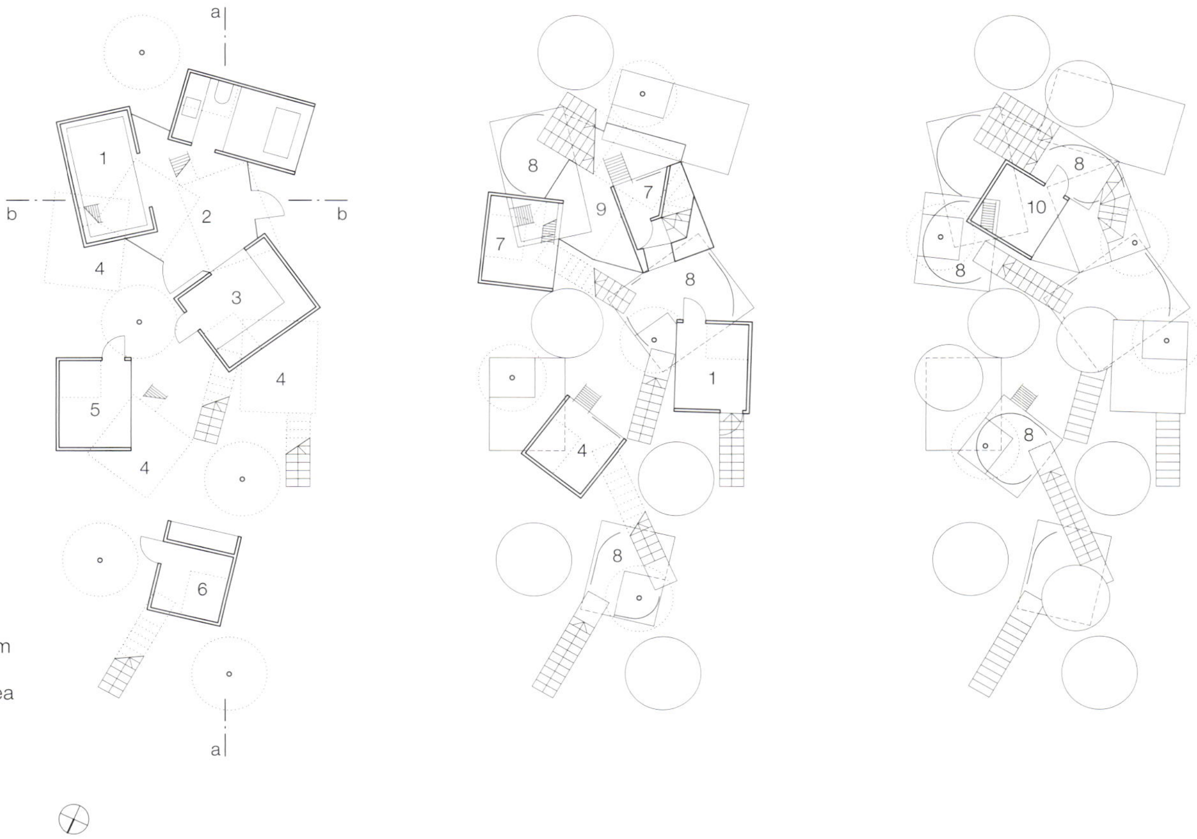

Section • Floor plans
Scale 1:200

1 Bedroom
2 Living/dining room
3 Kitchen
4 Outdoor living area
5 Guest room
6 Storage room
7 Gallery
8 Terrace
9 Void
10 Study

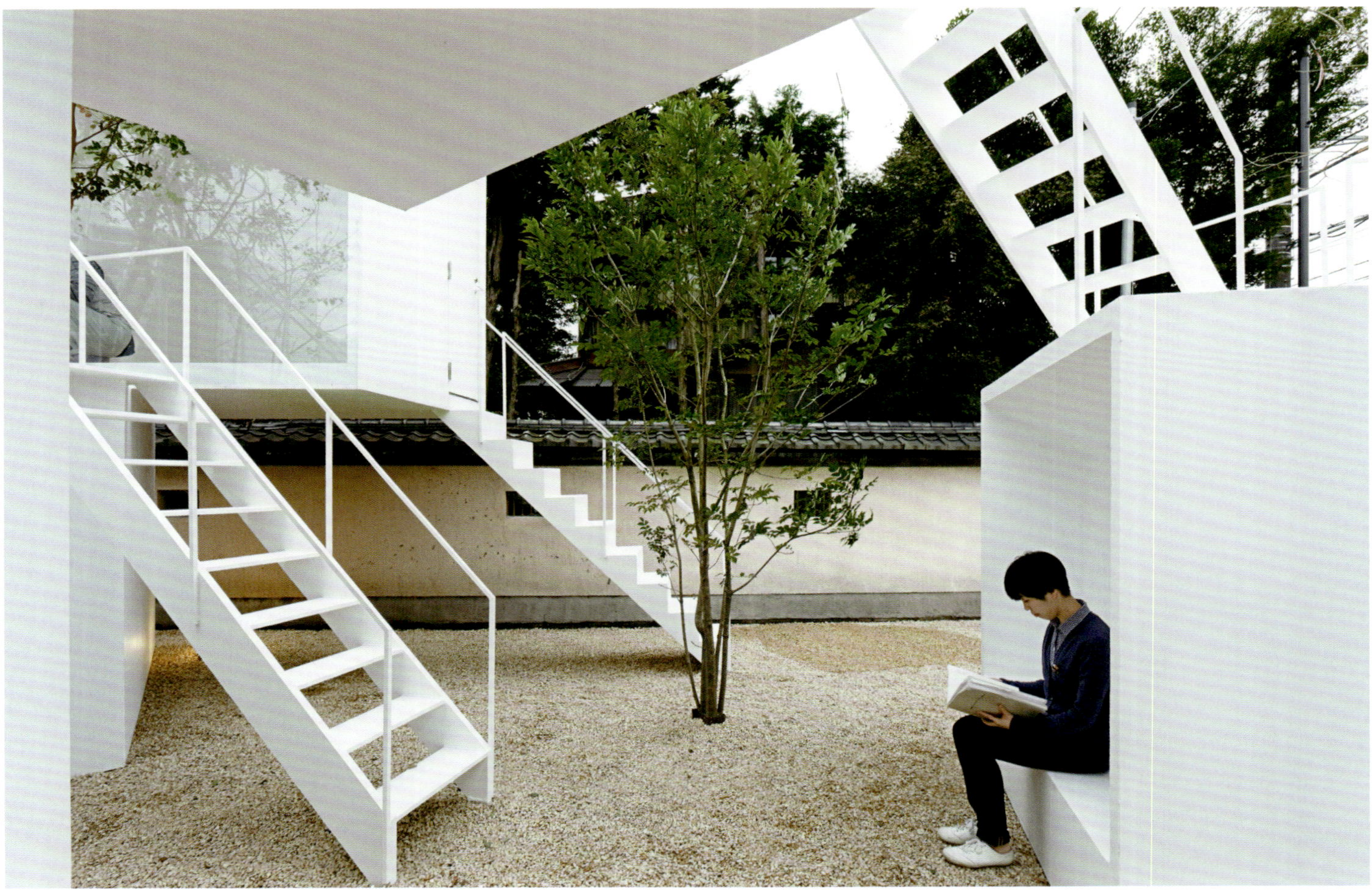

Vertical section
Scale 1:20

1 Double glazing in wooden frame:
2× 5 mm toughened glass +
15 mm cavity
2 3.2 mm welded steel sheeting
with polyurethane coating
30 mm rigid-foam insulation
between
5/50 mm steel flats
12/45 mm battens point fixed
with 50/50/6 mm steel angles
12.5 mm plasterboard
3 Step: 4.5 mm steel plate with
polyurethane coating
4 100/100/7 mm steel angle
5 3.2 mm welded steel sheeting
with polyurethane coating
30 mm rigid-foam insulation between
5/75 mm steel flats
15/45 mm battens point fixed with
50/50/6 mm steel angles
9.5 mm plasterboard
6 5 mm toughened glass
7 12 mm spruce parquet, waxed
12 mm plywood
45/45 mm battens point fixed with
50/50/6 mm steel angles
with 45 mm insulation
50 mm rigid-foam insulation between
5/75 mm steel flats
3.2 mm welded steel sheeting
8 22/22 mm steel profile balustrade
9 Planting box:
50 mm bed of gravel
Lightweight granules
root-resisting layer
100 mm granular drainage bed
Sealing layer; 3.2 mm welded steel
sheeting with polyurethane coating
10 150/50/2.3 mm welded sheet-steel
run-off
with polyurethane coating
11 12 mm spruce parquet, waxed
12 mm plywood; 45/45 mm battens
45 mm insulation
210 mm reinforced concrete slab

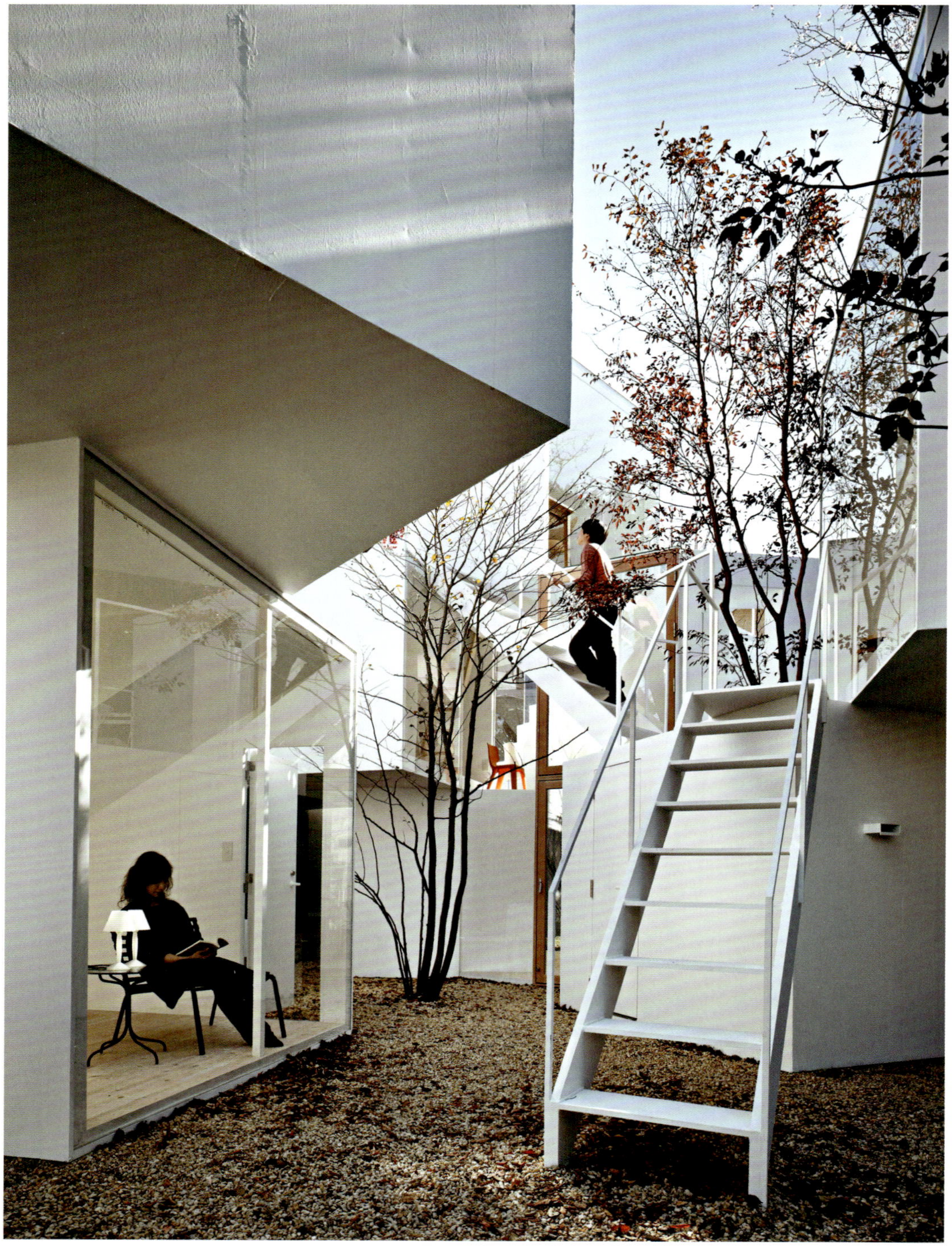

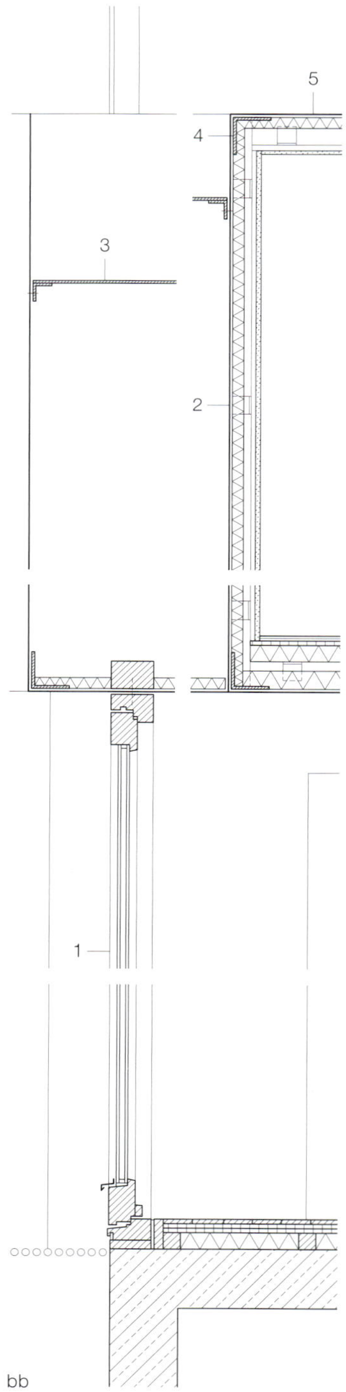

bb

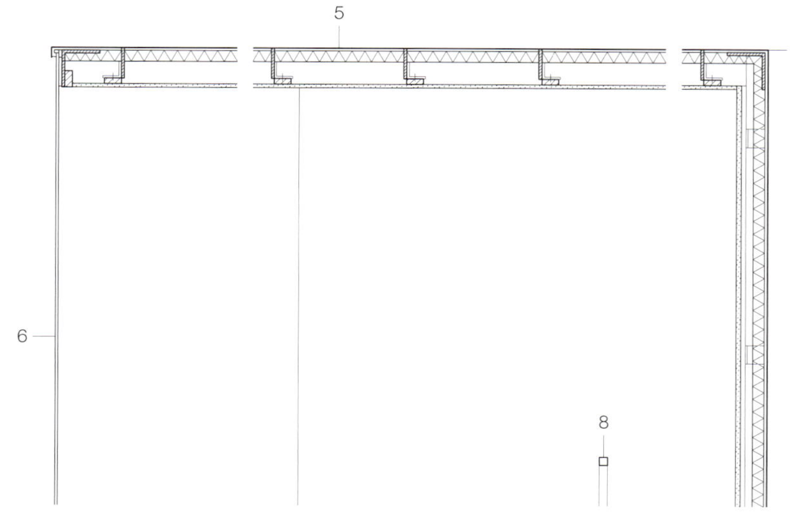
5
6
8

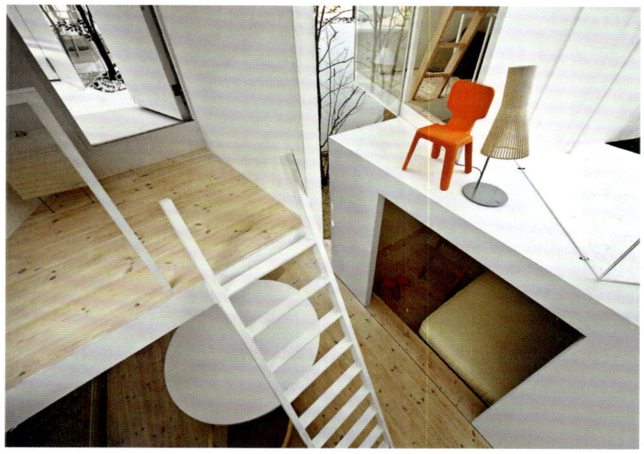

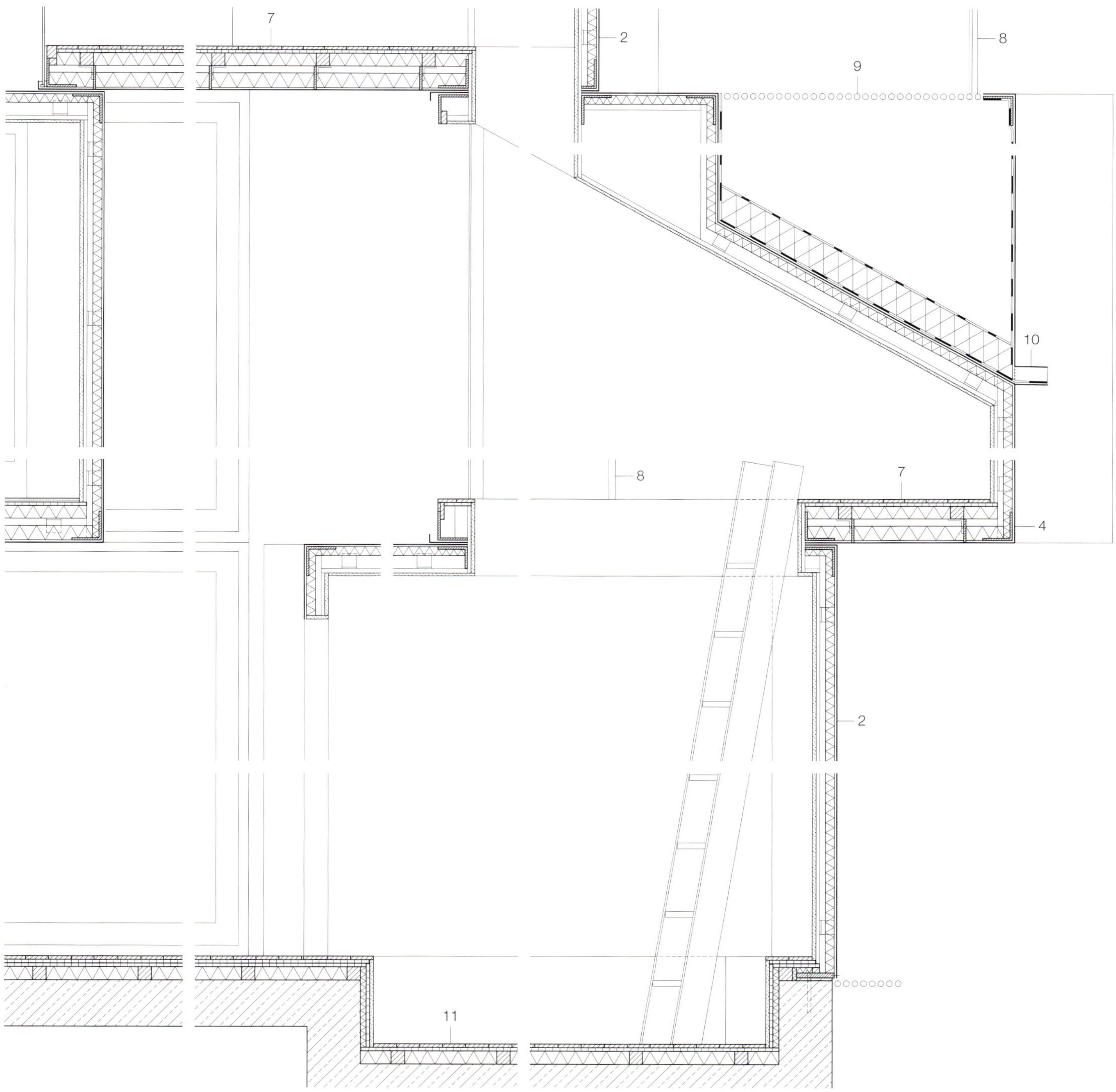
7
2
8
9
10
8
7
4
2
11

Centre for senior citizens in Maienfeld

Architects: Arbeitsgemeinschaft Isler Gysel / bhend.klammer architekten, Zurich

As the focal point of the building, the atrium creates visual links which connect the residential areas and provides the residents with a venue for social contact.

The Alterszentrum Bündner Herrschaft can accommodate approximately 60 residents in 54 care rooms and four sheltered apartments. It is situated in the immediate vicinity of the historical town centre of Maienfeld, directly beneath Brandis Castle, and enjoys open views of the surrounding area on all sides. With its projections and recesses and multi-angular layout, the intricately structured building has a scale to match the surrounding green areas of the Chur Rhine Valley as well as providing patios for the residents on all floors.
On entering the centre for senior citizens, visitors find themselves in the three-storey central entrance hall. This polygonal atrium, which is illuminated by several large skylights in the corners, determines the internal layout of the building and serves as a central point of orientation within its spatial structure. The cafeteria is located here, providing a meeting point for social interaction. However, it is also used for various events, and in addition serves as a day care centre with lunchtime dining facilities for the adjacent school. This is a way of attracting outside life into the building and encouraging interaction between the different generations. On three sides, the central space opens onto the park-like grounds through intimate recesses. From the entrance hall, vertical access to the building is provided via two centrally situated stairwells with lifts. In the event of a fire the building is not evacuated via the central hall, so it can also absorb fire loads, which is what makes it possible to use it to its full potential. In the event of a fire, fire protection doors isolate the stairwells from the hall and so lead the escape routes directly to the emergency exits.

Spatial diversity

The care stations each extend over one floor and are organised as large "apartments" with their own kitchen and dining area. Instead of long corridors, a peripheral circulation route runs right round the floor in the form of a semi-public access space which repeatedly opens out into areas which serve as common areas and also features sheltered niches for residents who prefer a quiet retreat. From here, residents can enjoy views out of the building towards the surrounding landscape as well as glimpses of the hall through internal windows in various formats. These create interesting vistas across the central open space of the hall. The spatial diversity and choice of materials creates a pleasant atmosphere, without any "hospital" feel. Residents and staff appreciate how the inner organisation of the centre for senior citizens creates a lively surroundings.

Site plan
Scale 1:4000

1 Centre for senior citizens
2 School
3 Brandis Castle
4 Neighbouring housing

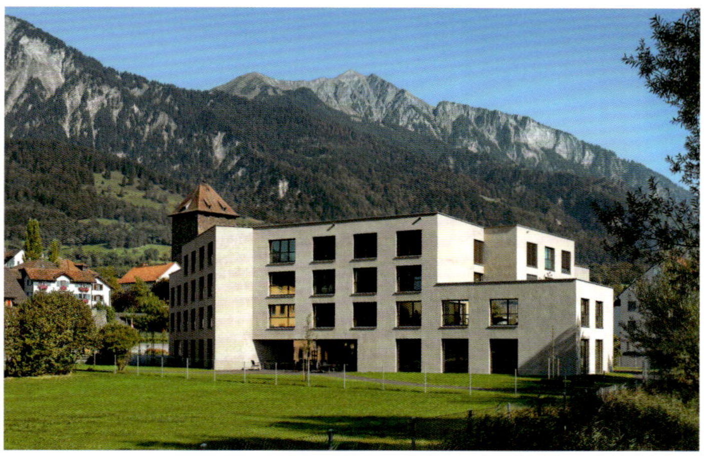

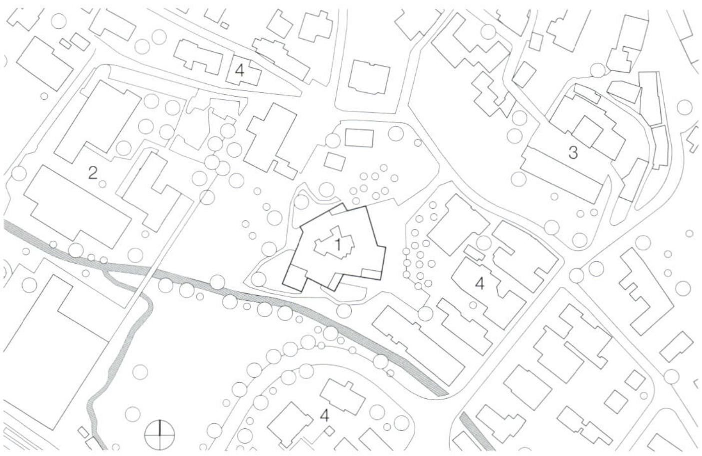

Project data:

Use:	Home for senior citizens and care home
Access:	Central hall, stairwells, lifts, peripheral corridors with common areas
Number of floors:	4 + 1 basement level
Clear room height:	2.84 m (ground floor), 2.54 m (upper floors)
Gross volume:	25,200 m^3
Gross floor area:	7,300 m^2
Year of completion:	2011
Construction period:	21 months

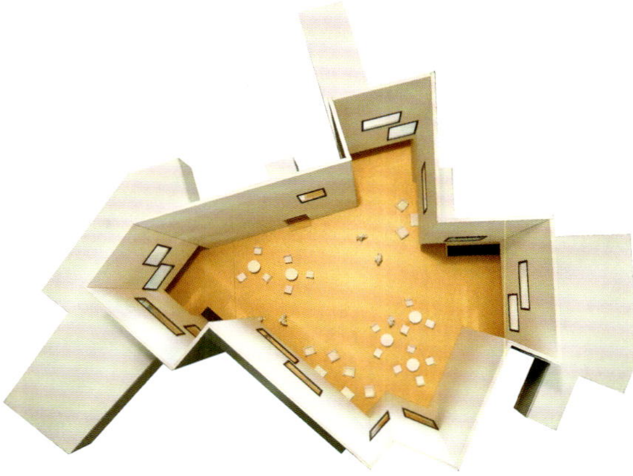

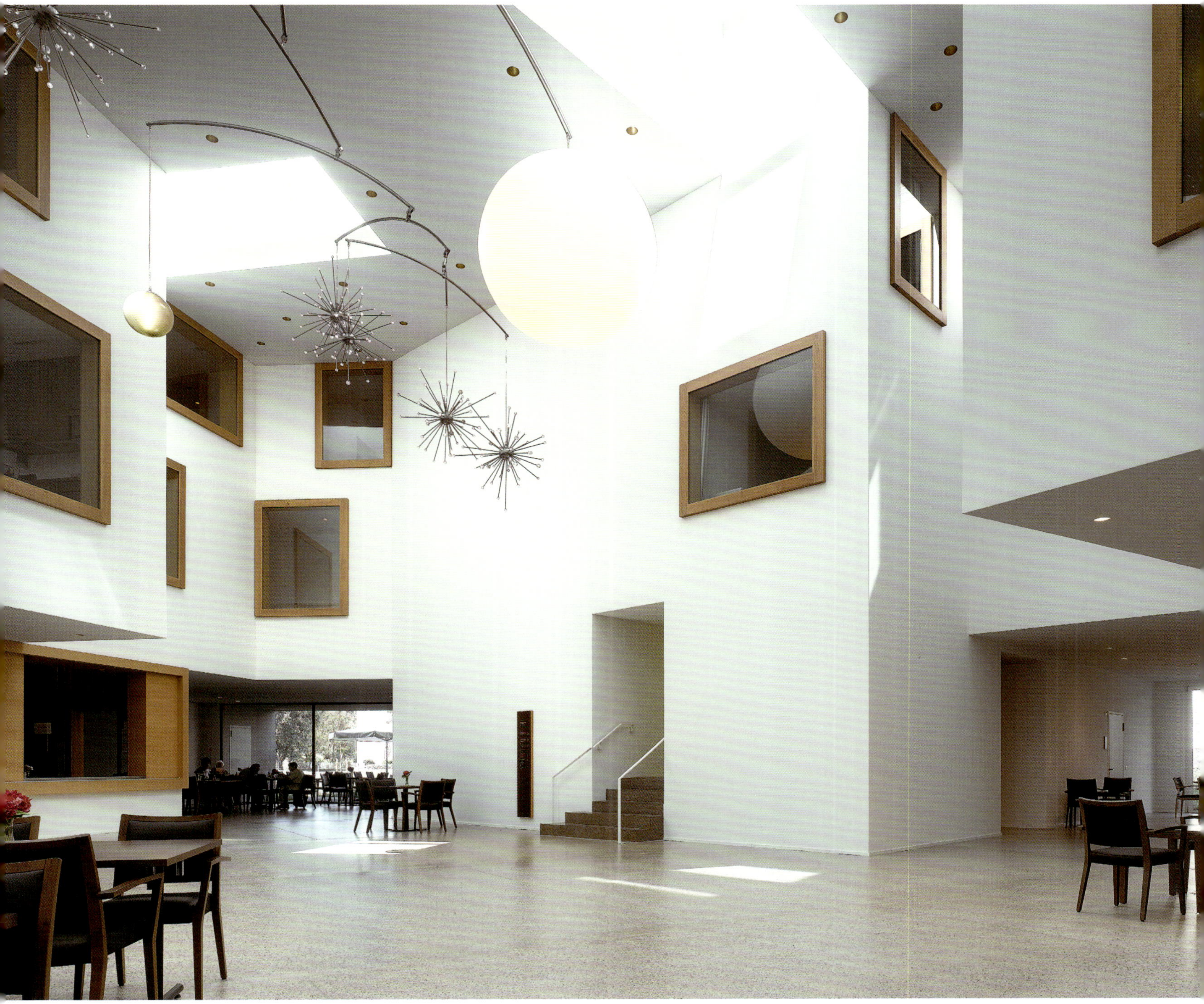

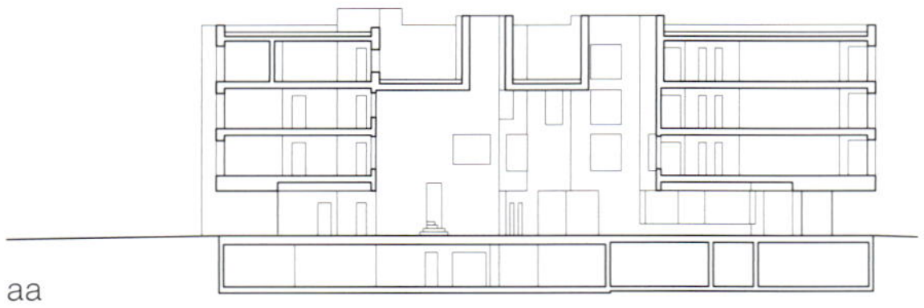

aa

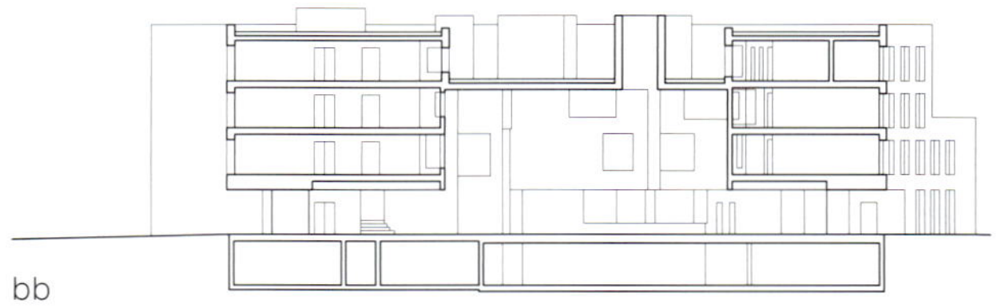

bb

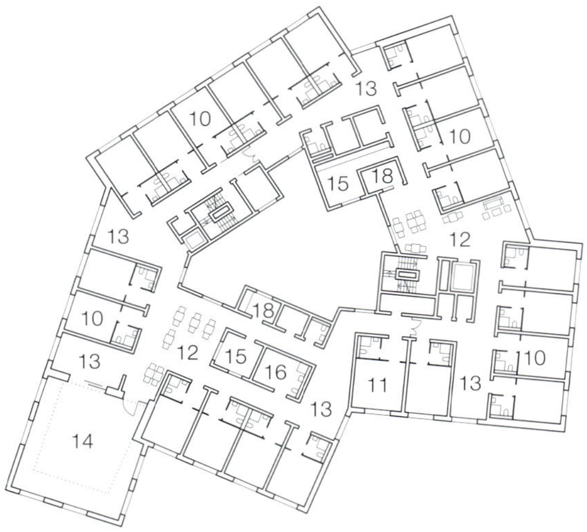

1st floor

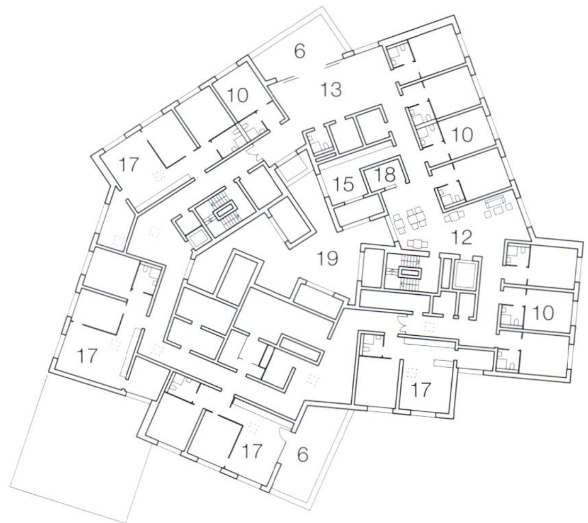

3rd floor

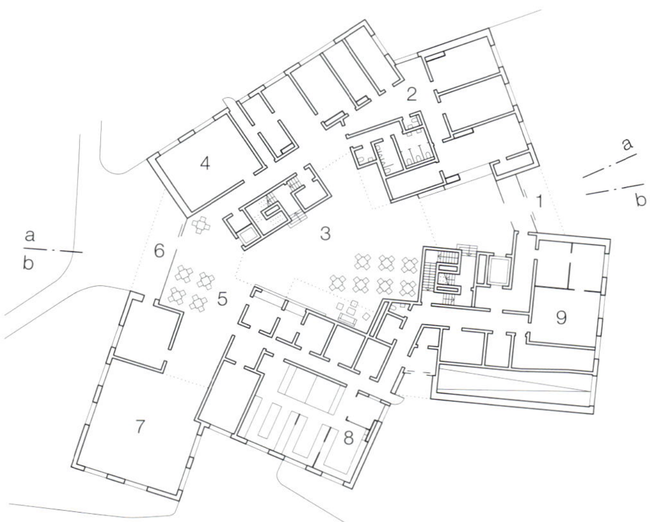

Ground floor

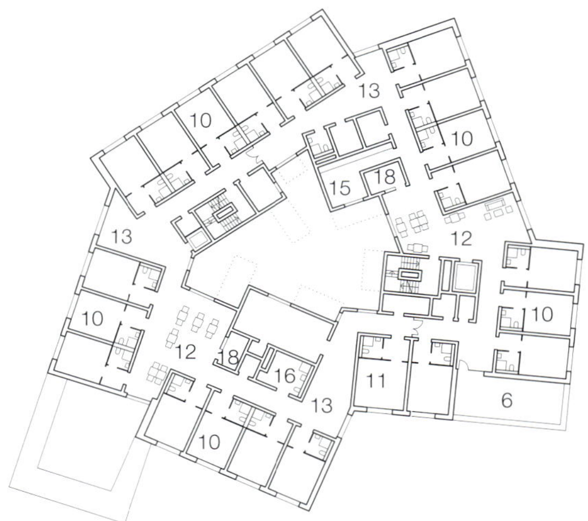

2nd floor

Sections • Floor plans
Scale 1:750

1 Entrance
2 Administration / "Spitex" care station
3 Hall
4 Occupational therapy room
5 Cafeteria
6 Terrace
7 Events space
8 Kitchen
9 Laundry room
10 Single room
11 Double room
12 Communal space
13 Quiet corner
14 Sheltered garden terrace for "forgetful" residents
15 Nurses station
16 Care bathroom
17 Sheltered accommodation
18 Ward kitchen
19 Roof garden (not accessible)

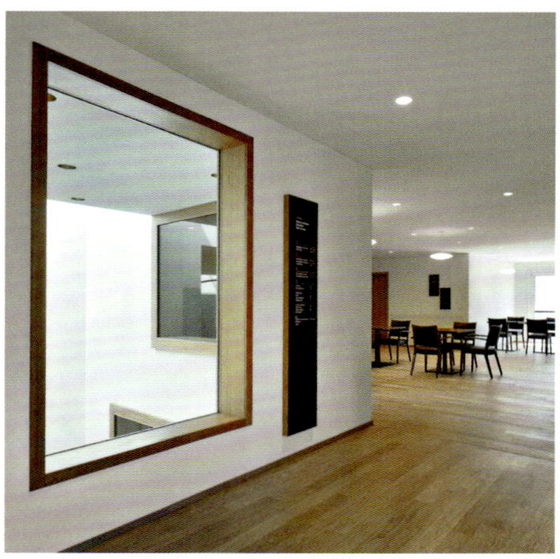

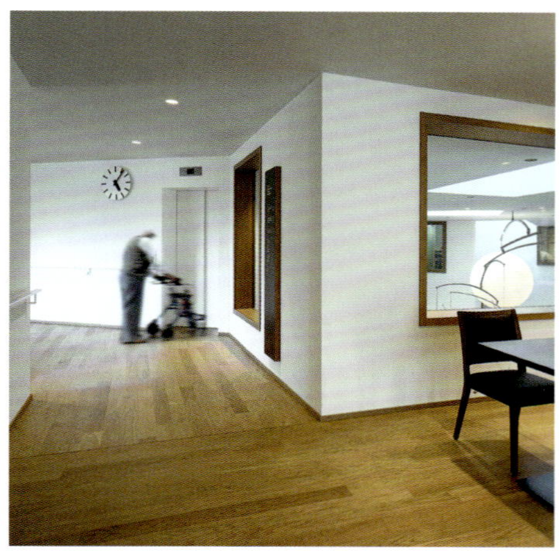

Apartment mound in Copenhagen

Architects: BIG – Bjarke Ingels Group, Copenhagen;
JDS Architects, Copenhagen

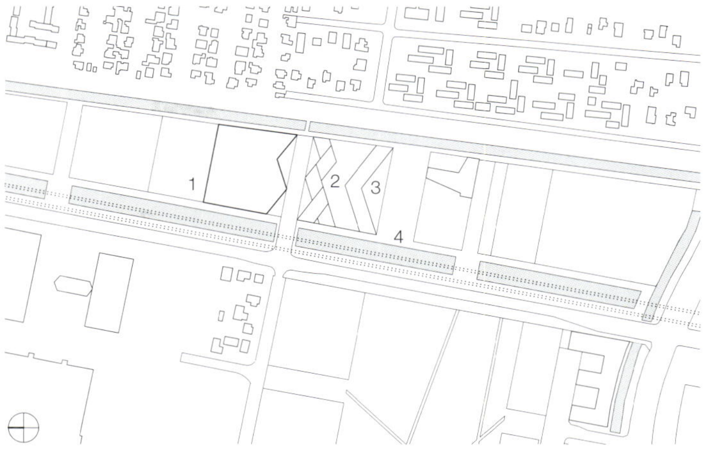

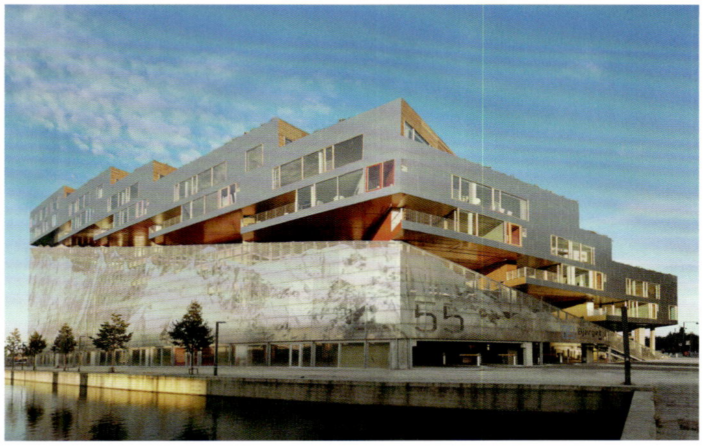

A colourful cathedral-like multi-storey car park forms the base for the terraced apartment complex, in which a sloping elevator provides access to the different floors.

Construction work on the new Ørestad City district, situated between the centre of Copenhagen and the airport, has been under way since 1992. The "Mountain" project, the third development phase situated to the north of the so-called House V and House M, represents an unconventional mix of multi-storey car park and terraced apartment complex. The client's brief required parking spaces on two thirds of the site and a residential development on the remaining third. Instead of placing two buildings next to each other, the architects combined the two functions. Thus, the multi-storey car park with its 480 parking spaces completely covers the site and serves as the base for 80 apartments, as well as offices arranged facing the street. The building descends in steps over ten storeys towards the south in generously dimensioned terraces offering views of the surrounding countryside. In contrast to the timber-clad facades of the apartments, the parking levels are clad in perforated aluminium sheeting depicting a rastered image of Mount Everest.
The parking levels are connected by a ramp. Pedestrian access to the apartments from the parking spaces is provided via steel stairways running along the edges of the building, several stairwells with lifts as well as a central sloping elevator. As it ascends, this runs beneath the internal walkways, glazed in different colours on each floor, via which the residents reach their apartments. The cathedral-like parking levels, with ceiling heights reaching up to 16 m, constitute a clear departure from the usual expectation of a multi-storey car park. The boundary between private and public space extends deep into the "mountain" – even on the upper levels, residents are able to park right outside their apartment.

Site plan
Scale 1:8000
Access plan

1 "The Mountain"
2 House M
3 House V
4 Rapid transit line
5 Stairwells with lifts
6 Main stairwell
7 Sloping elevator
8 Connection between sloping elevator and corridor
9 Corridor leading to the apartments
10 Entrance to underground car park
11 Building entrance

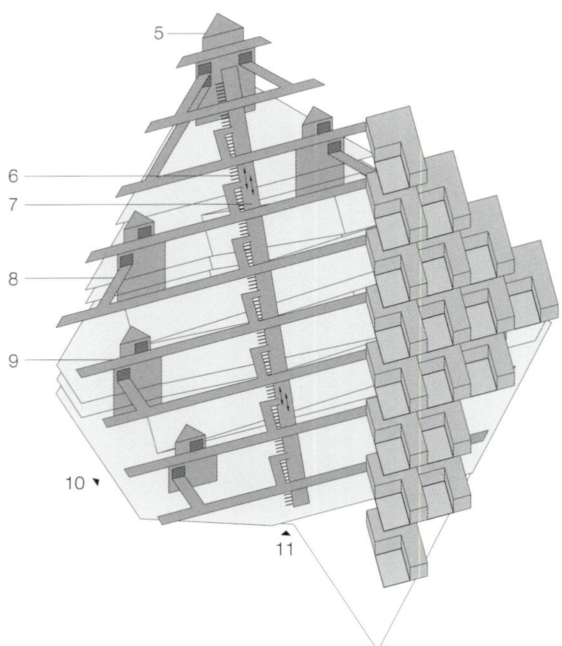

Project data:

Use:	Parking, residential
Access:	Sloping elevator, stairwells with lifts, connecting walkways leading to the apartments
Floors:	10 + 1 basement level
Clear room height:	3 m
Gross volume:	99,000 m³
Gross floor area:	33,000 m²
Year of completion:	2007

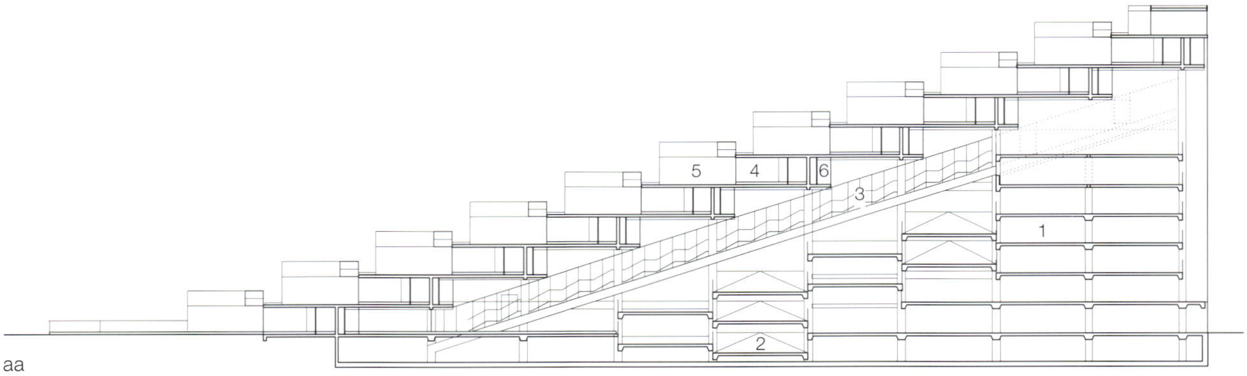

aa

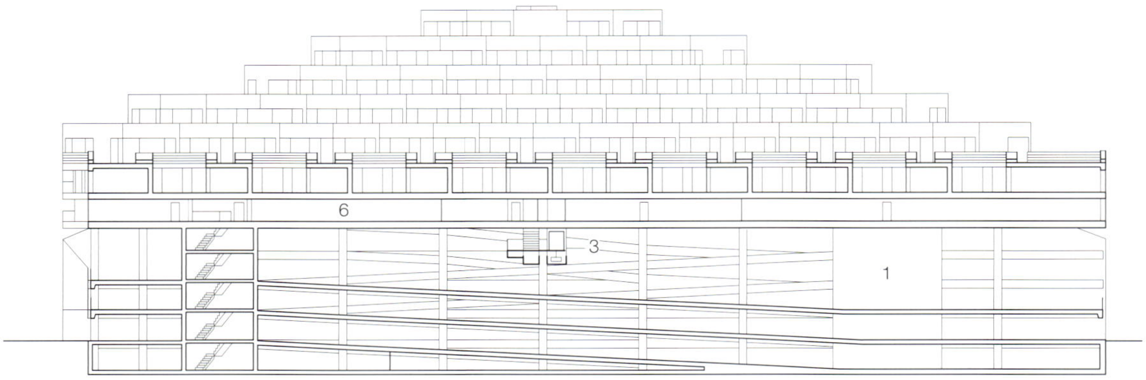

bb

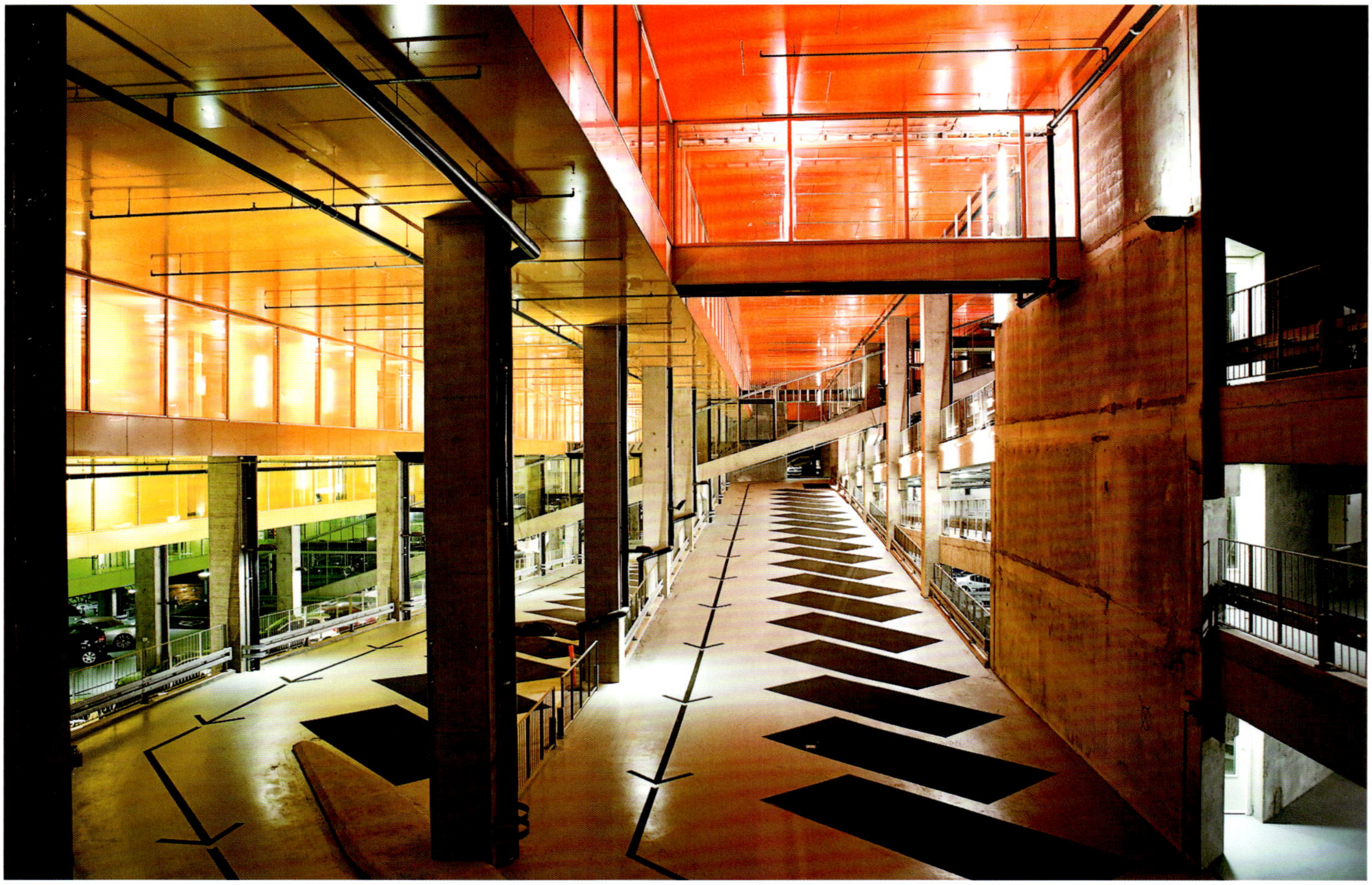

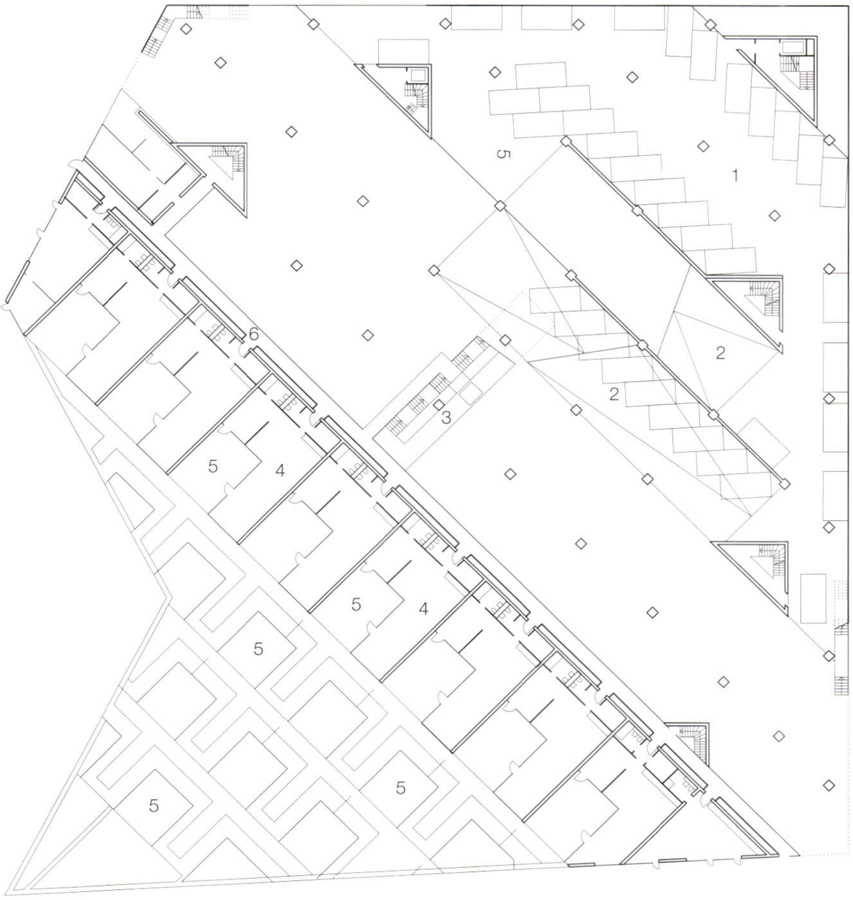

3rd floor

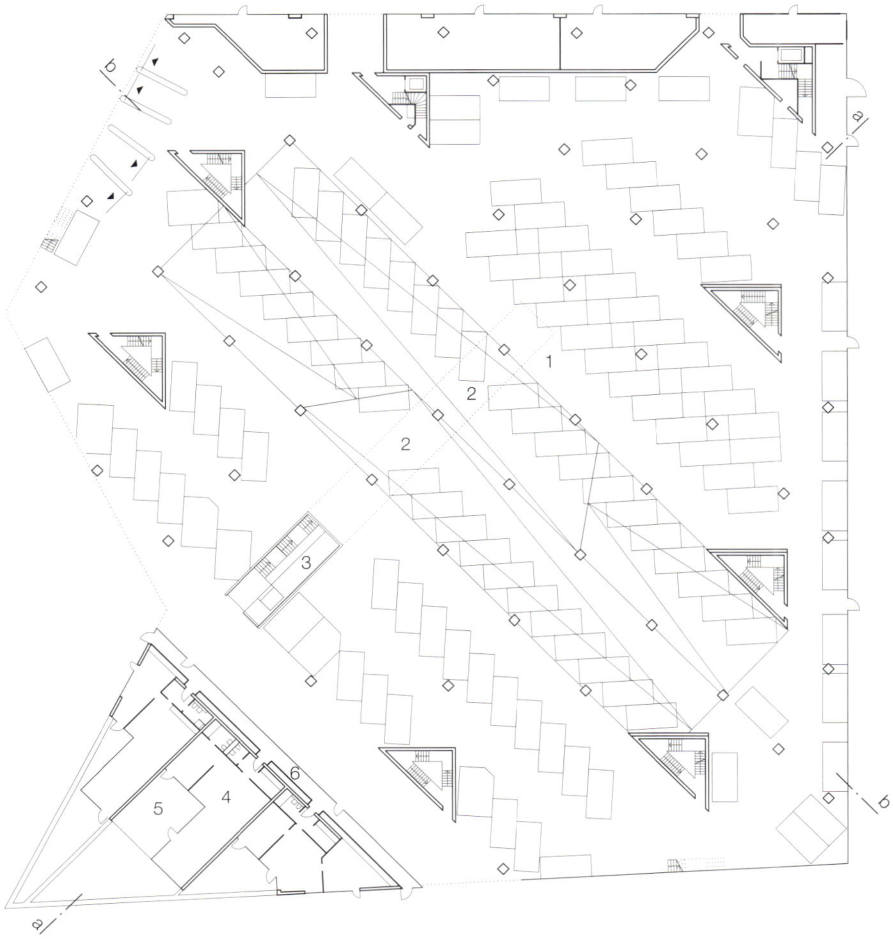

Ground floor

Sections • Floor plans
Scale 1:750

1 Multi-storey car park
2 Ramp
3 Sloping elevator
4 Apartment
5 Terrace
6 Corridor leading
to the apartments

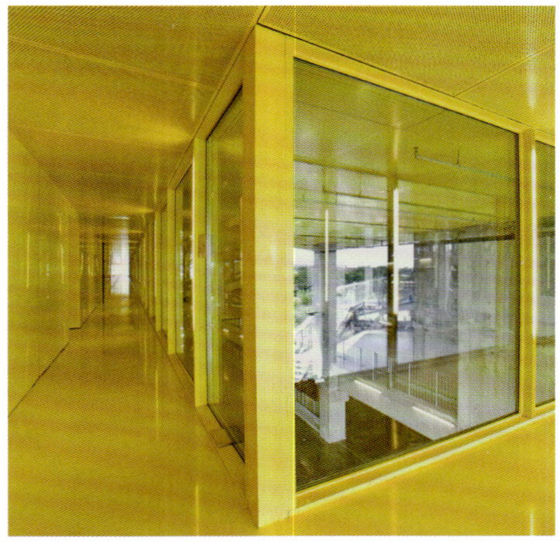

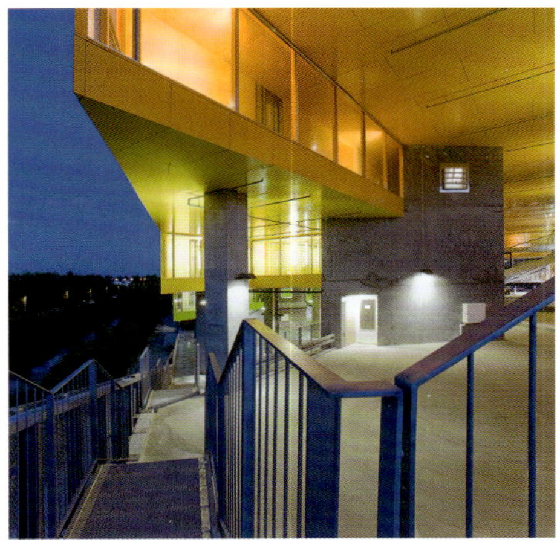

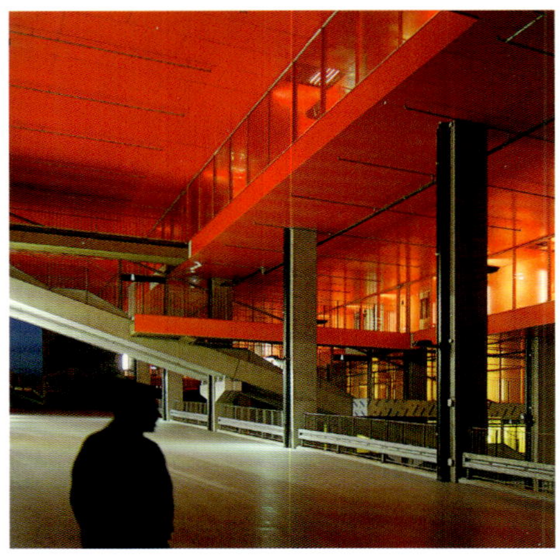

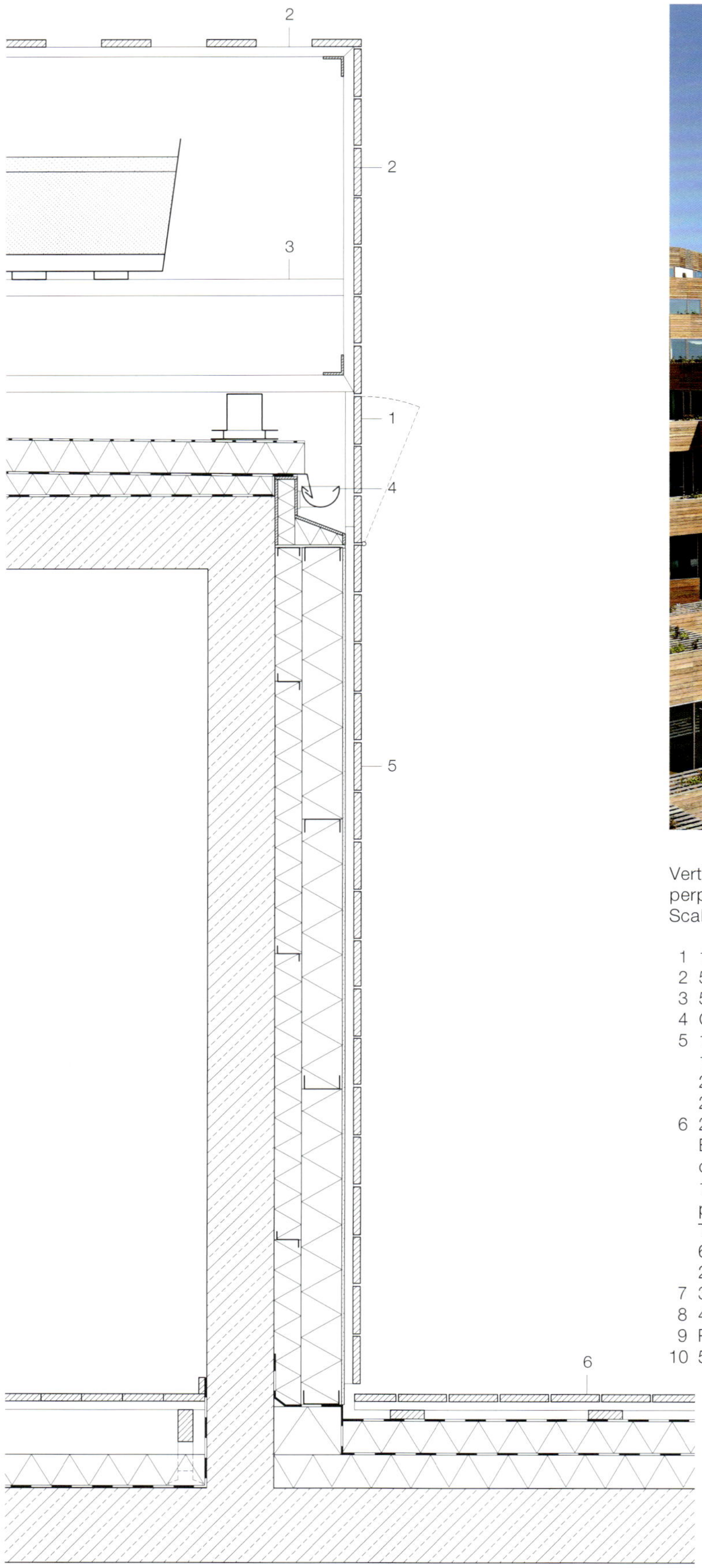

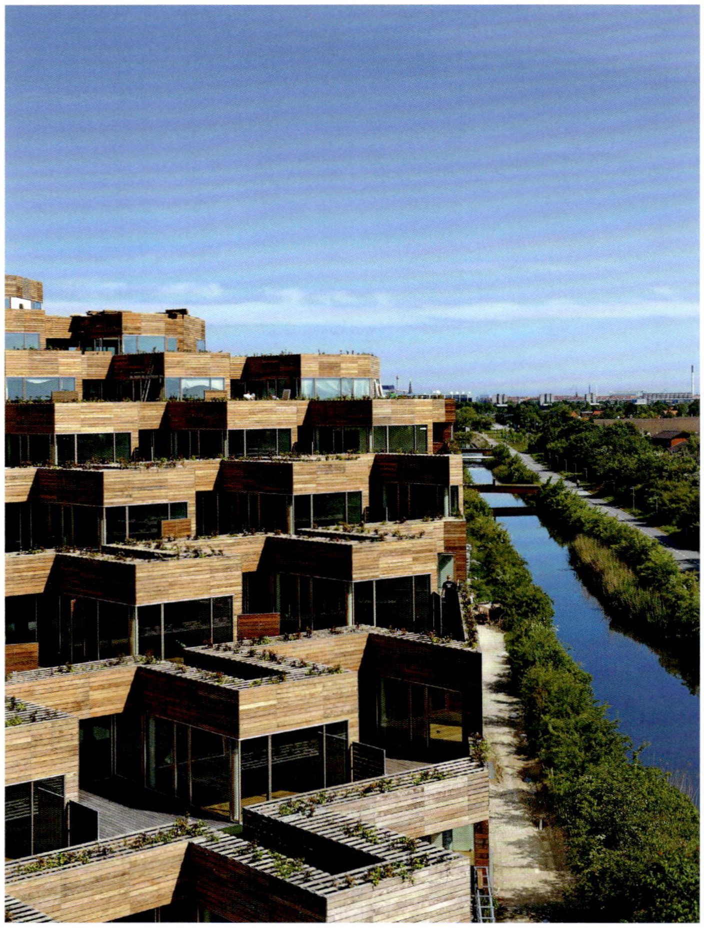

Vertical section through roof terraces
perpendicular to slope
Scale 1:20

1 145/22 mm timber cladding, Ipe
2 50/30/2 mm tubular steel, galvanised
3 50/30/4 mm tubular steel, galvanised
4 Gutter, galvanised
5 145/22 mm timber cladding, Ipe
10 mm fire-resistant panel
200 mm thermal insulation
200 mm reinforced concrete, painted
6 22 mm Ipe decking, untreated
Battening, 25/120 mm on
counter-battening; fireproof sealing
100 mm thermal insulation extruded
polystyrene
Two-layer bituminous sheet sealing
60–100 mm foam glass on slope
220 mm reinforced concrete
7 30/50/4 mm tubular steel, galvanised
8 4 mm aluminium cladding
9 Plastic planter
10 50/50/4 mm tubular steel, galvanised

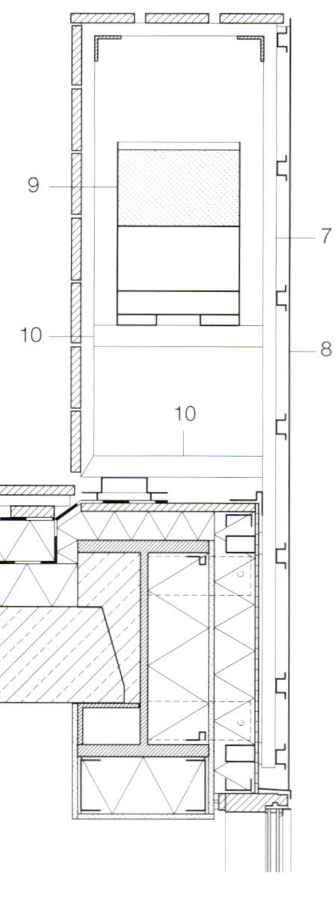

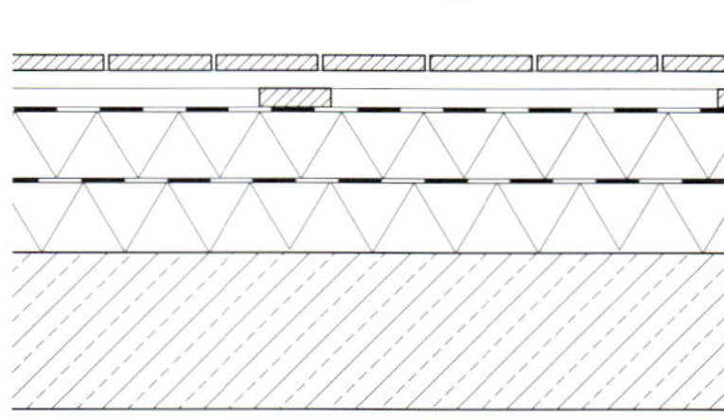

Hostel in Split

Architects: STUDIO UP, Zagreb

The windows of the hostel facing the square glow invitingly in bright yellow. The corridor layout and zoning of the rooms follow the pattern of the facade.

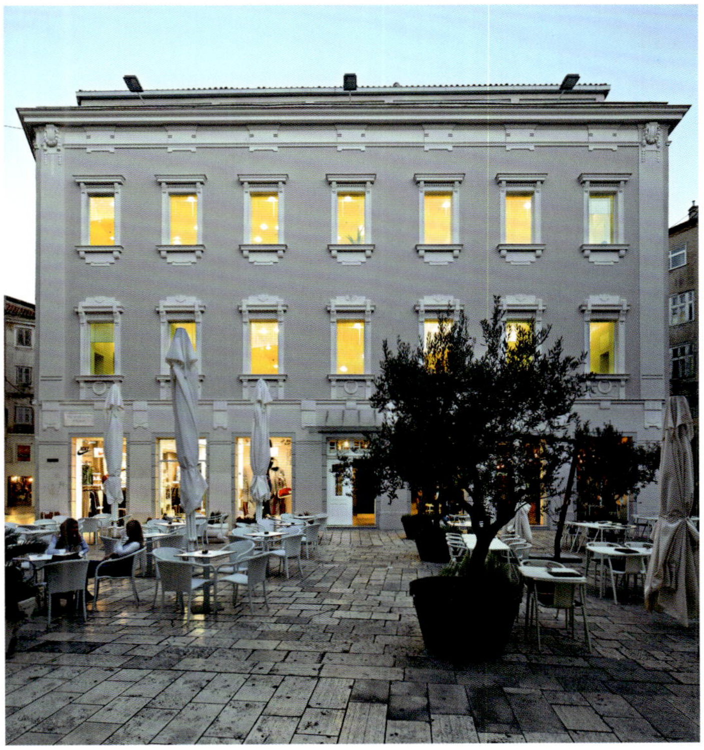

The hostel occupies a three-storey, mansion-like townhouse situated in the historical city centre of Split. Prior to the conversion, the building, built at the end of the 19th century, was for ten years home to a department store. The panoramic lift and escalator date from this time, as does the stairwell, which in 2011 was remodelled using simple means to create a social space within the building: the balustrades were covered with plasterboard panels, the reinforced concrete floors were simply coated with epoxy resin. The architects developed the interior layout of the three floors around this central access zone, making optimal use of the space. With their sleeping compartments, integrated lockers, waste bins and bathroom areas, the room-high fixtures constructed of white OSB or chipboard panels are reminiscent of Japanese capsule hotels. In total, 29 rooms with 138 beds were accommodated in the 3.50 metre high former sales floors, which contain virtually no large fixtures or partition walls. The conversion only took 100 days. Whereas on the first two floors short side corridors provide access to the dormitories holding up to eight 8 sleeping compartments, the attic floor contains apartments with sleeping galleries which are reached by means of simple ladders. The rooms on the second floor are somewhat more spacious; the few rooms only containing a double bed are also located on this floor.

Pictograms used for orientation

Clearly understandable signage on floors and walls provides information about the city and guides guests through the building. Pictograms on the floor identify the functions of the fixtures in the rooms. Super-sized numbers in the stairwell identify the different floors. In contrast to the white room fixtures, the stairwell, including all balustrades, floors and ceilings, as well as the media room with its seating steps are finished in an intense brilliant yellow – reminiscent of the city's famous sulphur baths. In order to guarantee natural lighting and ventilation, the zoning of the rooms and the access areas follow the window pattern of the original classical facade. Thus, with its glowing yellow window openings the building presents an inviting face to the square outside. On the ground floor, guests are greeted by a large oval black counter fixture which serves simultaneously as reception desk, room divider and, on the rear side, as the bar of the in-house restaurant.

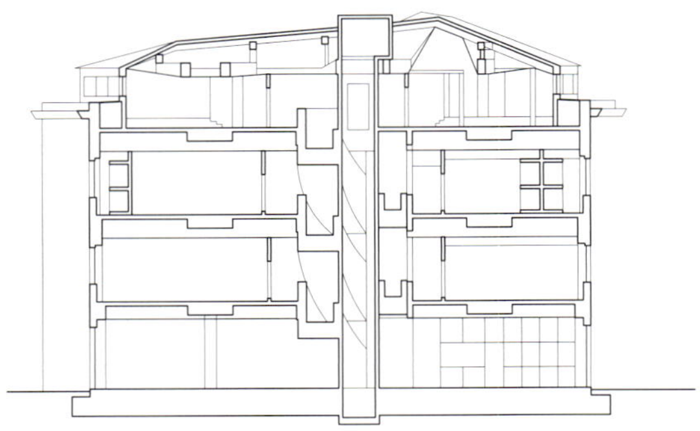

aa

Section Scale 1:500

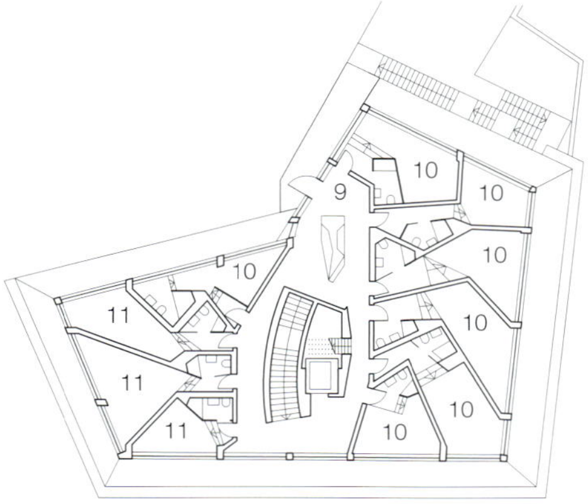

Attic floor

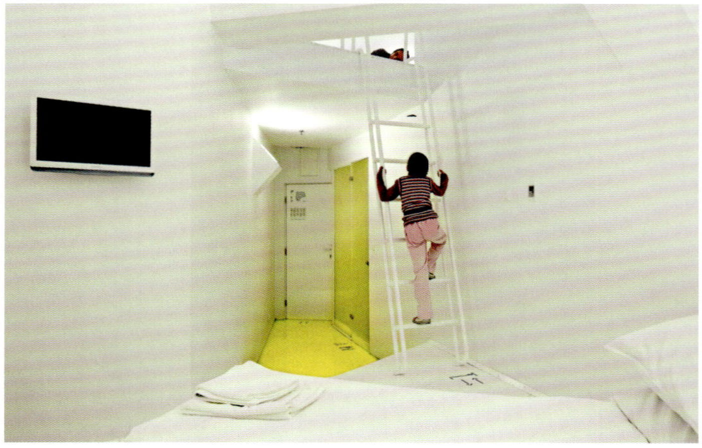

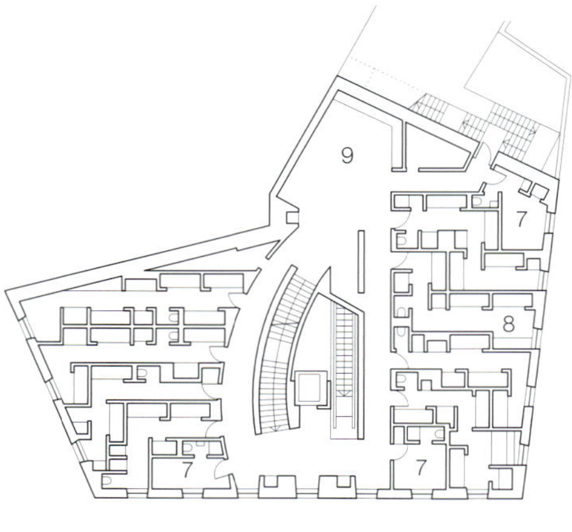

2nd floor

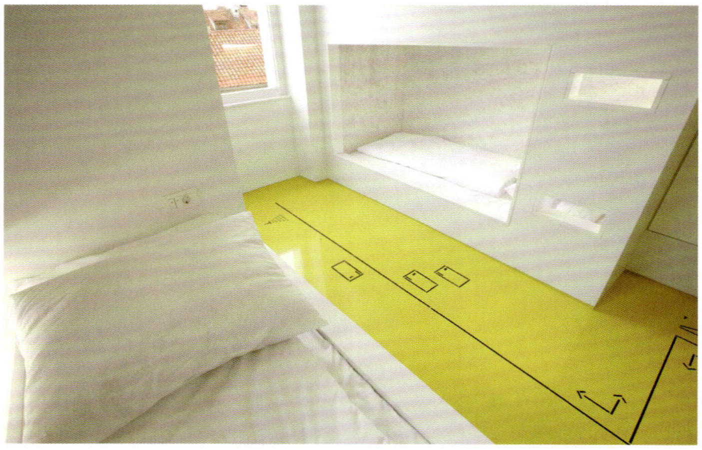

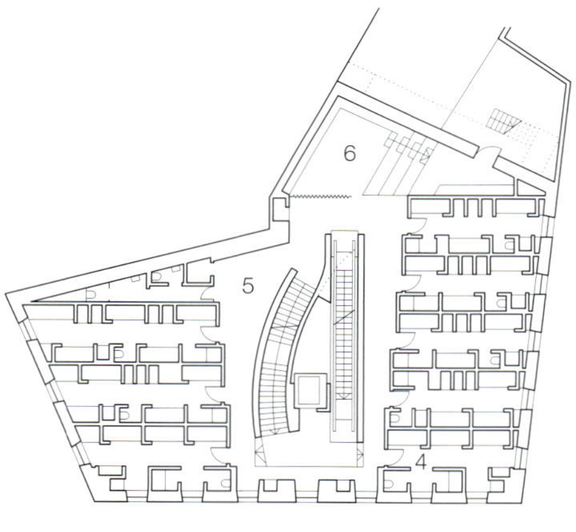

1st floor

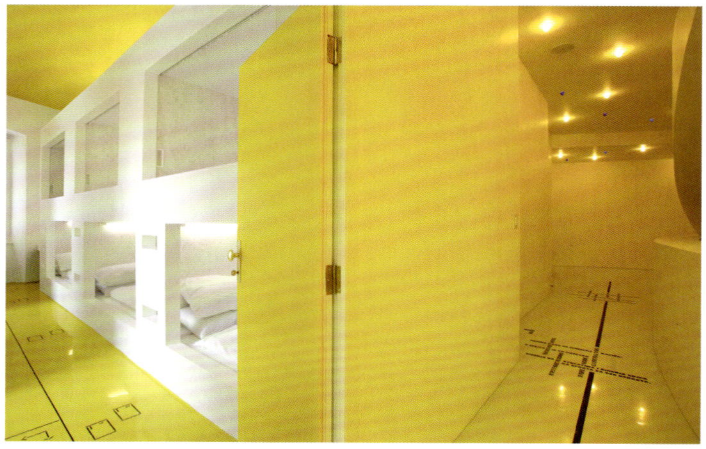

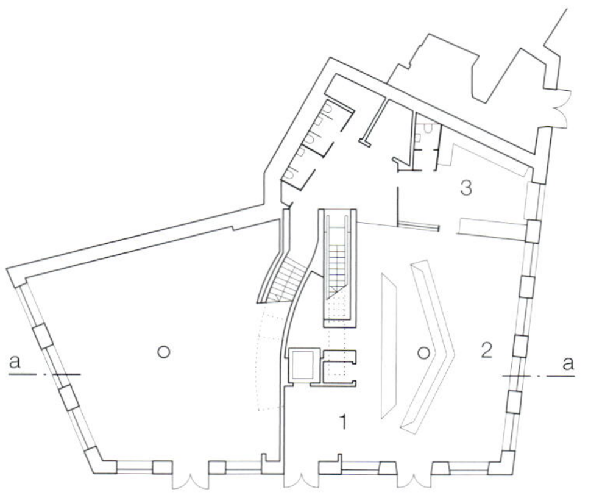

Ground floor

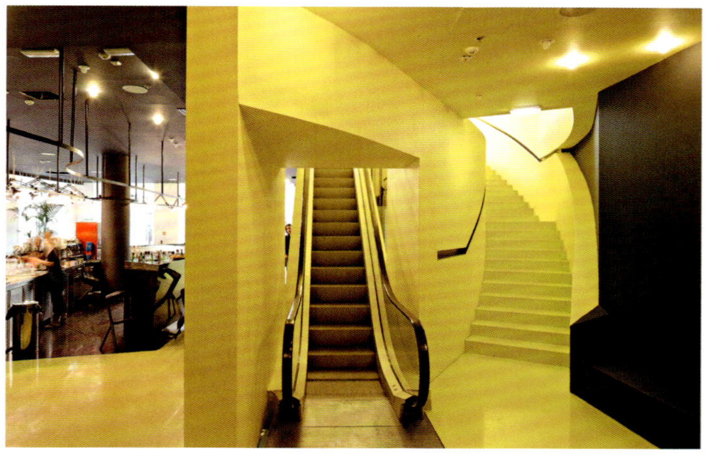

Floor plans
Scale 1:500

1 Reception
2 Bar/restaurant
3 Kitchen
4 6-bed dormitory
5 Seating area
6 Media room with seating steps
7 Single/double rooms
8 Multiple-bed room
9 Communal kitchen/dining area
10 Apartment (with gallery)
11 Apartment (without gallery)

Sectional detail of room fixtures
Sleeping alcoves and washing area
Scale 1:20

12 Wall structure:
2× 12 mm OSB, painted white
50/50 mm steel U-section
100 mm insulated cavity
12 mm OSB, painted white
13 Suspended ceiling:
2× 12 mm OSB, painted yellow
50/50 mm square timber
50/50 mm steel U-section
14 19 mm MDF as locker front,
painted white with snap closure
Cupboard fittings made
of 18 mm chipboard
15 Slatted frame for mattress
16 Curtain rail for fabric curtain
17 Fluorescent tube, fixed to OSB
18 12 mm OSB in washing area and side-
boards, painted white, rubber coated
19 Waste bin mounted on MDF front
20 Epoxy resin coating, coloured yellow,
applied to reinforced concrete floor
(part of original fabric)

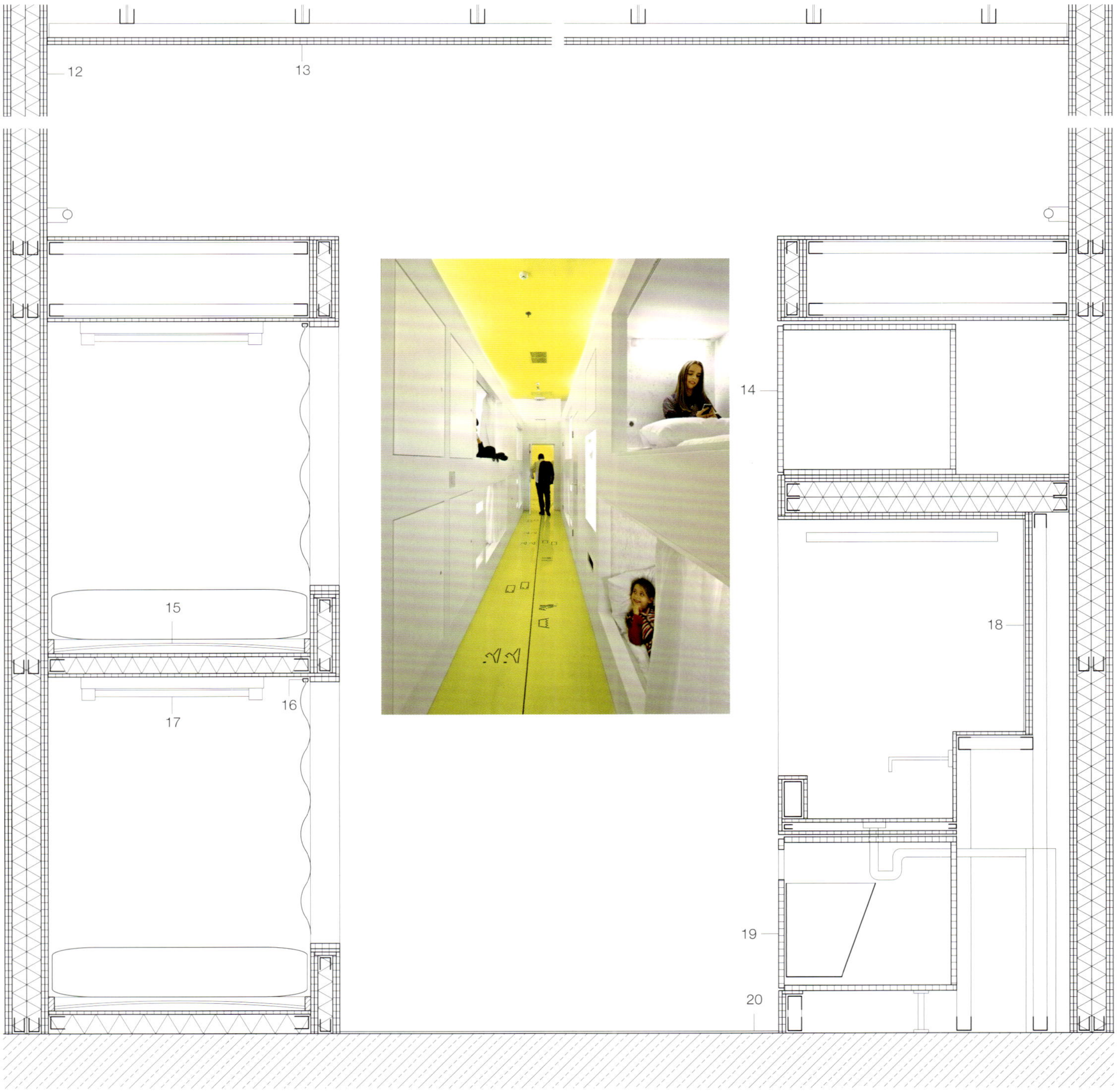

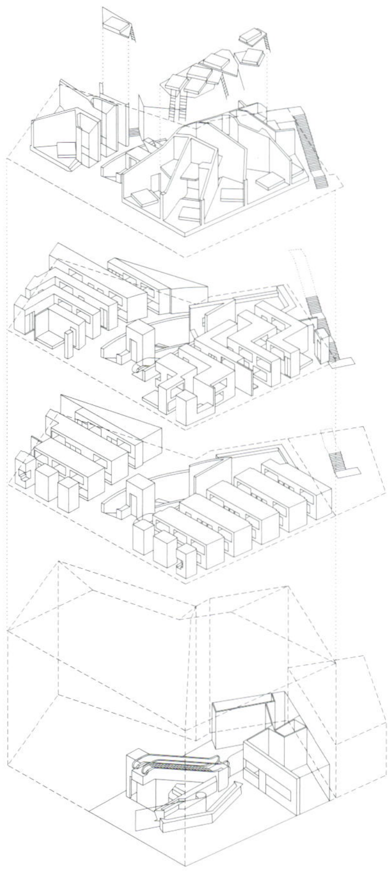

Project data:

Use:	Hostel
Access :	Panoramic lift, escalator, stairs (part of original fabric)
Number of floors:	4 + 1 gallery level
Clear room height:	3.50 m
Gross volume:	5,050 m^3
Gross floor area:	1,360 m^2
Year of completion:	2010
Construction period:	100 days

Vertical section through stair balustrades Scale 1:20

1 12.5 mm plasterboard, painted yellow
2 Supporting structure 50/25 mm steel profile
3 Handrail inset at a height of 85 mm, lit by LED lights
4 10 mm black MDF
5 70 mm balustrade (part of original fabric)
6 Yellow epoxy resin coating, applied to reinforced concrete floor (part of original fabric)

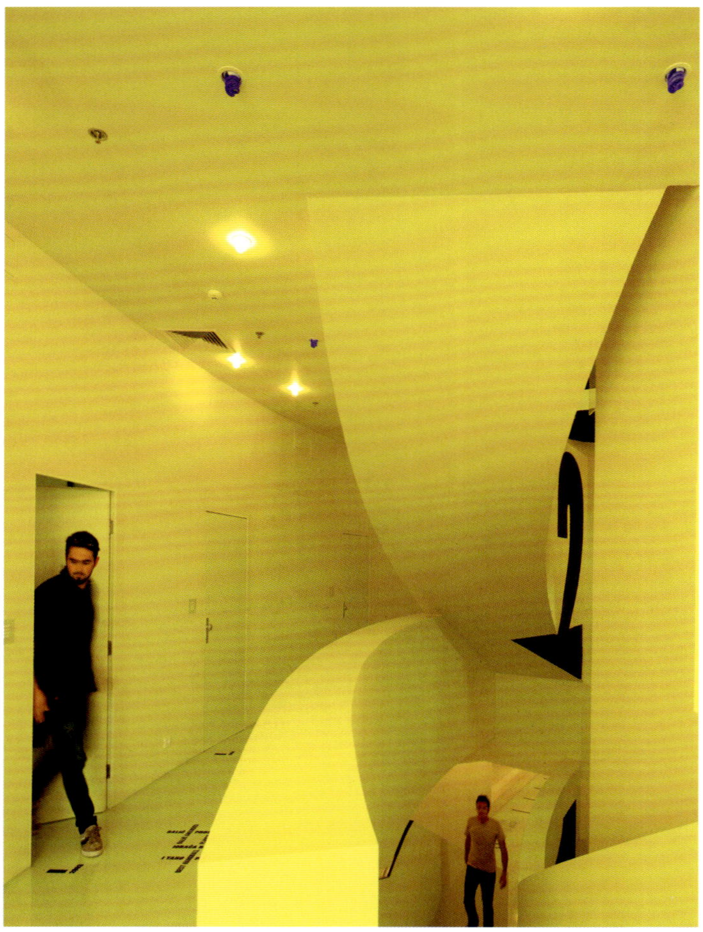

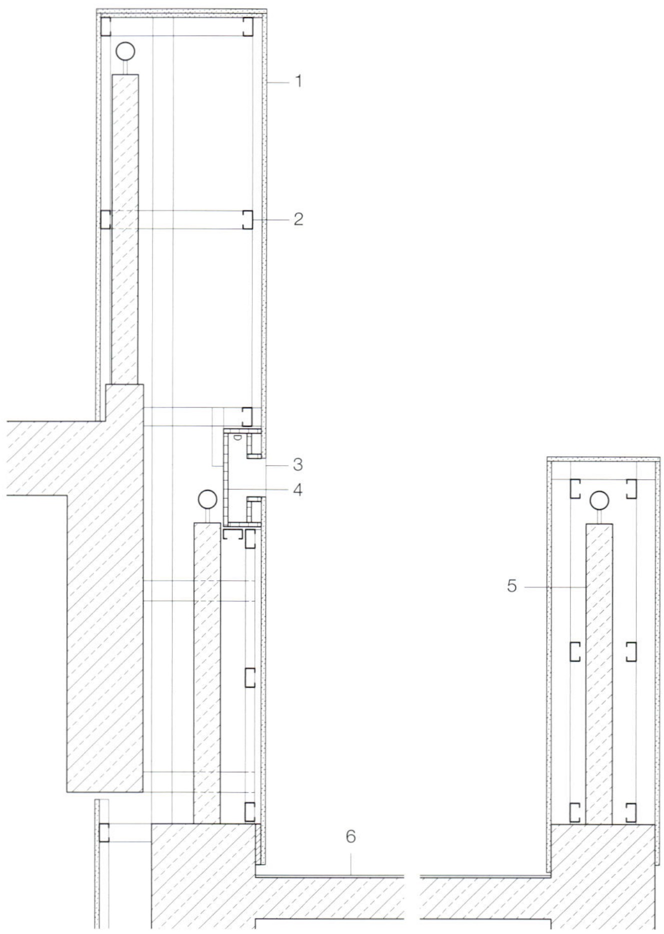

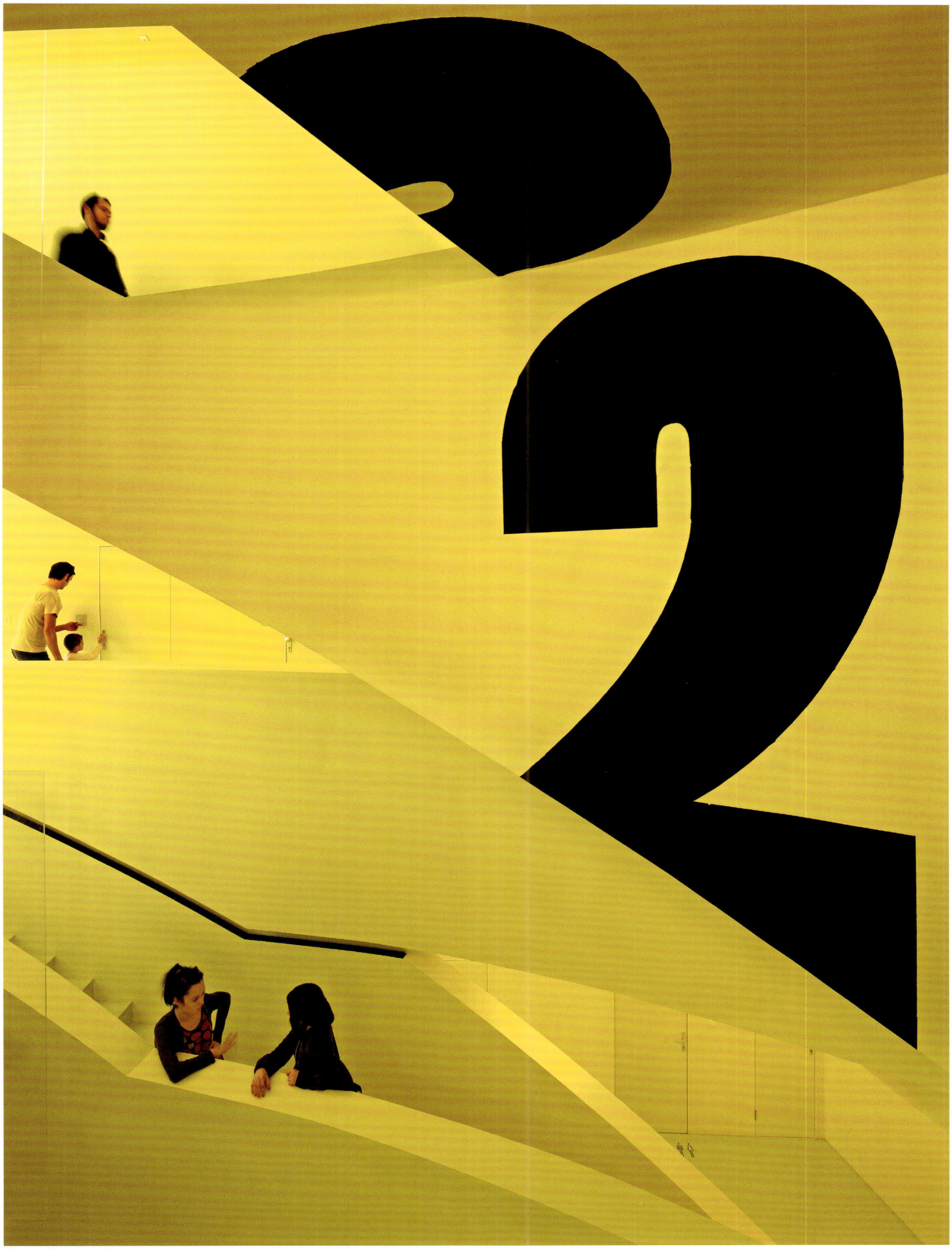

Hotel am Domplatz in Linz

Architects: hohensinn architektur, Graz

The full-height atrium, flooded with natural light and lined with balconies providing access to the guest rooms, forms the core of the hotel.

The Hotel am Domplatz is situated directly next to the neo-Gothic Mariendom cathedral, the largest church in Austria, in a puristically redesigned square in the centre of Linz. The architect's brief included not just the design of the square but also the construction of an underground car park and the revitalisation of two baroque buildings on the south-west side. A 16 cm shadow gap makes the building, with its clear-cut cubic lines, seem to float above the square. In contrast to the darkened natural stonework and stained glass windows of the cathedral, the five-storey skeleton structure of the hotel consists of pale beige fair-faced concrete and storey-high glazing with narrow larch wood frames. The building responds to the surrounding buildings with slight kinks in the facade, which slopes back as it ascends.

Interiors and materials used

The interior of the hotel is dominated by the atrium which runs through all the floors and provides natural light to the centre of the building from skylights. The slightly wedge-like form of the intervening space prevents narrow, dark corridors and lends the access area a feeling of spaciousness and transparency. This space is surrounded on the upper floors by access balconies leading to the 72 rooms. The parapet elements, which vary in height, are covered with stretched white fabric. LEDs integrated in the particle board panels periodically cause an art installation with lettering to appear on the parapet. Glass balustrades are only used on the short end sides. One of the parapet elements is continued down to around 1 m above the floor on the ground level and functions as a sight screen between the reception lobby and the breakfast area without interfering with the unity of the space which runs from facade to facade. The appearance of the hotel is characterised through the use of a warm, low-key colour palette of pale beige and brown tones for the furnishings and high-quality materials such as oak for the wall claddings and floor coverings. The room-high windows and even the partially glazed bathroom areas in the guest rooms provide a permanent sense of contact with the immediate surroundings – many of the rooms offer a close-up view of the cathedral. The recessed entrance, lift, stairwell and service rooms are located in the southern part of the building.

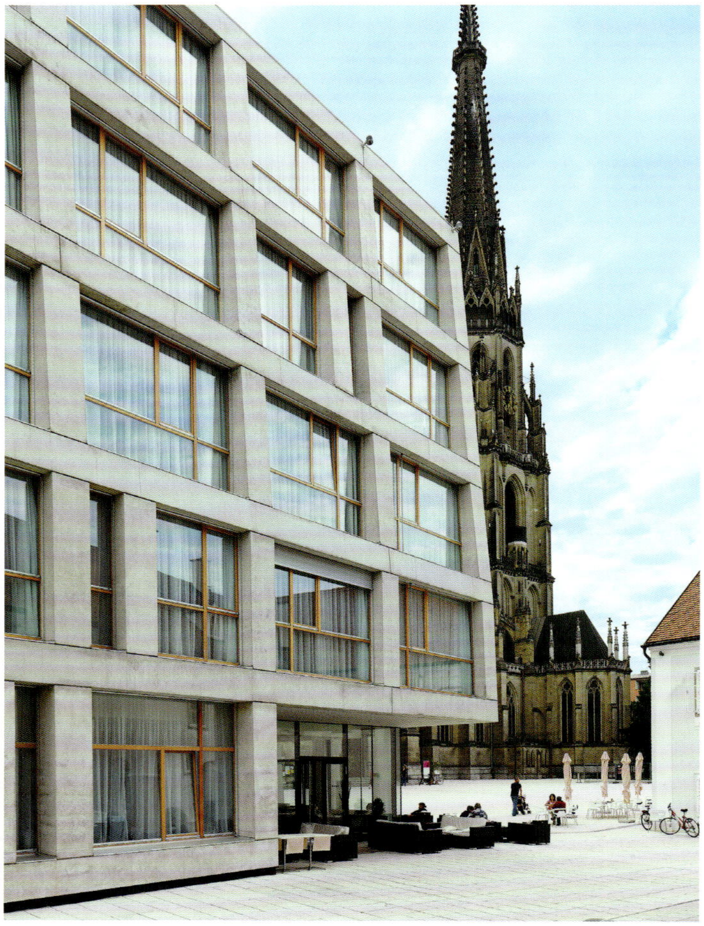

Section • Floor plans
Scale 1:500

1 Reception/entrance lobby
2 Bar
3 Breakfast area
4 Kitchen
5 Staff
6 Administration
7 Guest room
8 Void

Project data:

Use:	Hotel
Access:	Foyer/atrium, access balconies
Number of floors:	5 + 3 basement levels
Clear room height:	2.68 m
Gross volume:	14,871 m^3
Gross floor area:	4,191 m^2
Year of completion:	2009
Construction period:	24 months

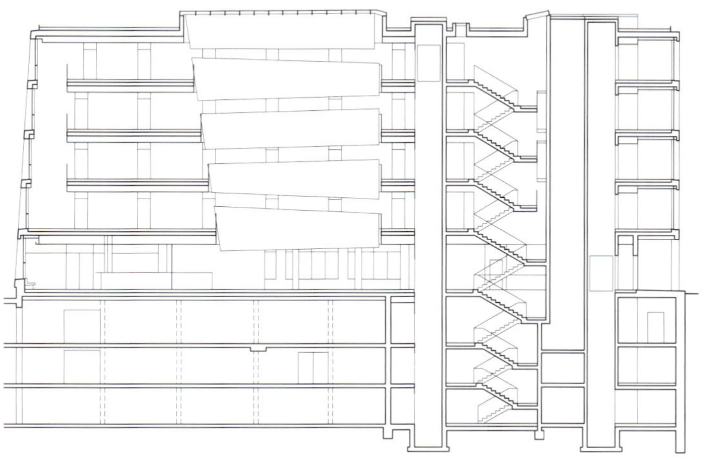

aa

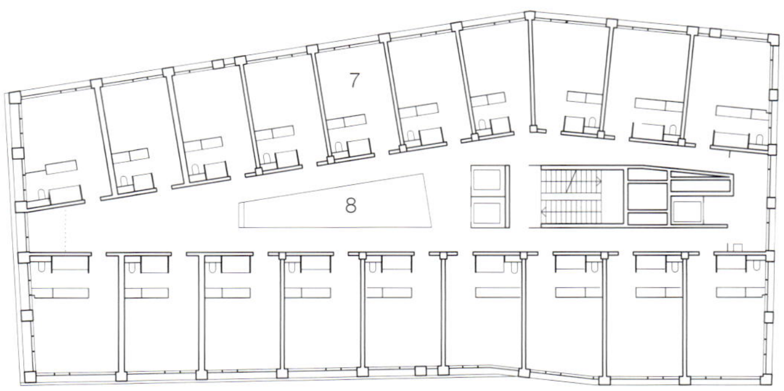

1st floor

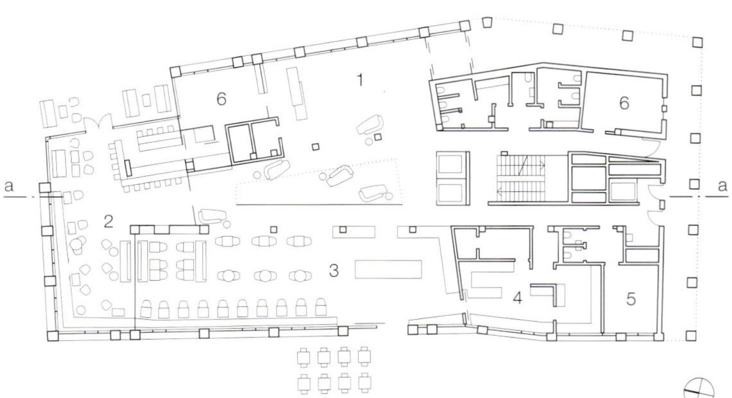

Ground floor

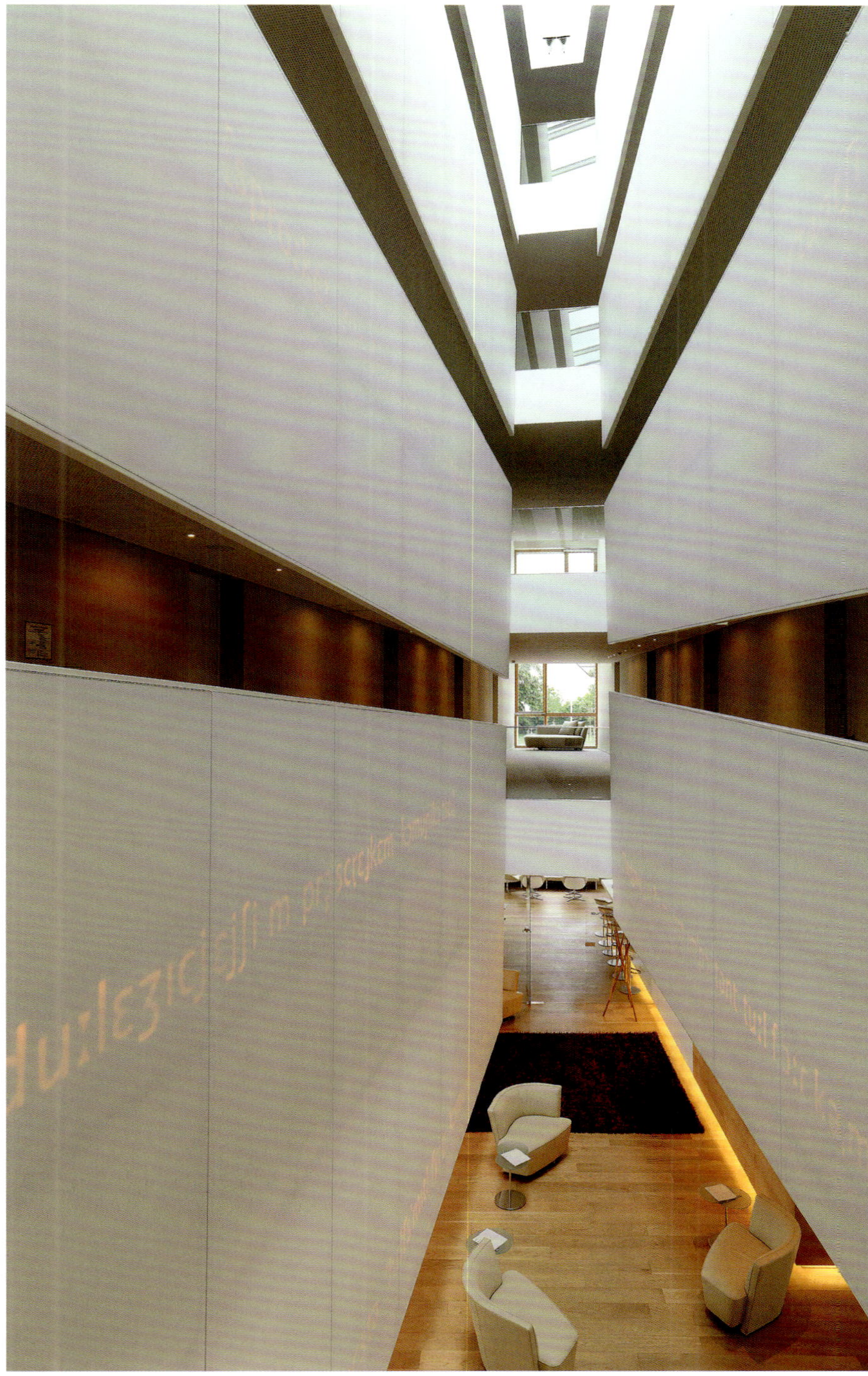

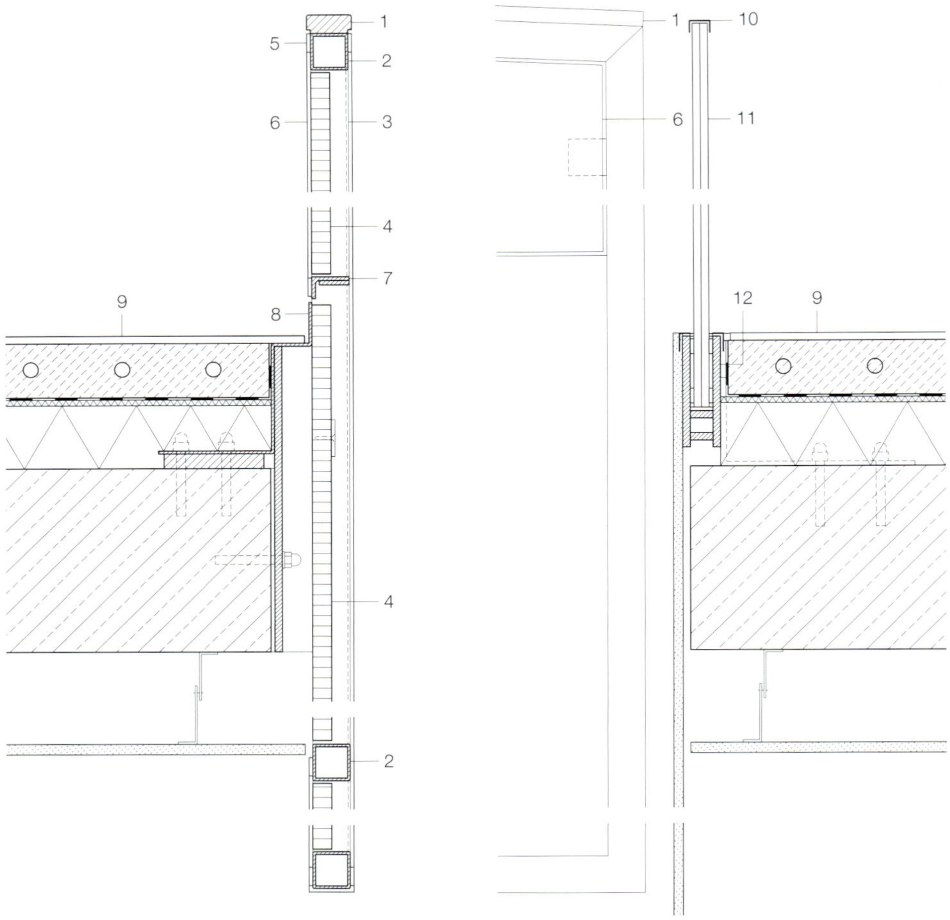

Vertical section
Balustrade • Glass balustrade
Scale 1:10

1 Handrail: 25/58 mm wood, painted white
2 50/50/2 mm hollow steel profile
3 50/50/3 mm hollow steel profile
4 25 mm chipboard, coated
in some places LEDs for art project
clamped on
5 5/25 mm surrounding plastic beading
6 Stretched white fabric, clamped in place
7 50/30/5 mm steel profile
40/6 mm flat steel
8 4 mm sheet steel, screwed
9 10 mm carpet
70 mm heated screed
PE sheeting
8 mm compression-resistant PE
footfall insulation
107 mm Styrofoam fill
250 mm reinforced concrete
17 mm plasterboard with edge protection,
perforated, suspended
10 1 mm stainless steel U-profile
11 20 mm laminated safety glass:
2× 10 mm toughened safety glass
12 2× 150/10 mm flat steel, primed, safety
glass in between

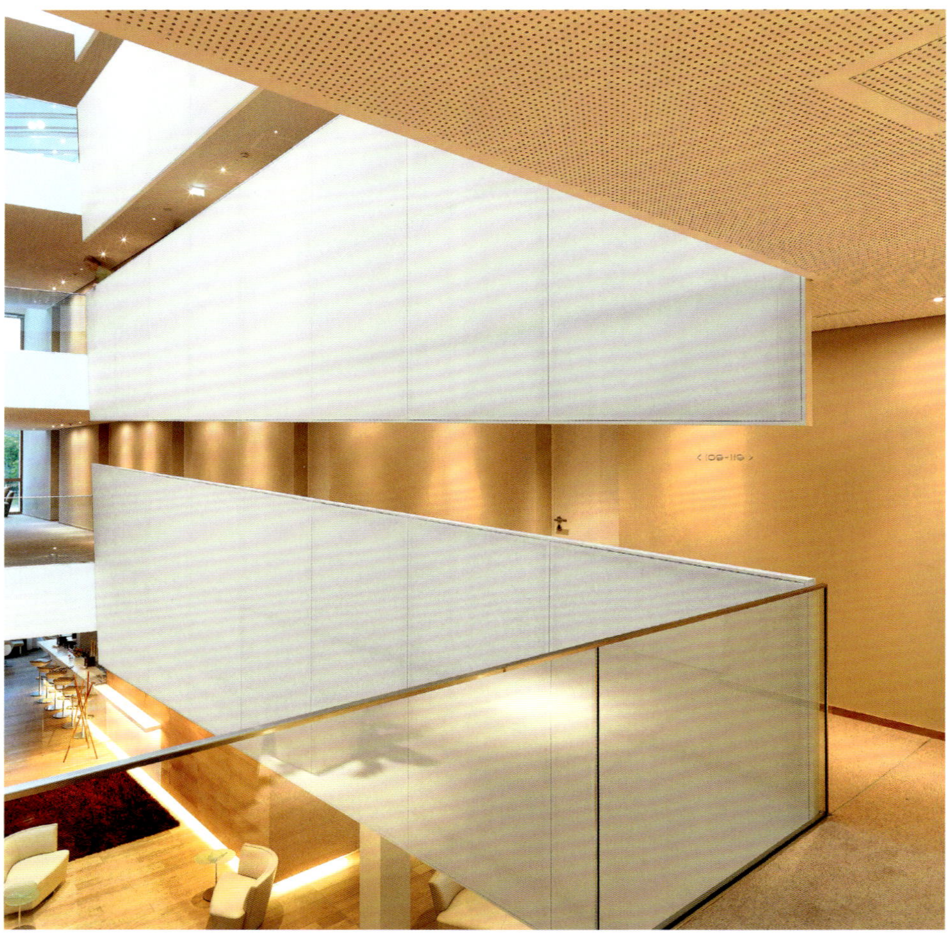

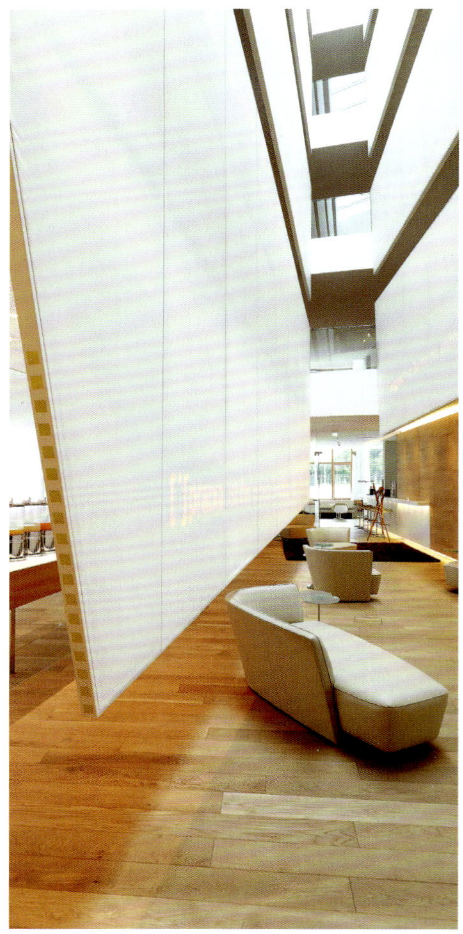

Children's hospital in Basel

Architects: Stump & Schibli Architekten, Basel

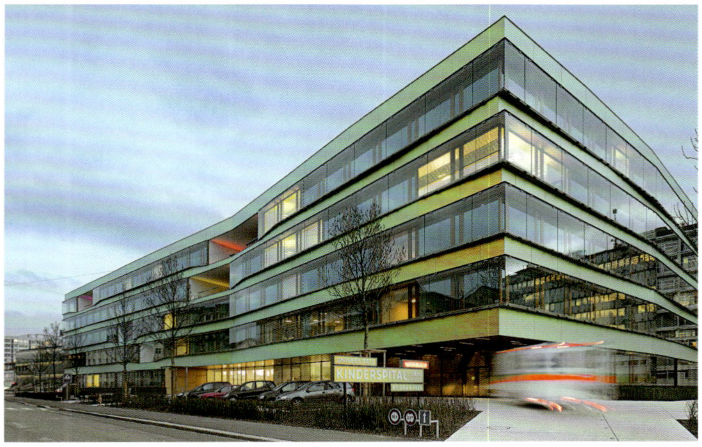

The centrally situated vertical access core with lifts and stairs makes it easier for patients and visitors to orient themselves within the building.

The new hospital building concentrates the facilities of the children's hospital, which were previously dispersed around different locations, in a single building. This required the creation of a distinctive building structure which asserts its presence among the detached buildings surrounding it. This was achieved through a striking facade design featuring parapet bands in changing colours which, depending on the angle of view and light conditions, appear to change from green through yellow to an intense glowing red, creating a surprising effect.
The colourful parapets continue into the planted courtyard, which all the patient rooms overlook. On the slightly kinked facade facing the Spitalstraße, set back from the street line, a spacious overhang shelters the entrance area which provides access to both the clinic and the accident and emergency unit.
In the interior, the building structure allows the implementation of different room typologies which are tailored to the specific functional procedures of the different departments and their widely differing requirements. The visitor enters the children's hospital via a bright and clearly laid-out entrance hall which in addition to containing the reception desk also provides access to the light-flooded stairwell and the visitor lifts. This central access area makes it possible for visitors to get their bearings quickly and facilitates orientation within the building. The departments with a high level of public traffic such as the accident and emergency unit, the polyclinic and therapy facilities are located on the open-plan ground floor, as is the cafeteria which faces the interior courtyard. Central functions such as the kitchen, laundry etc. are provided by an adjacent clinic which is connected with the new building at the first basement level via an existing service tunnel and a new pedestrian underpass.
The guidance system uses the material, size and colour of the information signs to differentiate between the different target groups – visitors, patients, staff. The lettering is either cut out of rubber so as to stand out three-dimensionally or is painted directly on the wall. All information is applied directly to the surfaces of the building, so that the signage becomes part of the architecture.

Corridors imitate villages

The challenge in the care departments was to combine a child-friendly, homely atmosphere with the functional efficiency of a modern medical care facility. In order to achieve this, low-key materials, surfaces and colour schemes were used throughout the building. For example, in the corridor zones of the inpatient wards reddish-brown rubber flooring, wood-veneered doors and wall coverings as well as white walls decorated with large-format illustrations and picture stories create a pleasant atmosphere. The corridors create connecting and circulation zones with a village-like ambience and, where they widen out into open spaces, provide relaxation areas furnished with colourful seating. The deeply inset loggias allow visual links to the city, as well as providing the circulation zones with natural light. They serve the young patients as play areas, recreational spaces and quiet corners.

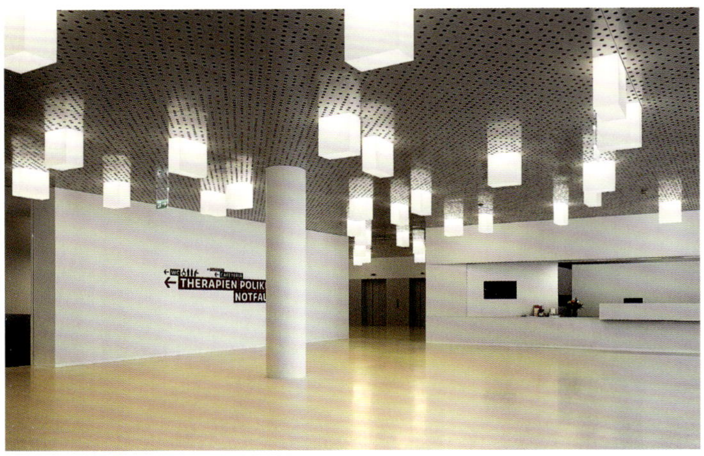

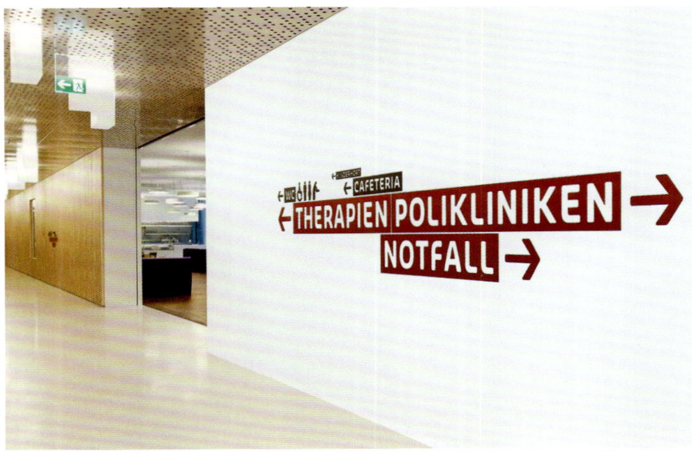

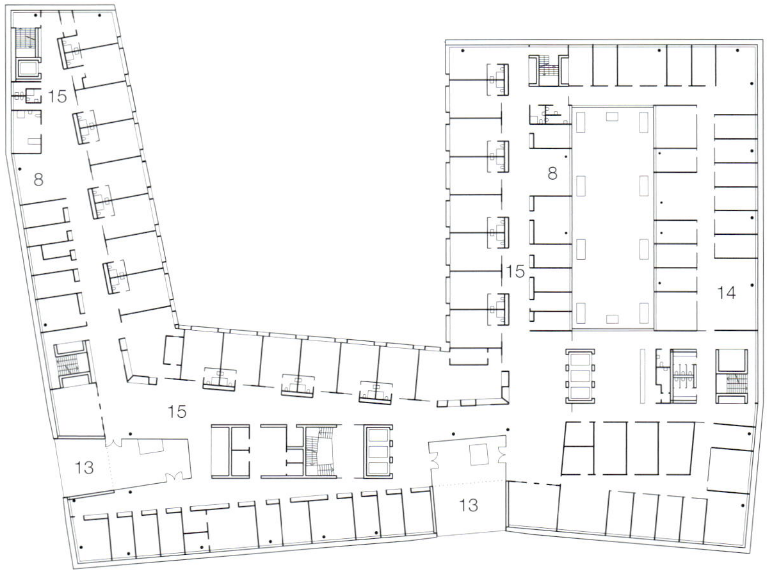

3rd floor

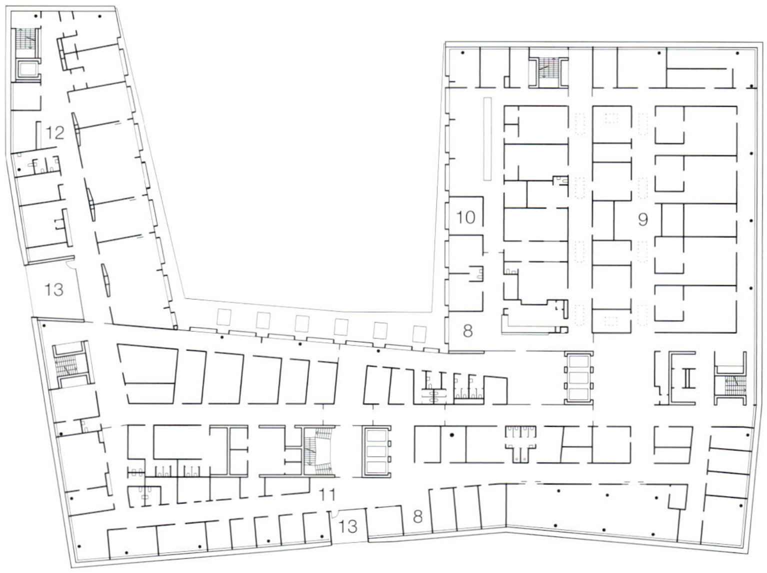

1st floor

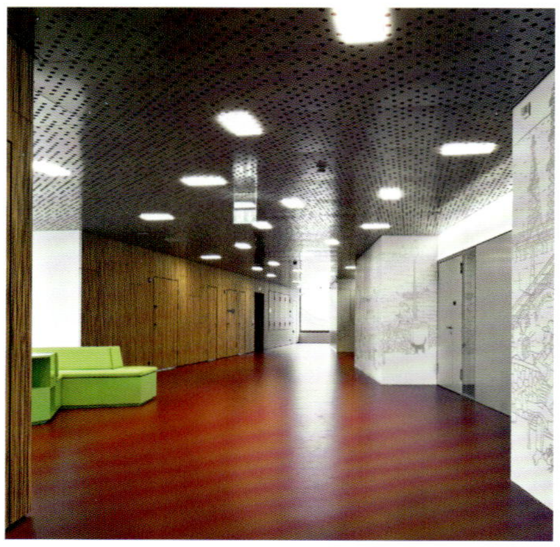
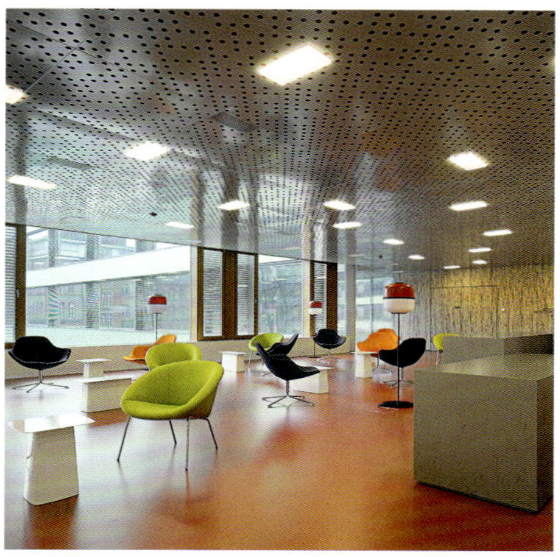
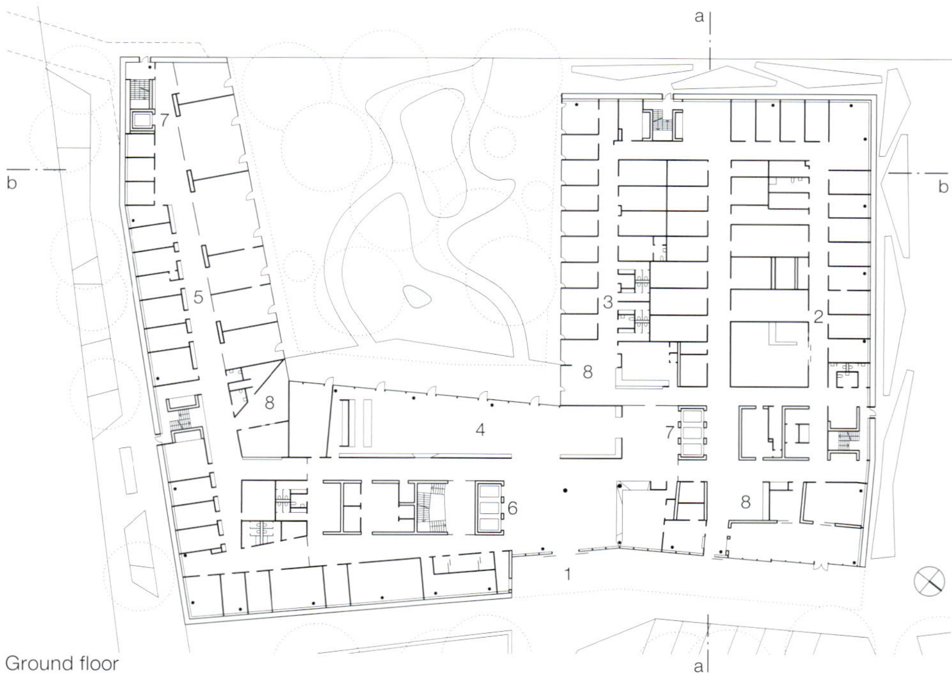

Ground floor

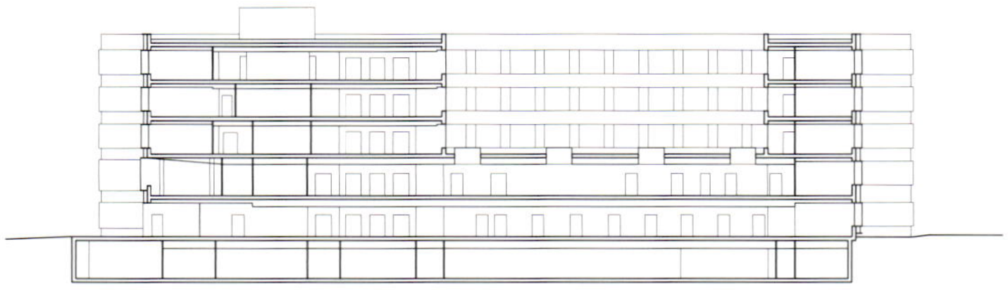

aa

Floor plans • Sections
Scale 1:1000

1 Main entrance
2 Emergency
3 Therapy
4 Cafeteria
5 Polyclinic
6 Visitors lift
7 Staff lift
8 Waiting area
9 Surgery
10 Radiology
11 Day unit
12 Intensive care unit
13 Loggia
14 Medical services
15 Nursing ward

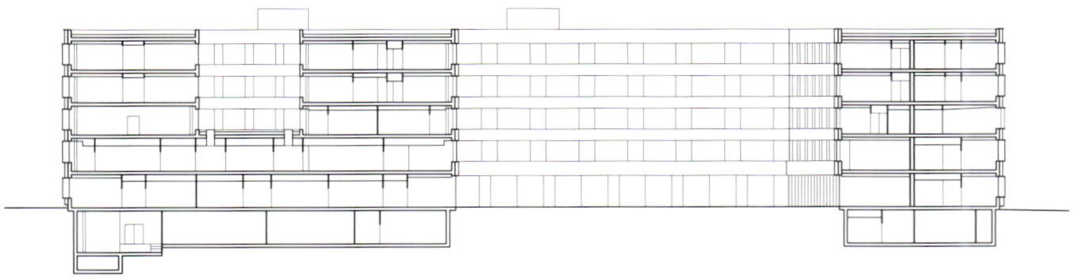

bb

Project data:

Use:	Hospital
Access:	Stairwells, lifts, corridors used as common areas
Number of floors:	5 + 1 basement level
Clear room height:	2.60–3 m
Gross volume:	116,000 m³
Gross floor area:	29,400 m²
Year of completion:	2009
Construction period:	36 months

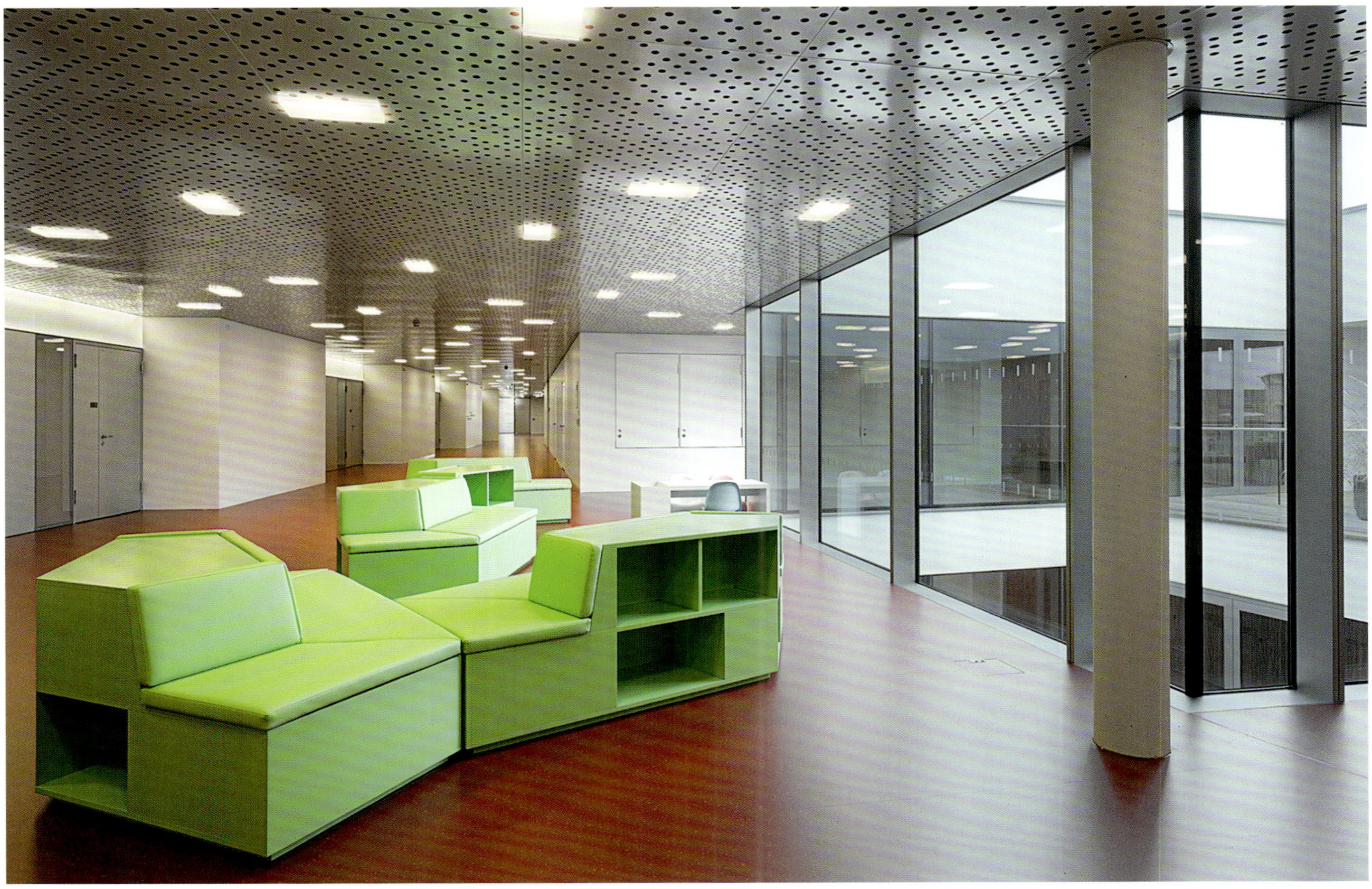

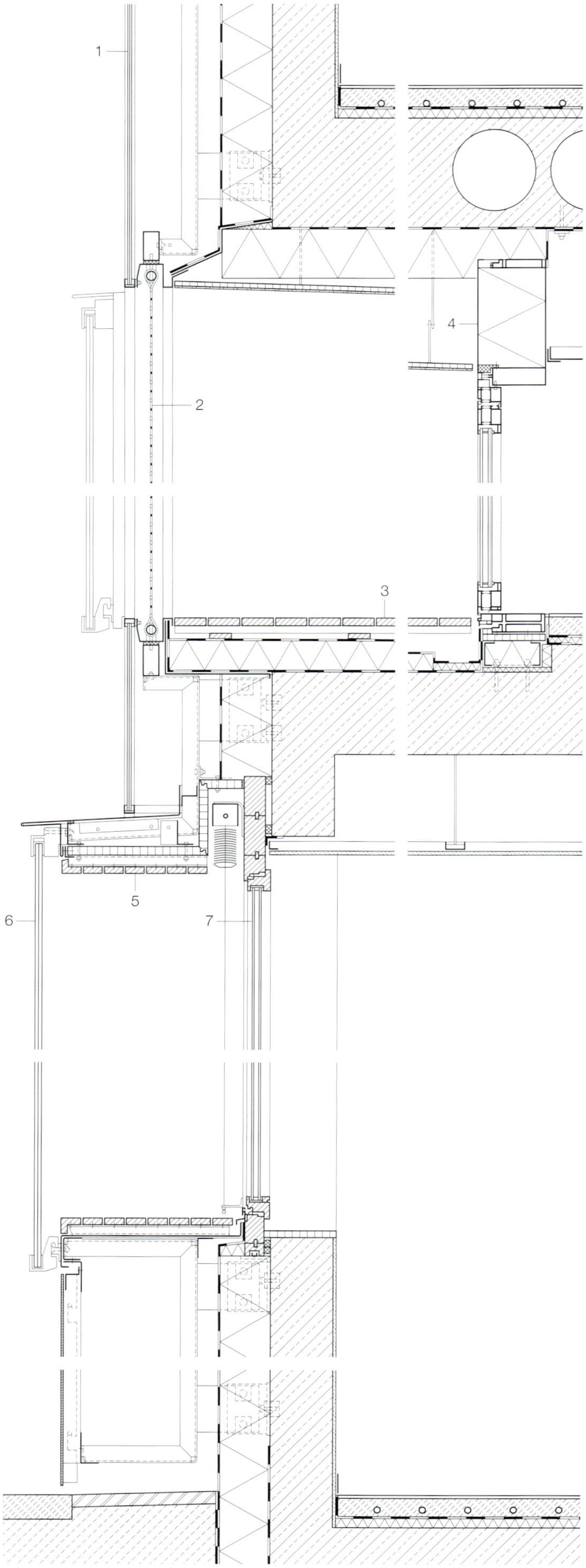

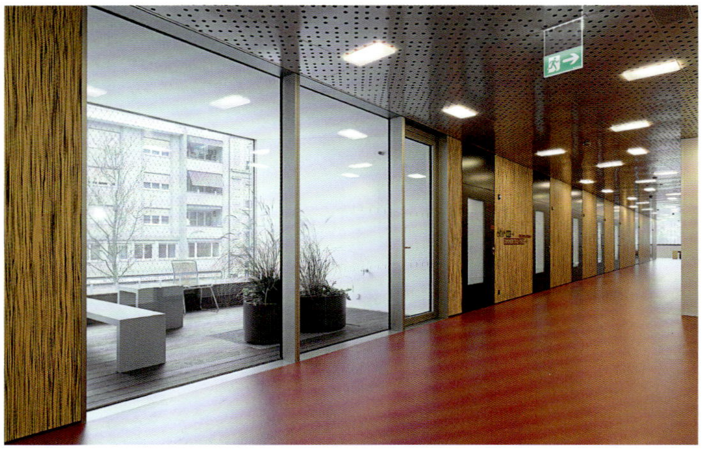

Vertical section of street facade
Scale 1:20

1 2× 6 mm laminated safety glass with multiple colourful siliconised-polyester interlayers
142–465 mm ventilated cavity
Fleece, dark tone
160 mm mineral-wool thermal insulation
200 mm reinforced concrete
15 mm rendering
2 Loggia fall protection: stainless-steel mesh
3 140/25 mm ash grating
25 mm battens
Separating layer: 6 mm recycled rubber mat
Bituminous seal, two layers
60–100 mm thermal insulation
Vapour barrier
270 mm reinforced concrete
4 2 mm aluminium sheet
200 mm thermal insulation
1.5 mm sheet steel
5 62/22 mm wood grating
39/39 mm aluminium angle
30 mm fibre-cement board
6 Laminated safety glass of 2× 10 mm float glass
7 Wood windows, insulation glazing, 8 mm toughened glass + 18 mm argon-filled cavity + 12 mm toughened glass, $U_g = 1.1\ W/m^2K$

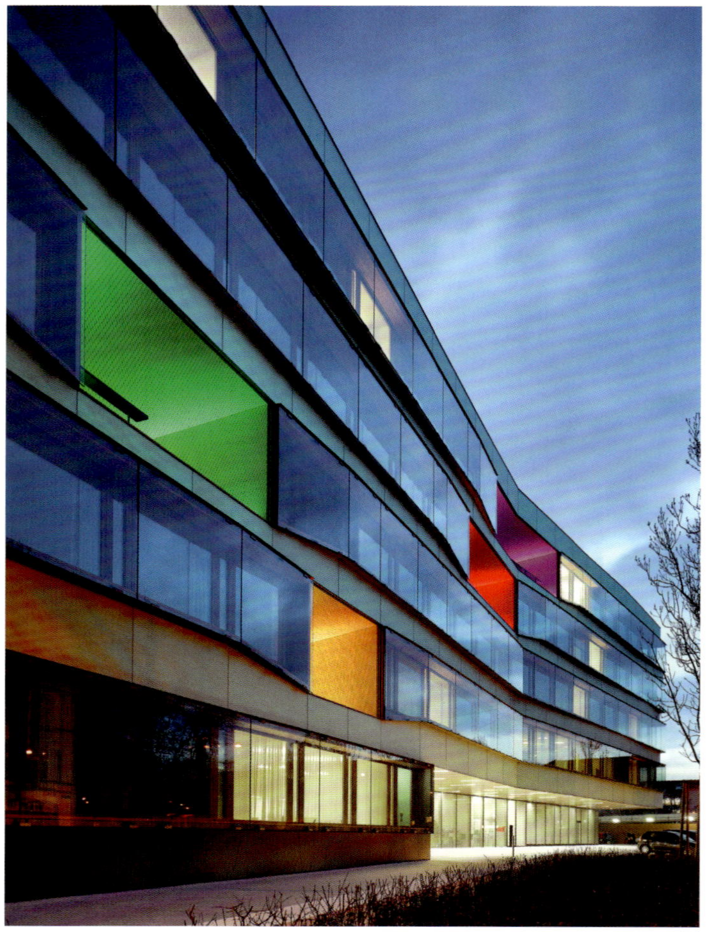

Cooper Union in New York

Architects: Morphosis Architects, Culver City

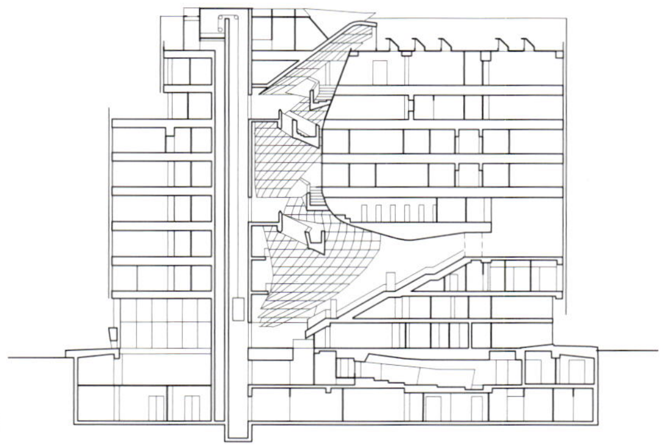

aa

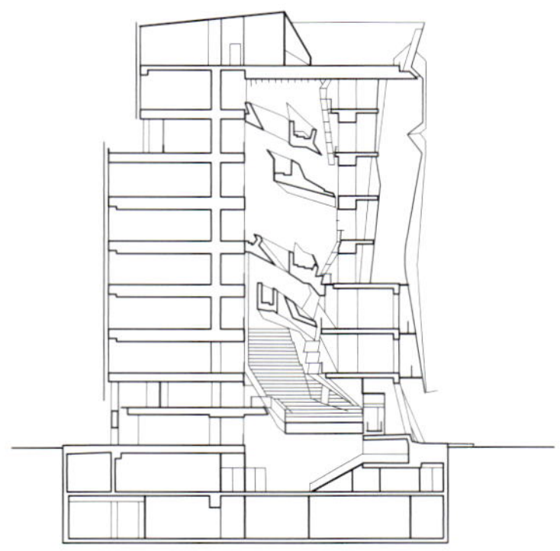

bb

Sections Scale 1:1000

An elaborately designed staircase forms the communicative centre of the university building.

Over its roughly 150 years of existence, the Cooper Union has developed into one of New York's most important intellectual and cultural centres. The new building brings together the faculties of Art, Architecture and Engineering Sciences, which had previously been housed in different buildings, under a single roof. With its sculptural form, it is intended to express the College's aspirations as an innovative institute of learning and encourage an interdisciplinary dialogue between the different faculties. The fractured building envelope of perforated stainless steel sheeting plays with light, shadows and transparency and allows the activities taking place within the building to shimmer through to the outside. The entrance to the building is marked by visually distorted lettering on the facade bearing the name of the college.

Full-height atrium

Inside, a central atrium rises the entire height of the building. As a centre for communication, this "vertical piazza" represents the heart of the building. From the two-storey entrance lobby with reception desk, a six-metre-wide staircase leads up four floors to a student lounge with views over the city. A wave-formed grid structure is suspended over the stairs, ascending up to the skylight which provides the stairwell with natural lighting. Leading off from this space from the fifth to the eighth floors are "sky lobbies" with seating, which are used by the students both for spontaneous socialising and also as work spaces, as well as meeting and seminar rooms grouped around the atrium. These areas are linked together by connecting bridges.

In order to encourage students and teaching staff to move around the building on foot and in this way promote chance encounters and informal dialogue, the main lift only stops on the ground floor and on the fourth and seventh floors. Two stairways with inclined and partially backlit balustrades wind through the atrium from the fourth and seventh floors, each running one floor up and one floor down. Additional lifts stop on all the floors in order to guarantee barrier-free access to the college and also allow heavy loads to be transported without any problem. The basement level contains public areas such as exhibition spaces and an auditorium, which are accessed via a broad stairway.

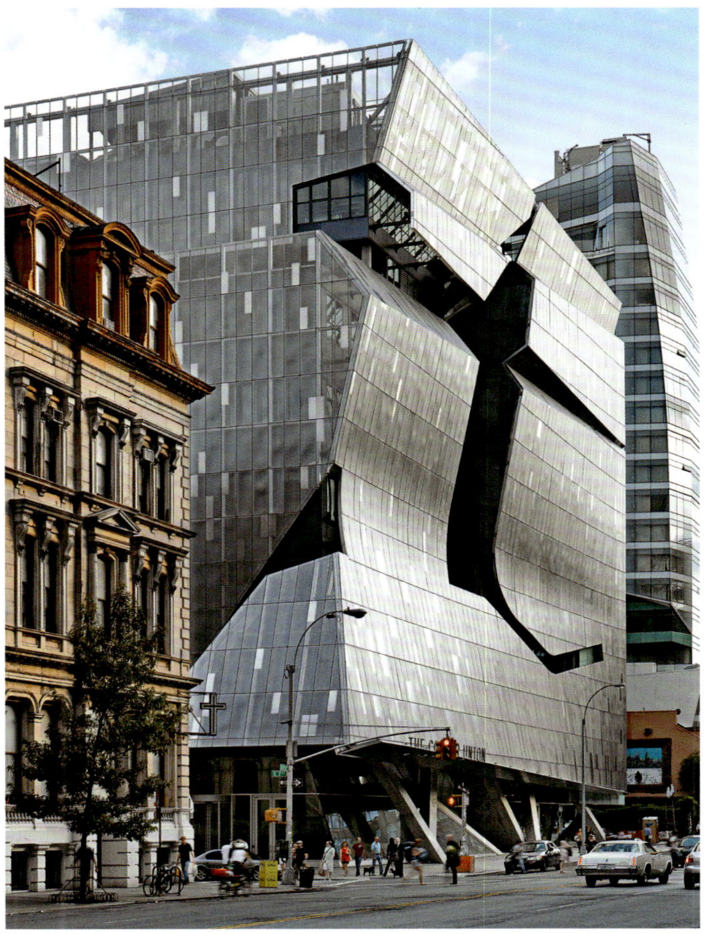

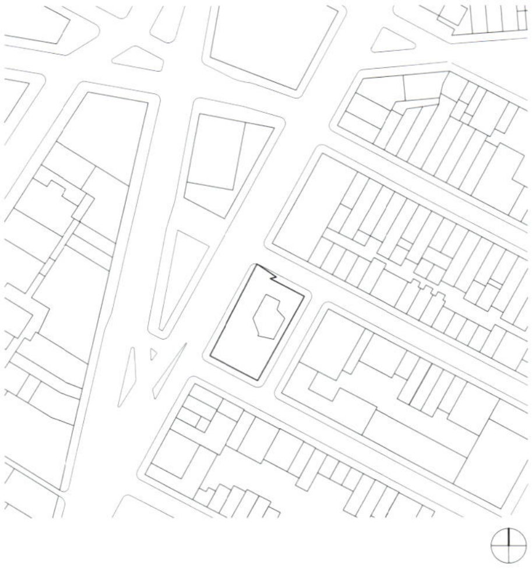

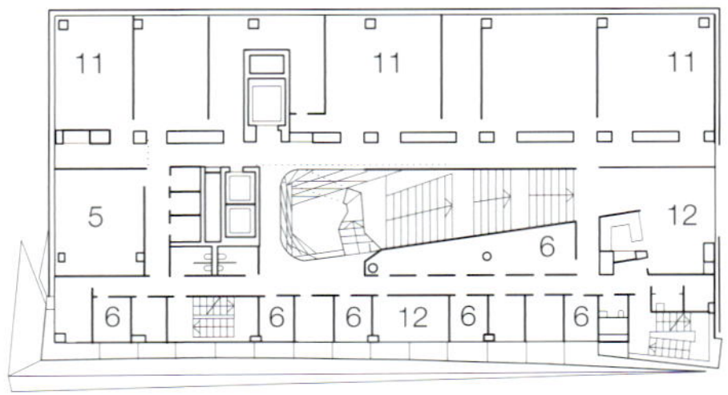

3rd floor

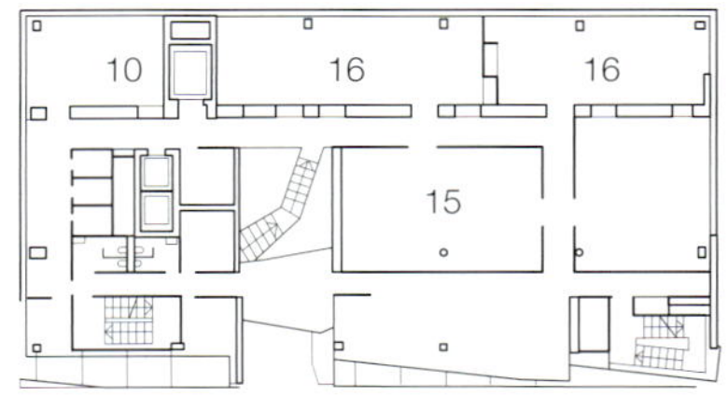

8th floor

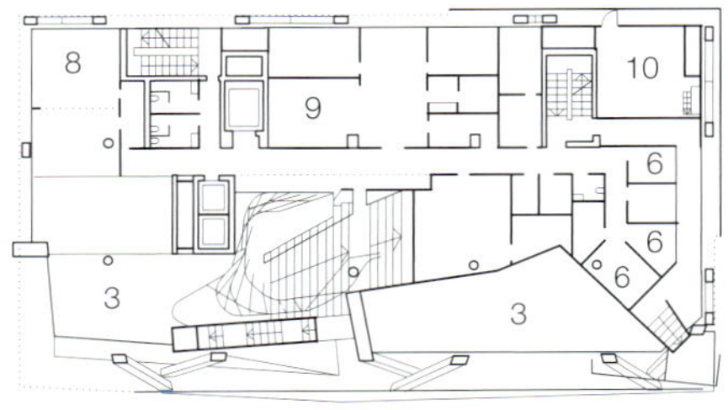

1st floor

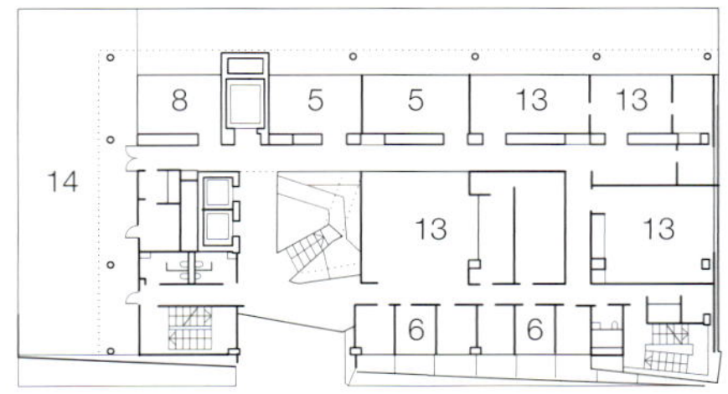

7th floor

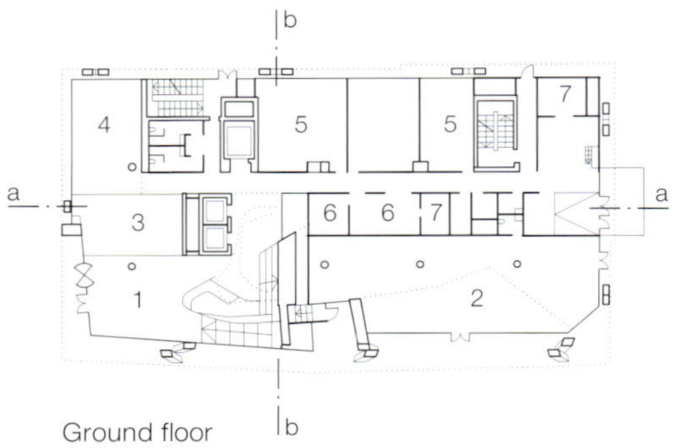

Ground floor

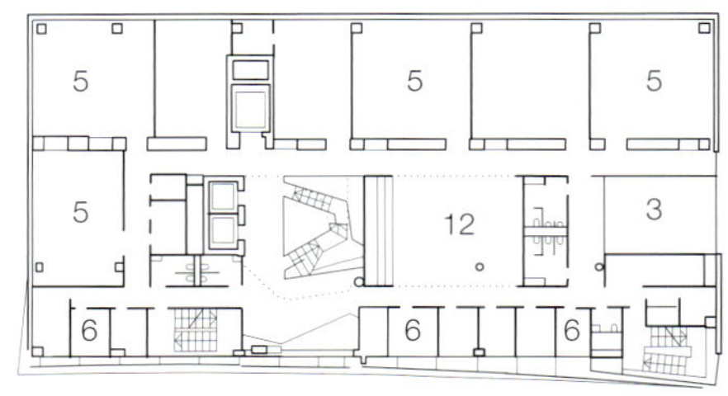

4th floor

Site plan
Scale 1:5000
Floor plans
Scale 1:1000

1 Entrance hall
2 Shop
3 Void
4 Seminar room
5 Classroom
6 Office
7 Storeroom
8 Meeting room
9 Dean's office
10 Technical services
11 Laboratory
12 Student lounge
13 IT rooms
14 Roof terrace
15 Student workshop
16 Art studio

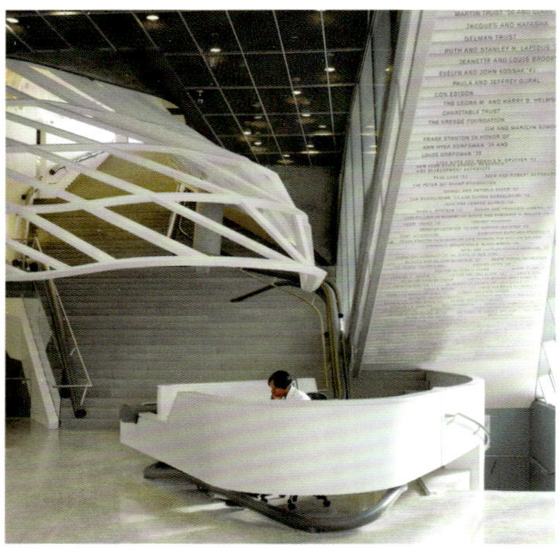

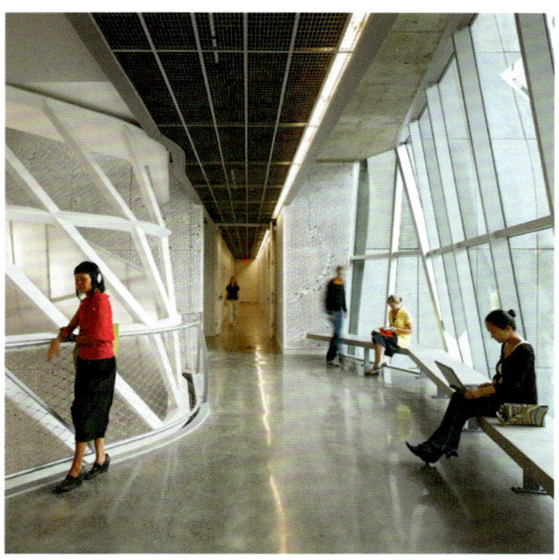

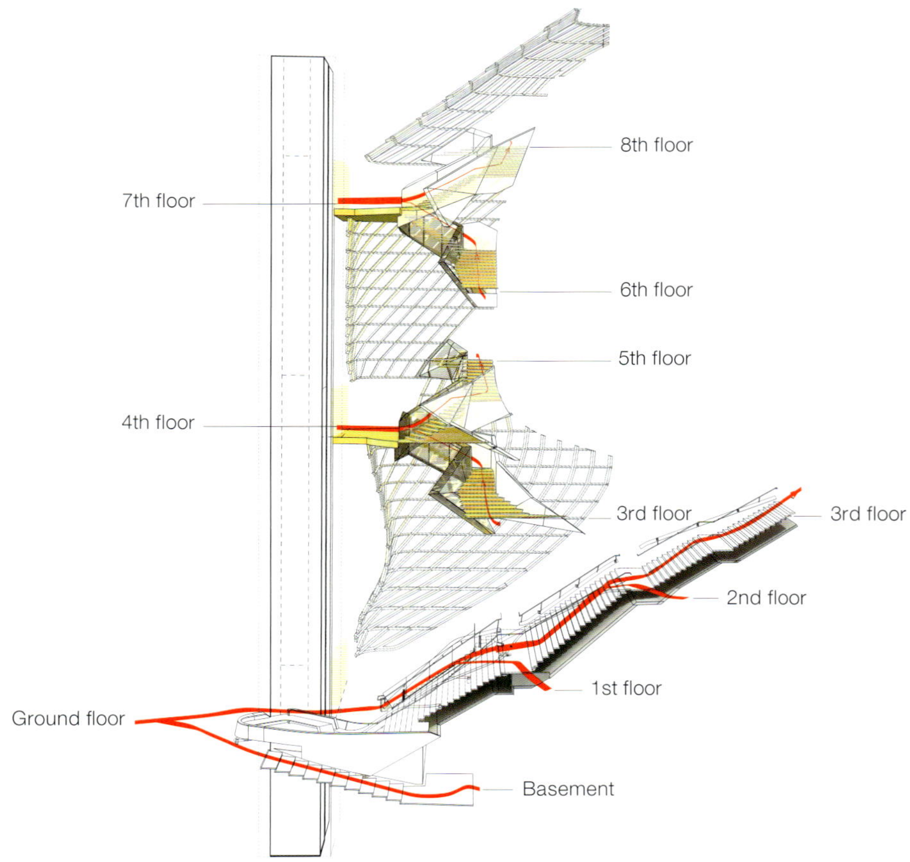

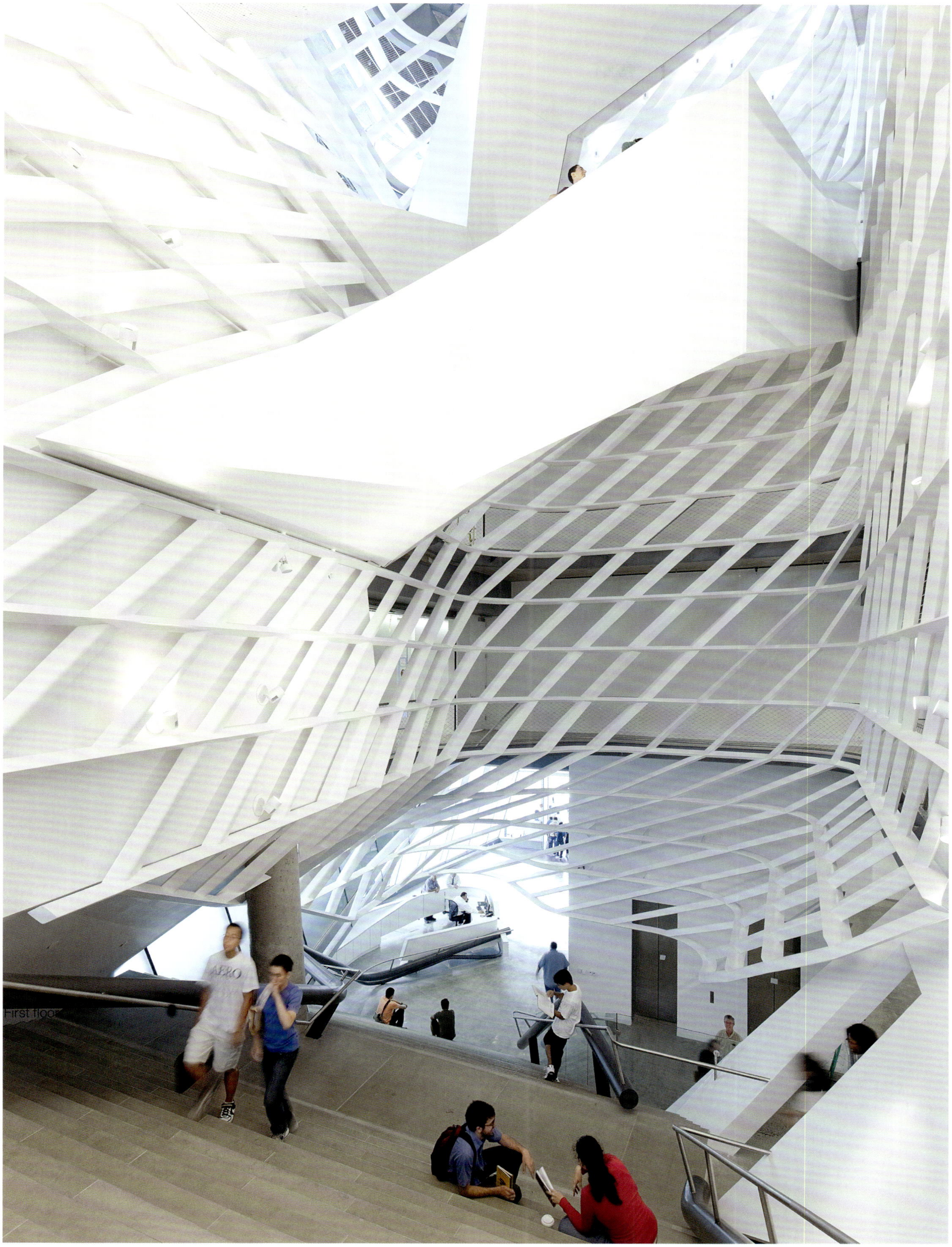

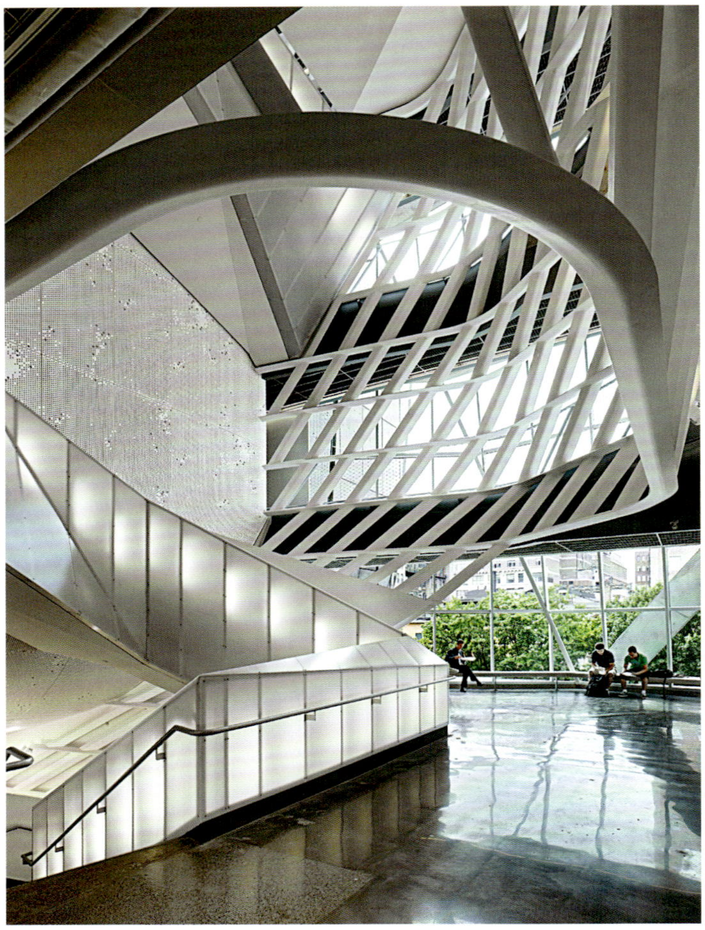

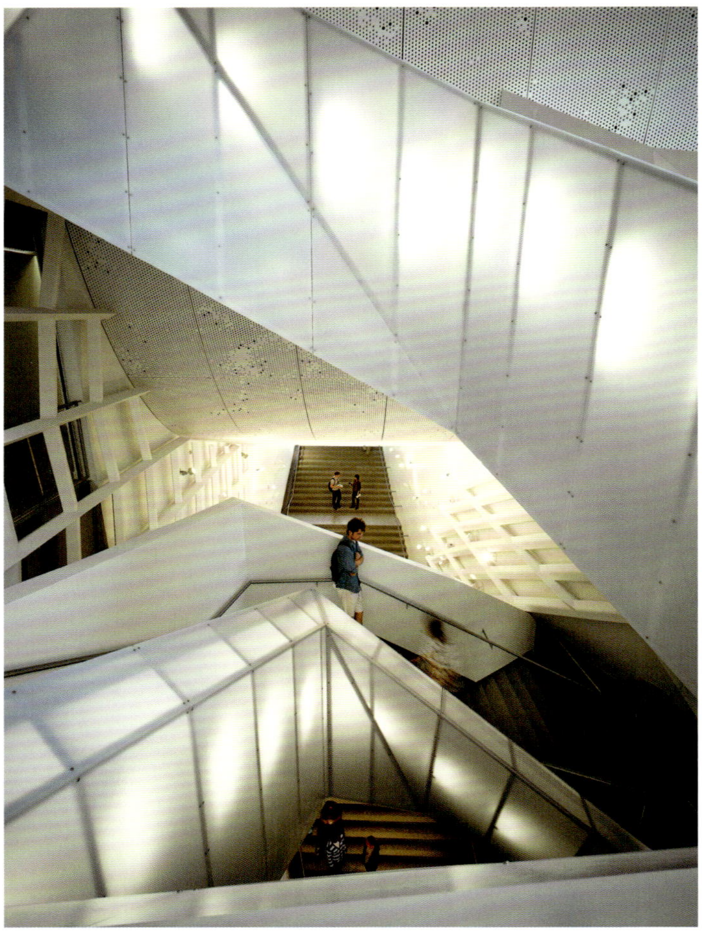

Sectional details of stairs
Scale 1:20
Axonometric view of stairs

1 76 mm square tubular steel
2 6.5 mm synthetic resin panel, screwed to supporting construction
3 Supporting construction: 25/25 mm angular frame
4 Handrail: 38 mm stainless steel
5 Fluorescent tubes
6 6.5 mm glass fibre reinforced plasterboard
7 6.5 mm sheet steel, painted
8 Landing: precast concrete element
9 Sprinkler
10 6.5 mm sheet steel, welded to stringers
11 Step: concrete on folded 6.5 mm sheet steel
12 6.5 mm plasterboard panel

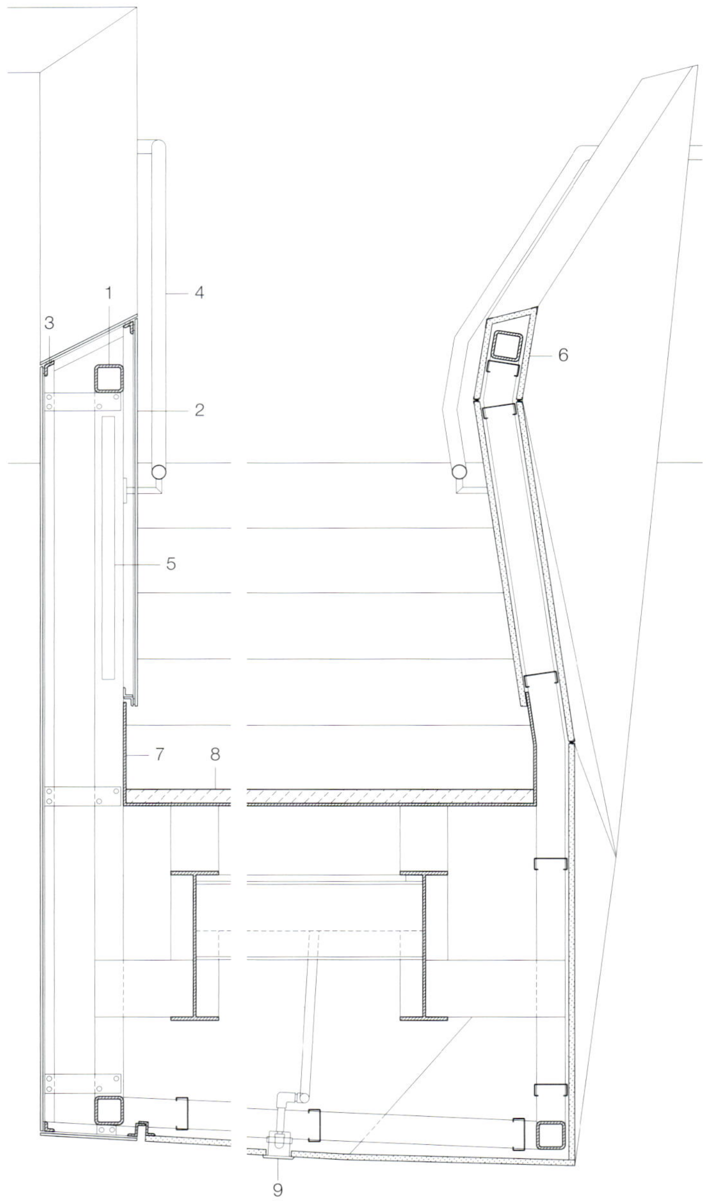

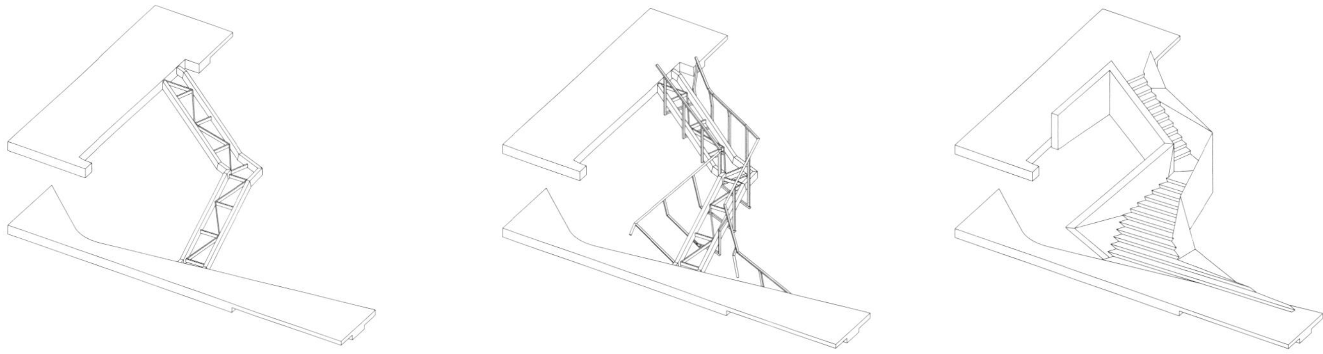

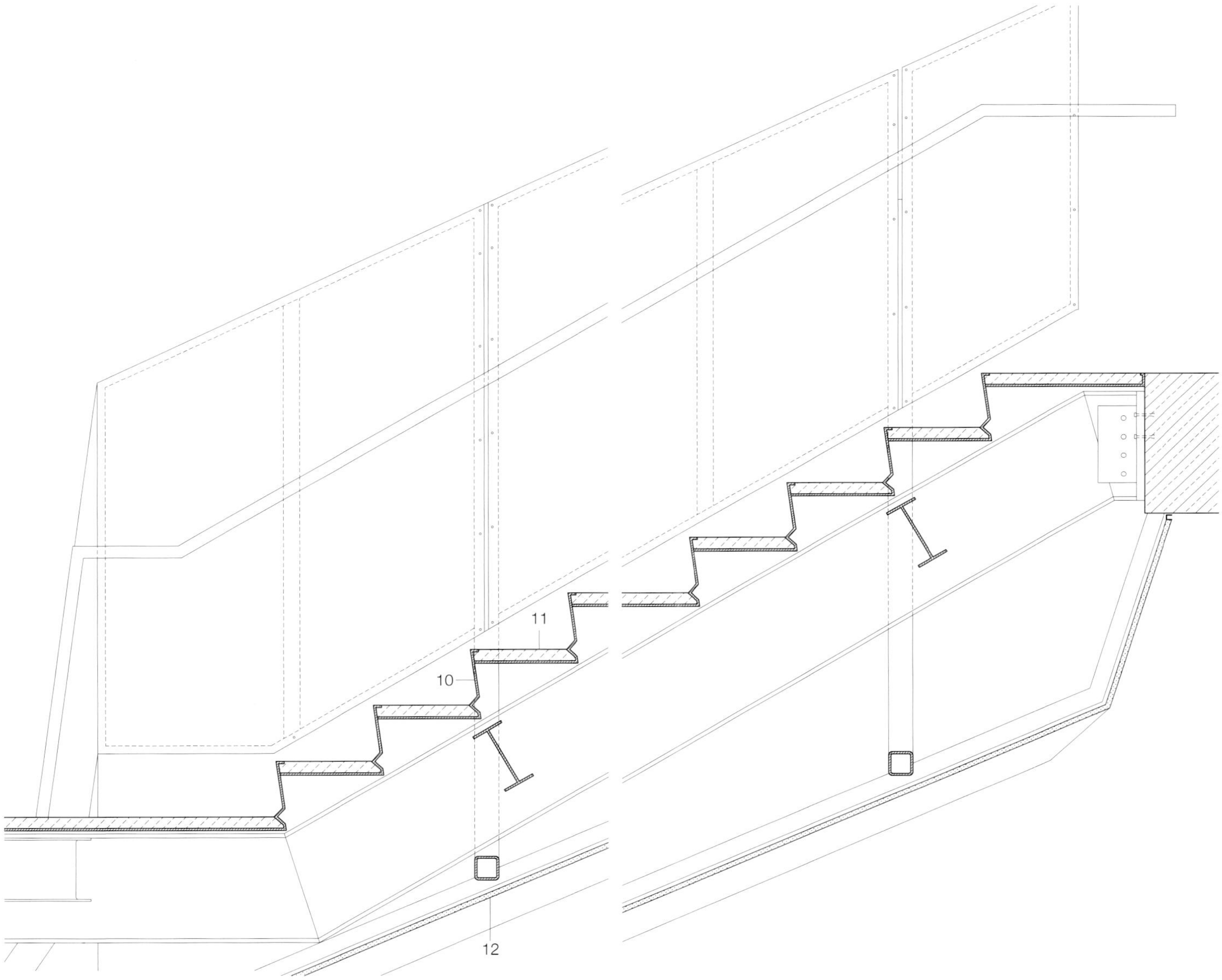

Project data:

Use:	Education/research
Access:	Stairs in atrium, elevators
Number of floors:	9 + 2 basement levels
Clear room height:	2.90 m (standard floor)
Gross floor area:	16,256 m^2
Year of completion:	2009
Construction period:	35 months

High school in Copenhagen

Architects: 3XN, Copenhagen

With its fluid spaces and sweeping stairs, the school is a spatial implementation of the recent reforms to Denmark's high school system.

Viewed from outside, the school building in Copenhagen's new district of Ørestad, which resembles something between an office block and a multi-storey car park, initially reveals little about its unusual interior layout that reflects the openness and transparency of the Danish approach to education. On closer inspection, it is perhaps the colourful, semi-transparent sunshades with numbers and letters printed on them on the facade of the cubic structure that first give a clue to the building's function as an educational institution, in which circulation and functional areas merge together on all levels to create a fluid, communicative space. Around the central spiral staircase – which leads from the basement level up to the fourth floor – segments of varying size and direction are cut out of the floors so that the levels are grouped around the staircase almost like extended platforms. The open space, slightly offset on each floor, reaches through the entire height of the building. Skylights and room-height glazing ensure that the building is flooded with daylight.

Spatial connections and visual relationships

Horizontal and vertical spatial connections and overlaps as well as varied visual relationships reflect the interdisciplinary approach of the learning concept. Each open-plan, boomerang-shaped floor level is assigned to a different area of knowledge. The pupils, aged 16 to 19, are grouped into study teams or make use of individual learning zones in the corners of the building. Only a few enclosed, classroom-like areas such as conventional specialist subject rooms, administration and music rooms are lined up along the facades. Instead, the picture is dominated by circular learning islands which are accessed via their own short flights of stairs and furnished with beanbags for group discussions and relaxation breaks. Lockers and sliding wall shelving elements provide storage space for learning materials and private possessions. In this spatial continuum, the three round cylinders which accommodate the emergency stairs, toilets and lifts also represent structurally necessary spatial constants and thus serve as orientation points. Acoustic ceilings and walls as well as noise-absorbent surfaces on the fixtures and fittings prevent a distracting level of background noise. Even the underside of the central staircase has been coated with acoustic plaster.

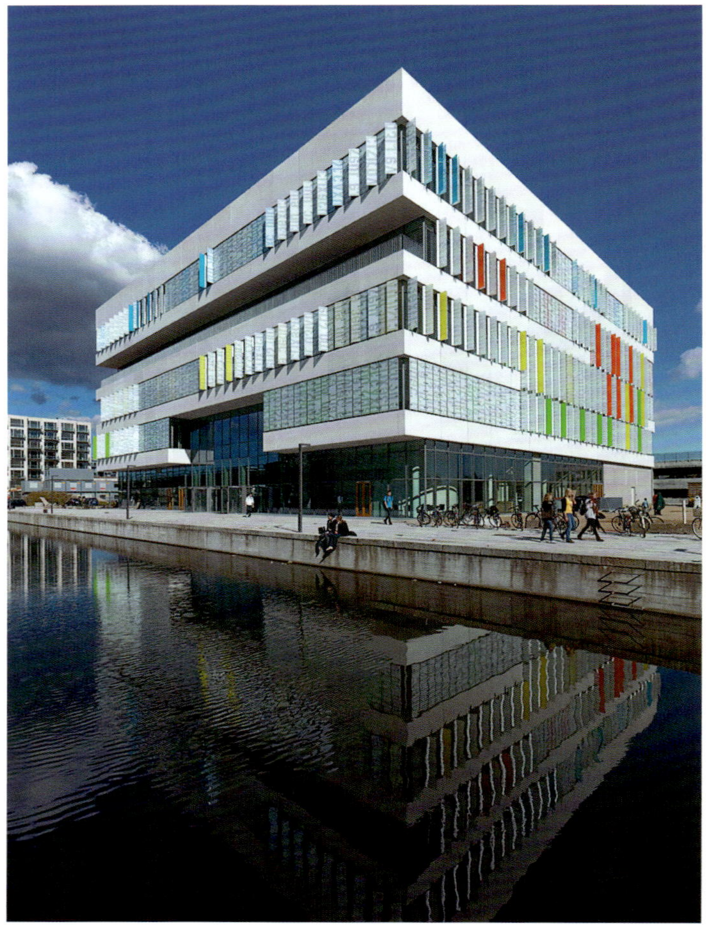

Project data:

Use:	School
Access:	Central spiral staircase, service core with emergency stairs, lifts
Number of floors:	5 + 1 basement level
Gross floor area:	12,000 m²
Year of completion:	2006
Construction period:	22 months

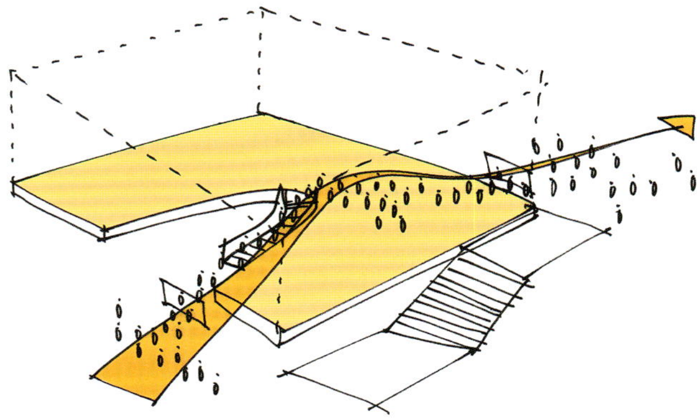

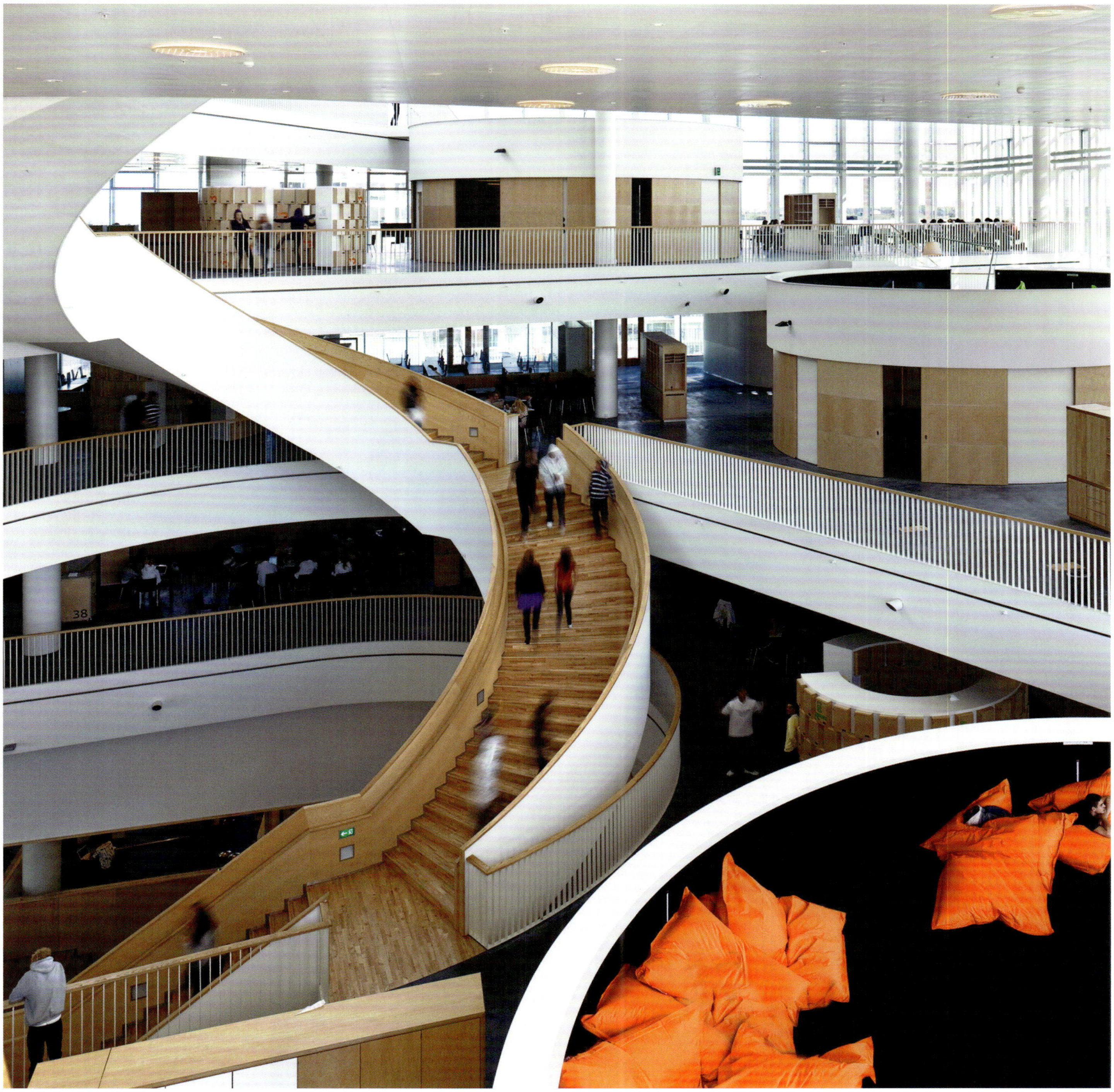

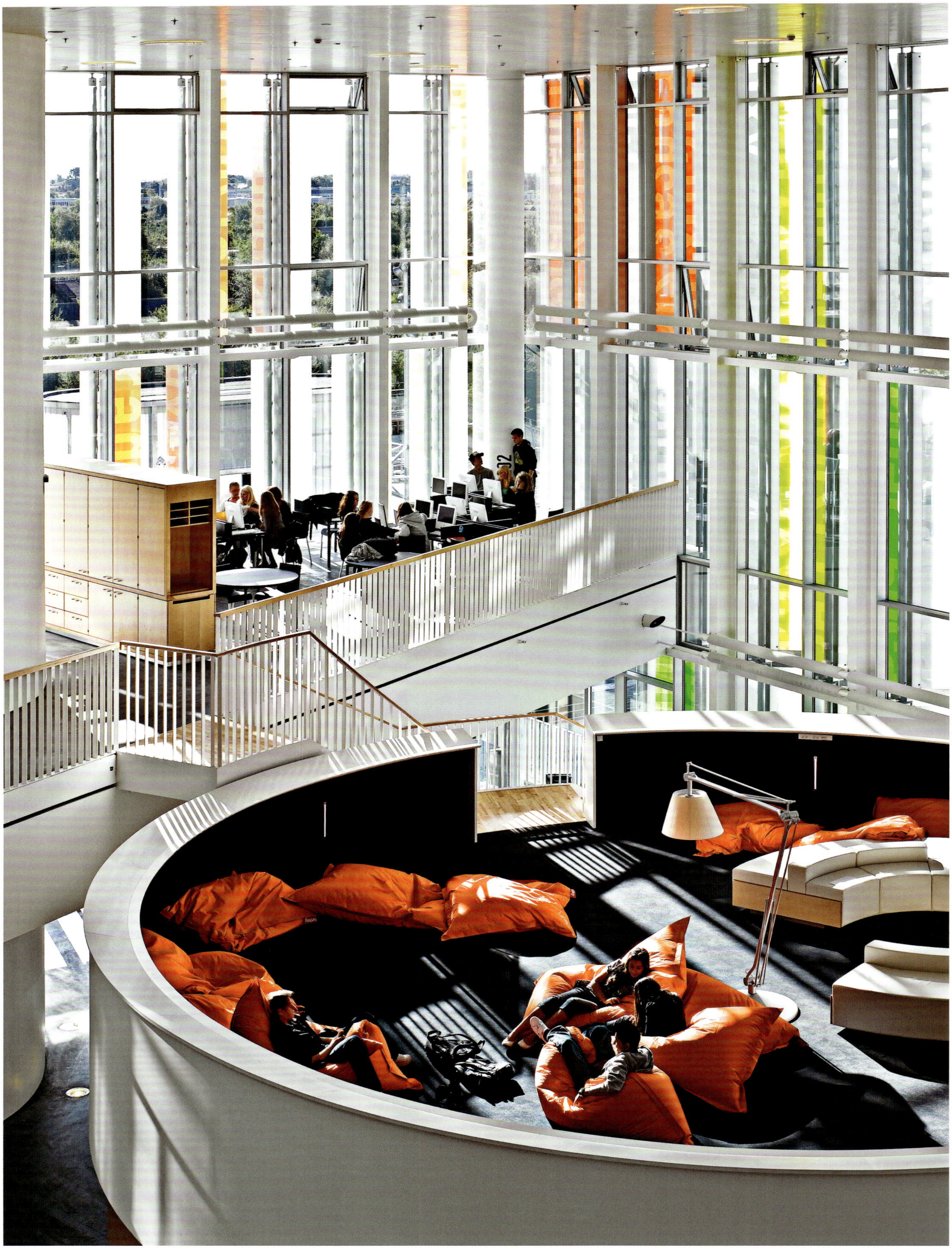

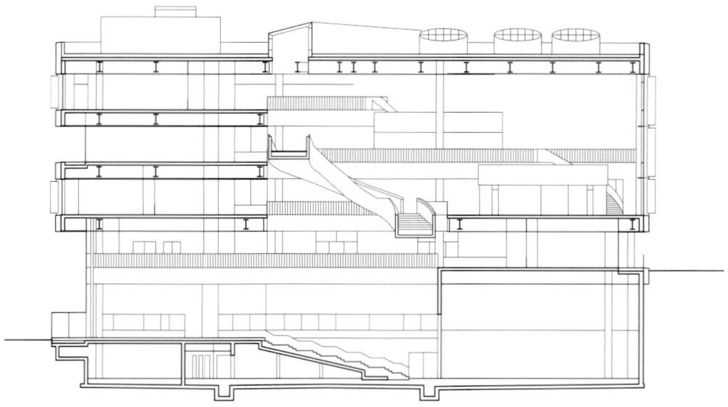

aa

Section • Floor plans
Scale 1:750

1 Foyer
2 Canteen
3 Administration
4 Void
5 Music room
6 Library
7 Atrium
8 Group room
9 Meeting space
10 Staff room

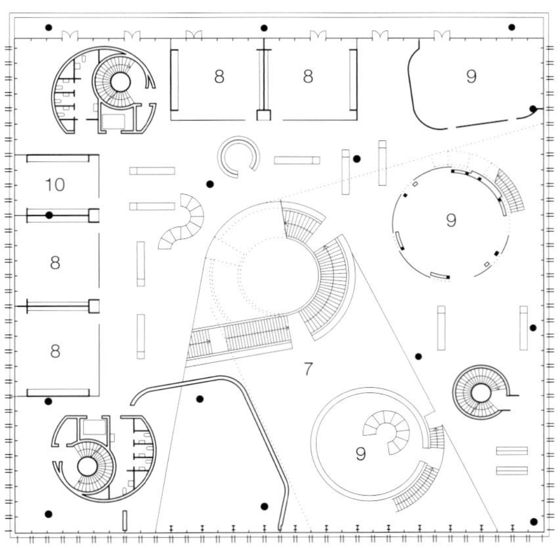

2nd floor

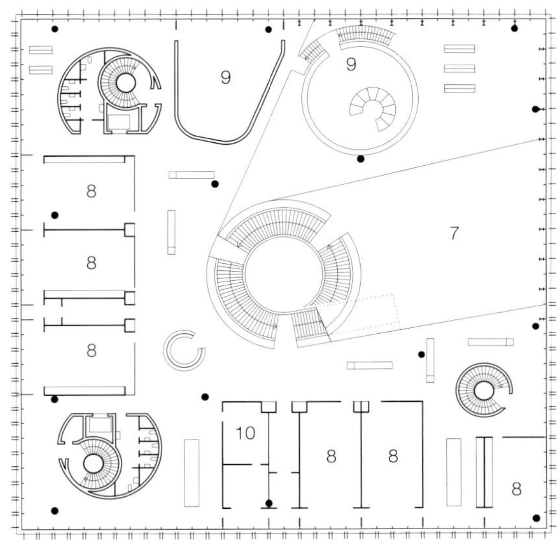

4th floor

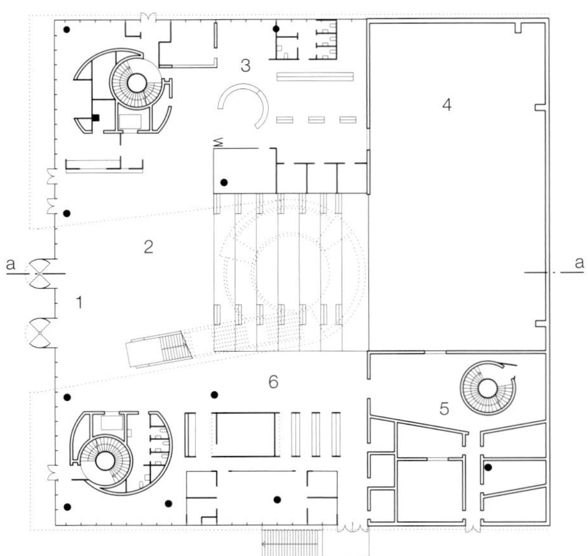

Ground floor

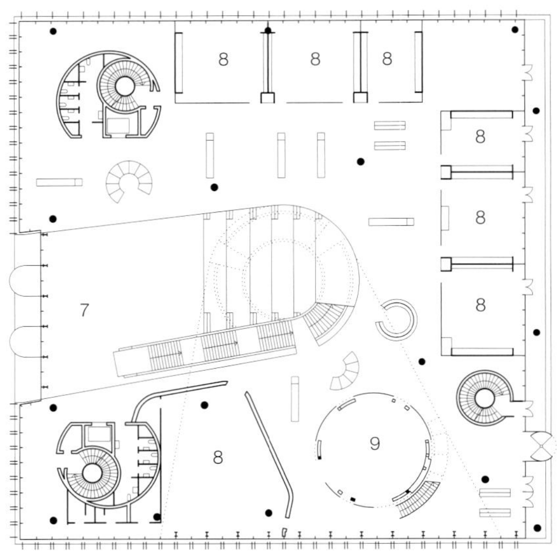

1st floor

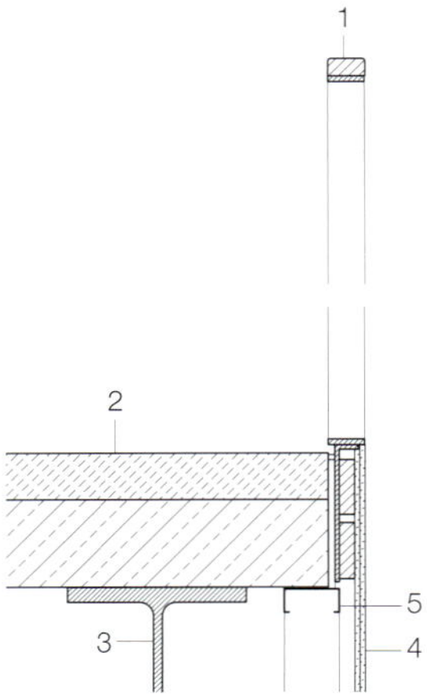

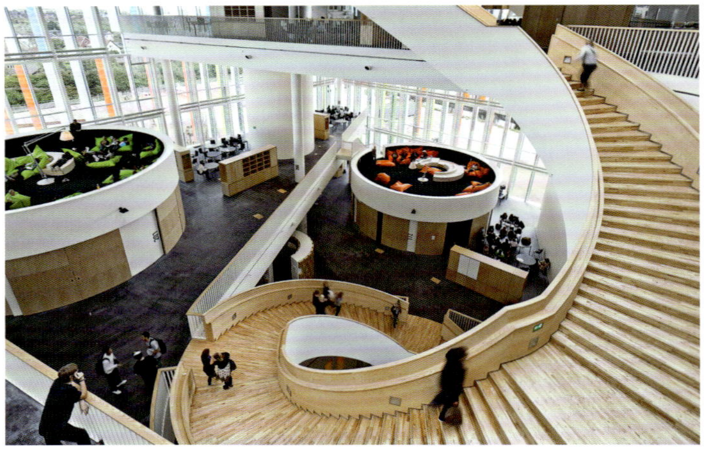

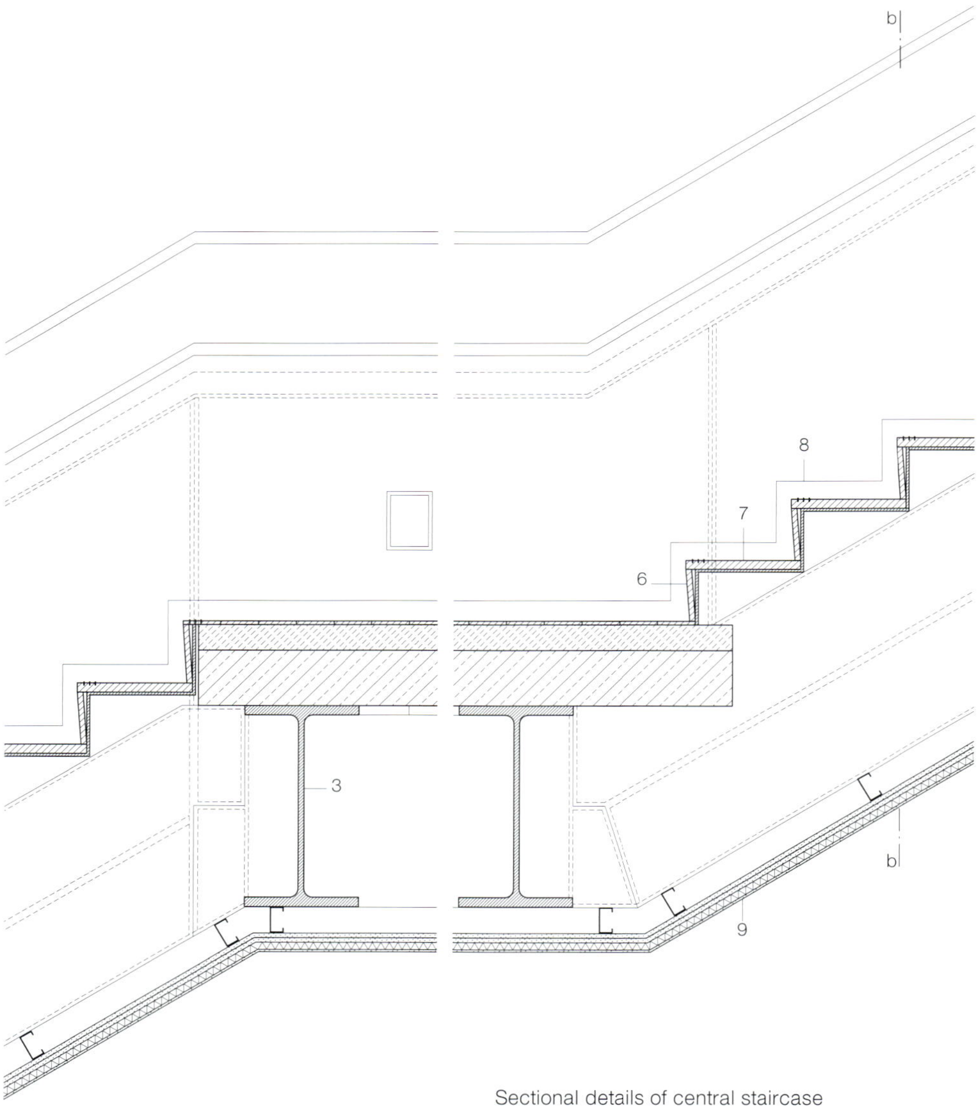

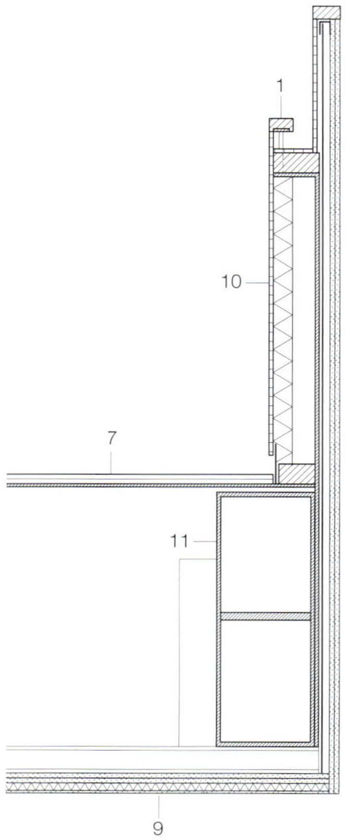

bb

Sectional details of central staircase
Scale 1:20

1 60/34 mm ash handrail
2 50 mm magnesite screed
  120 mm reinforced concrete
3 Steel H-beam 550 mm deep with
  fire-resisting coating
4 2× 13 mm plasterboard
5 95/20 mm steel U-section
6 15–22 mm ash riser
7 22 mm ash tread; 2 mm granulated cork
  6 mm sheet steel
8 2/80 mm stainless-steel section, satin polished
9 Acoustic plaster on 25 mm sheeting,
  2× 13 mm plasterboard
  70/20 mm steel U-section
10 10.5 mm composite ash boarding
  45 mm mineral wool; 55 mm cavity
  10 mm sheet steel
  20 mm trapezoidal section steel
  2× 13 mm plasterboard
11 250/560/10 mm hollow sheet-steel beam

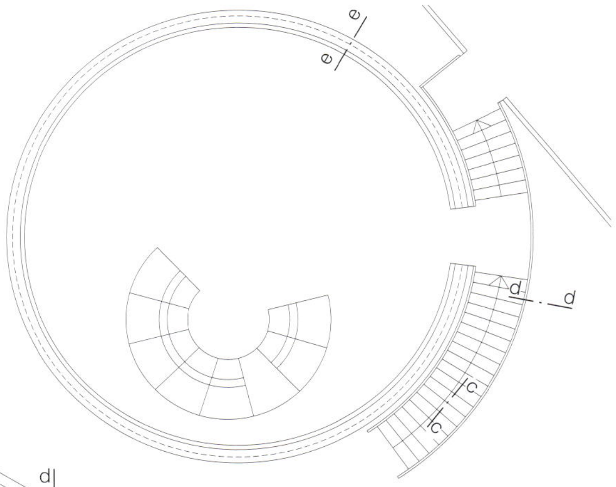

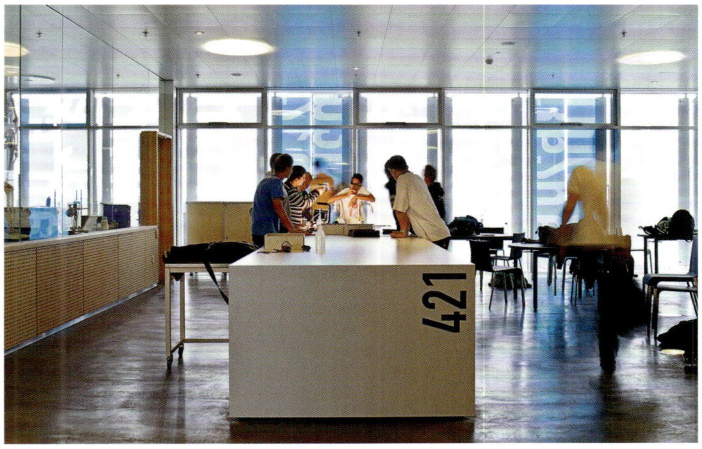

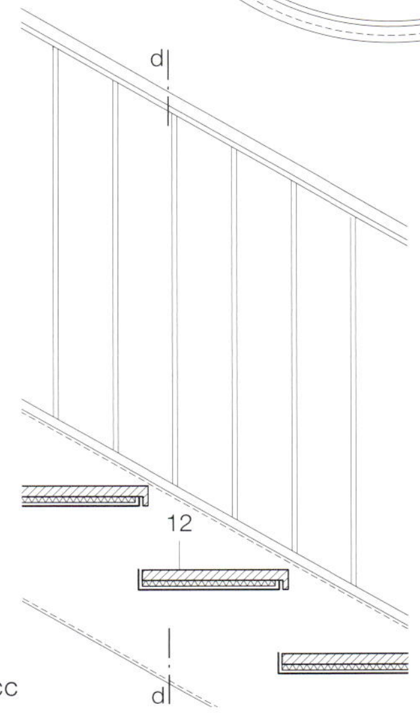

cc

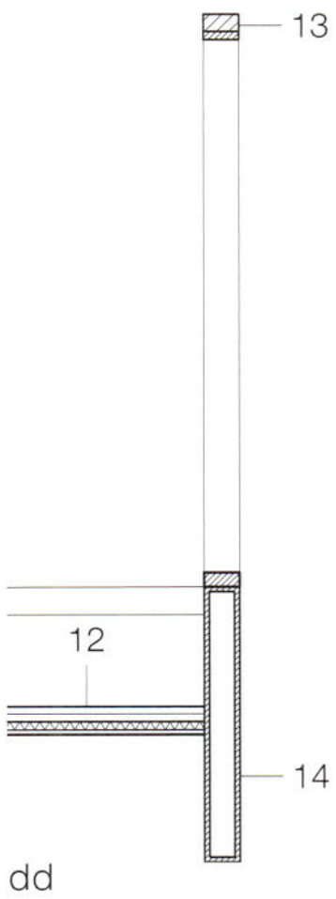

dd

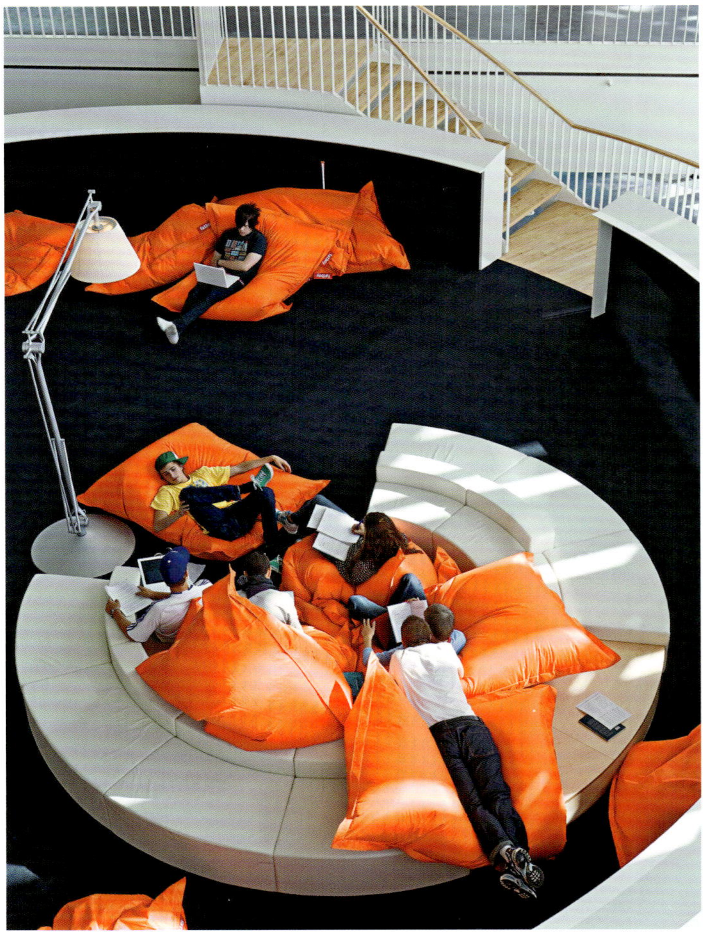

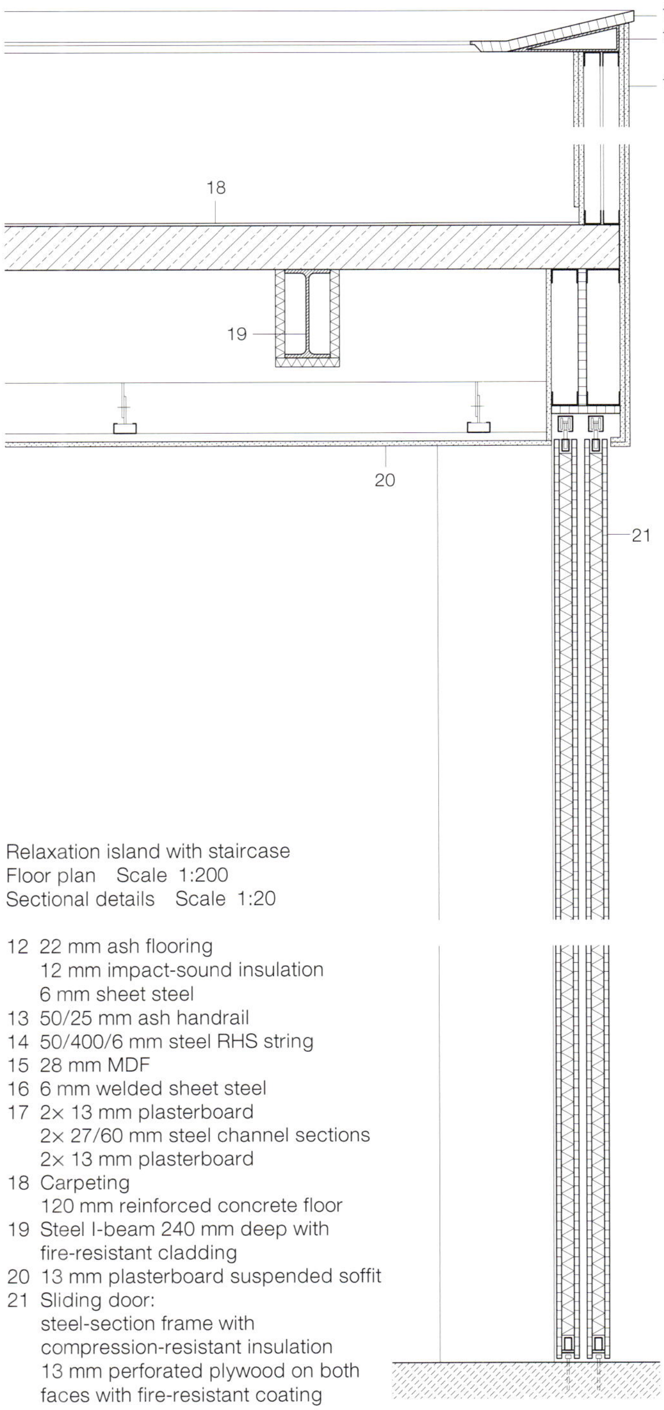

ee

Relaxation island with staircase
Floor plan Scale 1:200
Sectional details Scale 1:20

12 22 mm ash flooring
12 mm impact-sound insulation
6 mm sheet steel
13 50/25 mm ash handrail
14 50/400/6 mm steel RHS string
15 28 mm MDF
16 6 mm welded sheet steel
17 2× 13 mm plasterboard
2× 27/60 mm steel channel sections
2× 13 mm plasterboard
18 Carpeting
120 mm reinforced concrete floor
19 Steel I-beam 240 mm deep with
fire-resistant cladding
20 13 mm plasterboard suspended soffit
21 Sliding door:
steel-section frame with
compression-resistant insulation
13 mm perforated plywood on both
faces with fire-resistant coating

Municipal library in Stuttgart

Architect: Eun Young Yi, Cologne

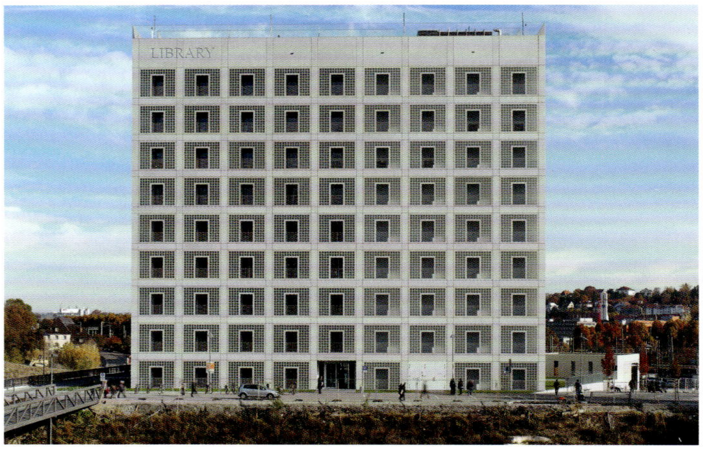

Stairs arranged around the reading room, which widens upwards like a funnel, lead to the different gallery levels.

The new city library is situated in the centre of the "Europaviertel" urban planning zone near Stuttgart's central station. Even though the west side has since established itself as the main entrance, the building was conceived in the form of a cube with four equal access points in the central axis. A foyer with self-service media return machines, visible sorting system, information desk and lifts surrounds on all sides the so-called heart, a 40-metre-high "negative monolith" inspired by the ancient pantheon room type which, according to the architects, is intended to serve as a place for meditative contemplation. This empty, also cube-formed space with its central skylight is flanked by stairwells which lead the visitor up to the fourth floor. From here, the funnel-shaped reading room, primarily housing literary works and works relating to literary studies, opens up, widening floor by floor as it ascends. The different reading gallery levels are accessed by stairs arranged around the central void. Additional media areas with shelf zones and workplaces branch off from the "heart" and reading room. Each level covers a different subject area, including a music library and children's library, but learning workshops and administration areas are also accommodated here. With their colourful spines, the books stand out clearly against the white shelving fixtures, stairs, floors and walls. Only the light blue upholstered seating elements add a different colour accent here. Free-standing digital information displays on each floor allow visitors to search for media as well as identifying their location within the building and having this displayed precisely on an enlarged floor plan. In addition, markings on the floor identify the shelf locations. The eighth floor houses a graphics library and a cafeteria. The viewing platform on the roof can also be reached from here. The basement level contains spaces which can be used for events.

Vertical sorting system

A unique vertical sorting system with 70 electric-motor-driven trolleys conveys all the returned media to the appropriate level. The library is open six days a week for 12 hours and also offers a 24-hour return service as well as the "library for insomniacs" – a small, changing selection of media available in the vestibule of the east entrance which can also be borrowed outside of opening times. Security gates on the four exits prevent media from leaving the building without being entered in the system.

Accessible facade passage

The building envelope is designed as a double facade: behind the 9 × 9 m square facade sections of light grey fair-faced concrete filled by frosted glass bricks lies the inner thermal envelope, a mullion/transom construction. The peripheral intermediate space between the facades is accessible and, like the roof terrace, offers views over the city. The entire building is designed to be accessible. Since, with a height of 40 m, it is already classed a high-rise building, the construction and surfaces in the emergency stairwells have to conform to special fire protection requirements. Due to abrasive additives contained in their coating, the fair-faced concrete steps comply with the R 10 slip-resistance rating.

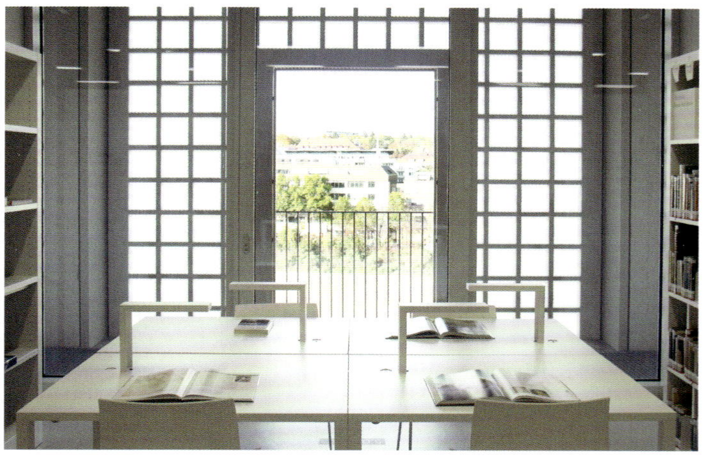

Project data:

Use:	Library
Access:	Stairs, lifts
Number of floors:	9 + 2 basement levels + roof terrace
Clear room height:	3.00 m
Gross volume:	98,249 m^3
Gross floor area:	20,225 m^2
Year of completion:	2011
Construction period:	24 months

Section • Floor plans
Scale 1:1000

1 Entrance
2 Foyer
3 “Heart”
4 Office
5 Sorting room
6 Media presentation
7 Workplaces
8 Group room
9 Reading room
10 Graphics library
11 Cafeteria

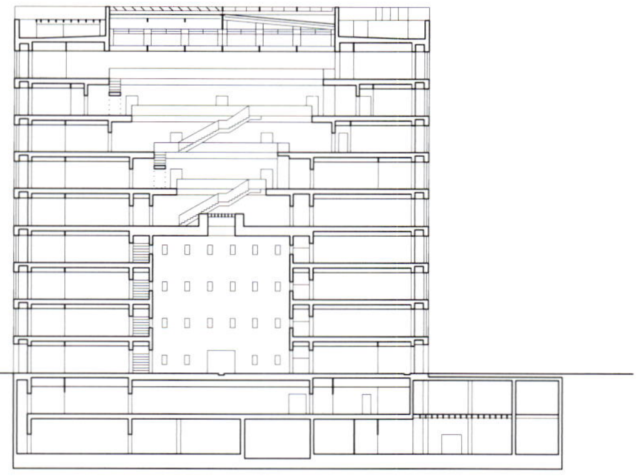

aa

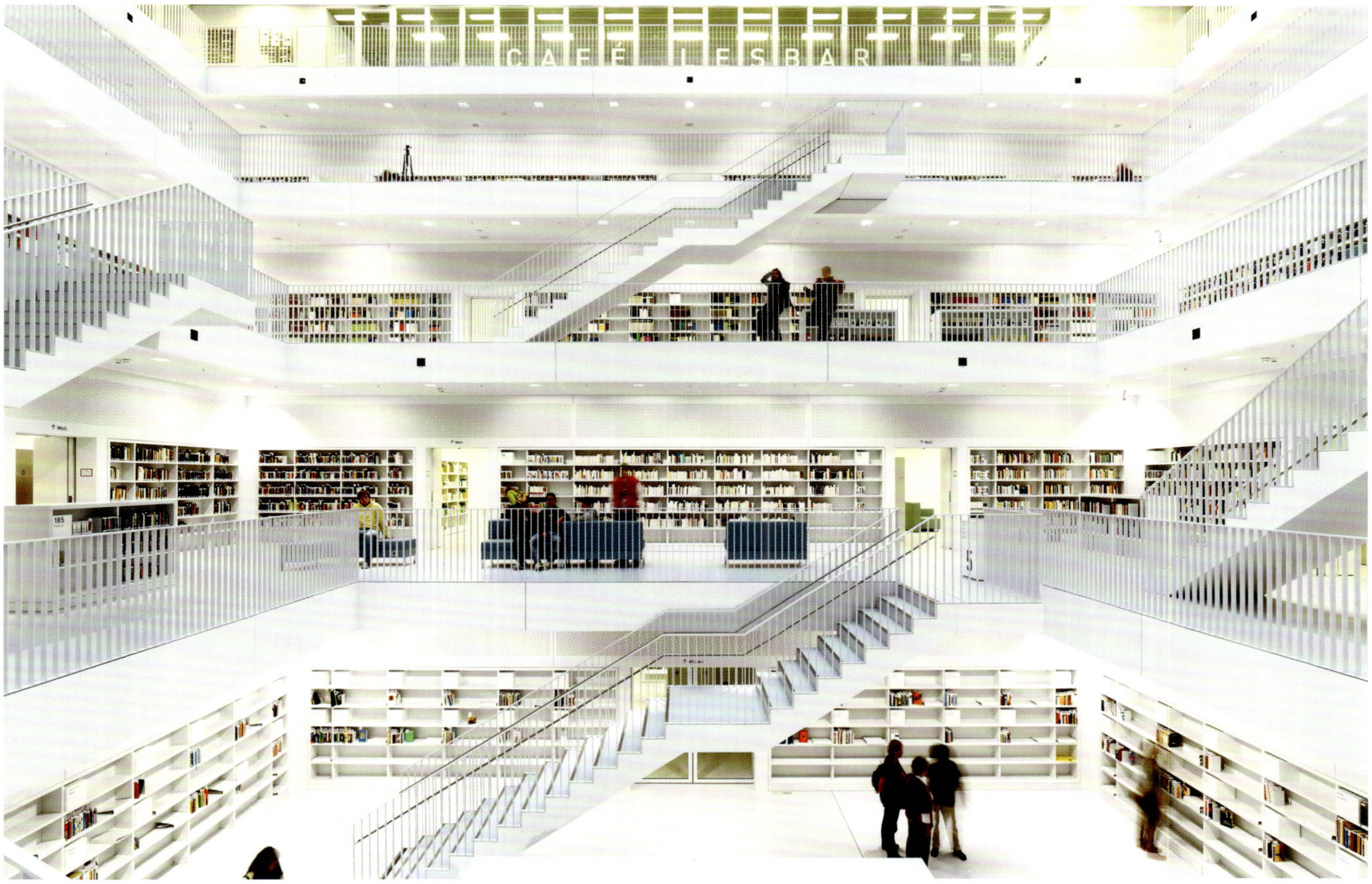

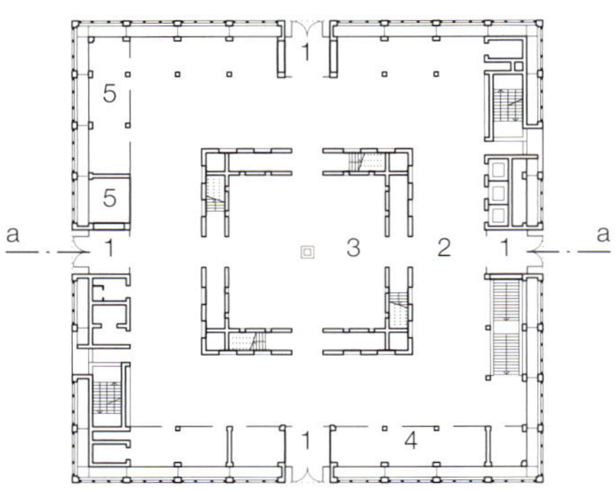

Ground floor

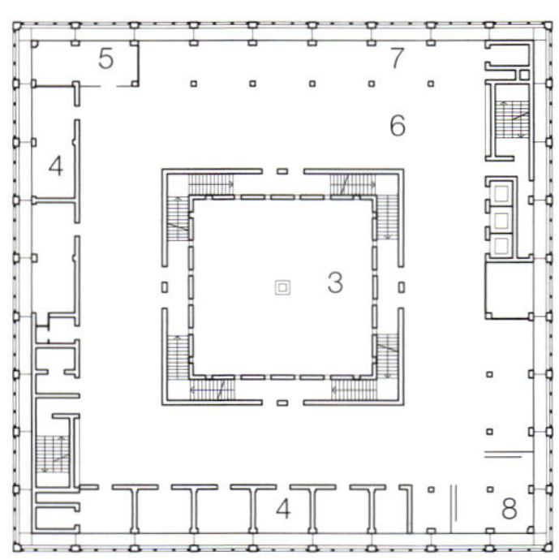

3rd floor

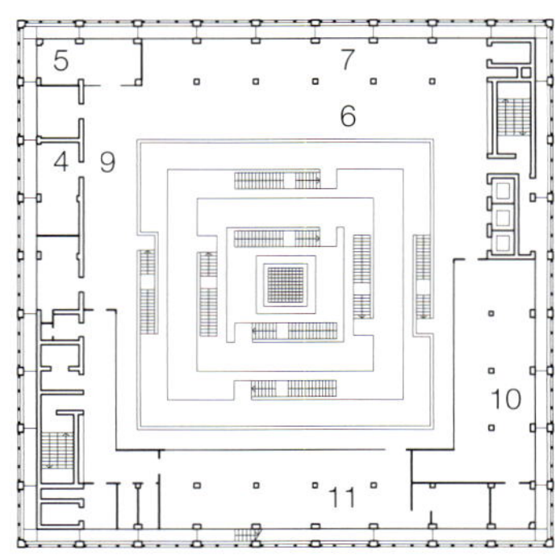

8th floor

Sectional details
Reading room stairs
Scale 1:20

1 Floor structure:
 6 mm linoleum, equalising layer
 89 mm heated screed
 separating polyethylene foil layer
 20 mm footfall insulation, sealing sheet
 85 mm equalising mortar layer
 300 mm reinforced concrete floor
2 Suspended ceiling:
 10 mm plasterboard, acoustic panel,
 perforated plastered, painted white
 40 mm insulation
3 180/90/10 mm angled steel profile
 shortened to 180/40/10 mm
 200/150/15 mm angled steel profile
4 Balustrade, hot-dip galvanised, sharp-edged:
 Top chord 40/15 mm flat steel
 Balustrade posts 40/10 mm vertical flat steel
 Bottom chord 40/10 mm flat steel
 Handrail: stainless steel ∅ 35 mm
 (height 85 mm)
5 Stairs: precast reinforced concrete, rendered
6 10 mm anti-slip linoleum (R10) with
 adhesive layer
 12 mm steel anchor plate, set in concrete in
7 Stairs underlay: flexible foam with supporting
 elastic cores for acoustic isolation

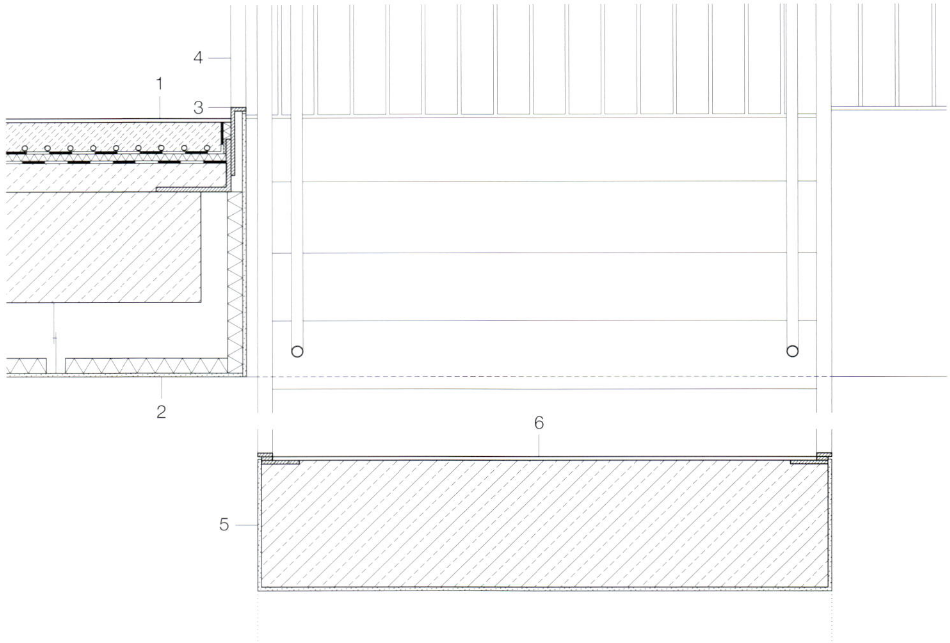

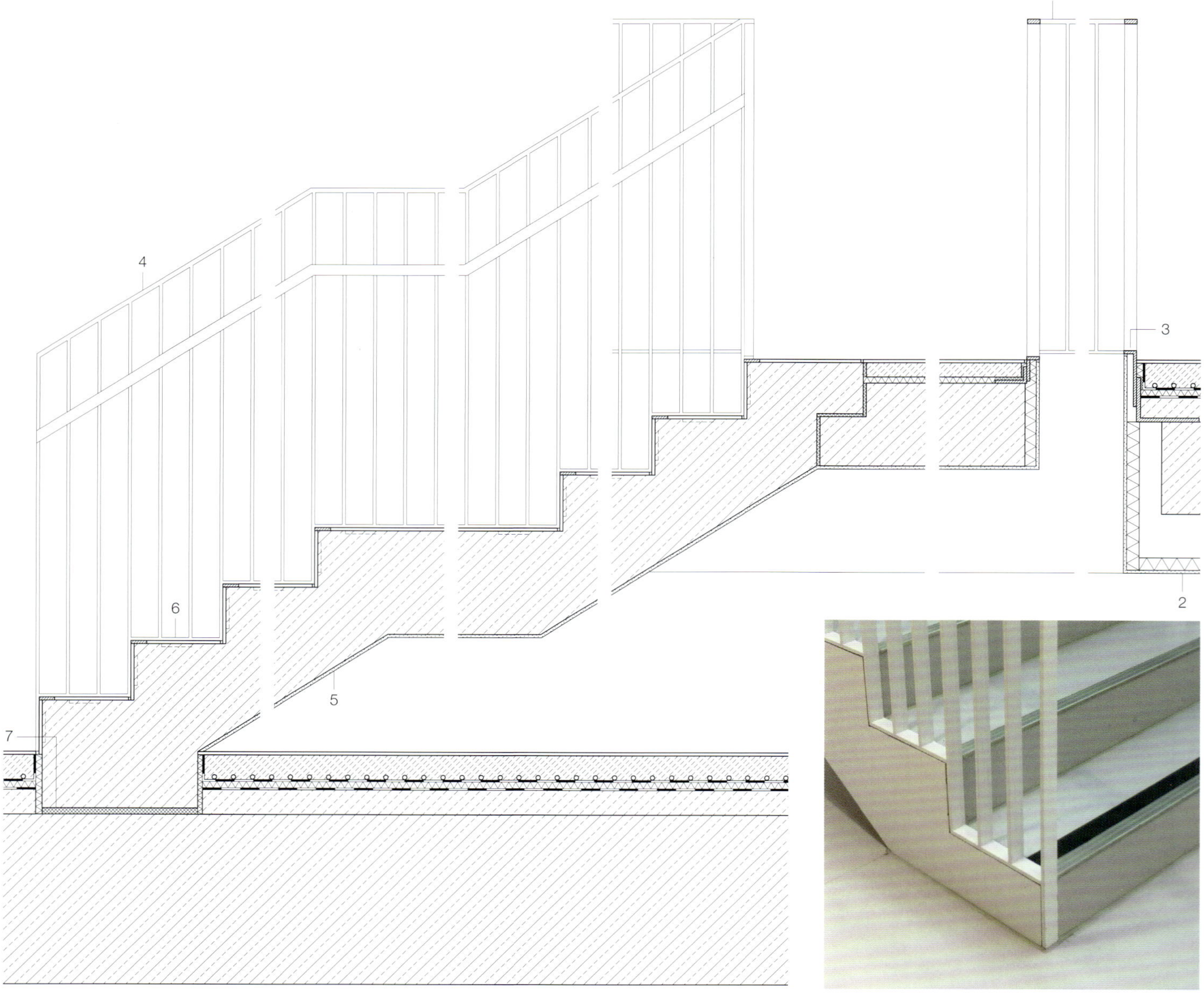

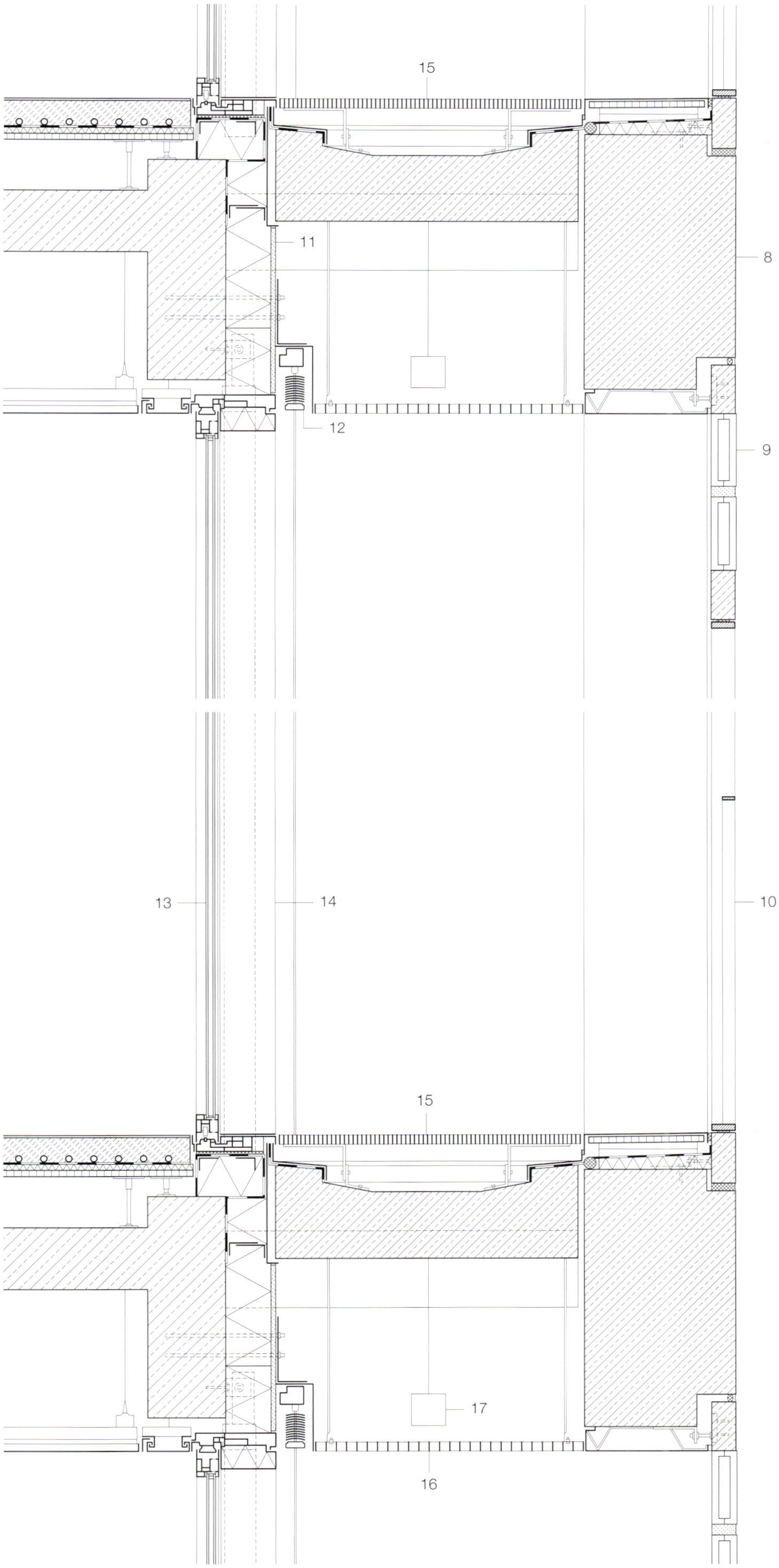

Vertical section, facade passage
Scale 1:20

8 Precast pillars and beams:
480 mm concrete reinforced, high white cement content
20 mm precompressed joint sealing tape anthracite grey, UV-resistant
9 Prefabricated glass brick element:
in total 4140/3320/80 mm
Concrete reinforced, high white cement content
Glass brick: white glass, low iron oxide content, frosted on outside, 240/240/80 mm surrounding precompressed joint sealing tape, anthracite grey, UV-resistant, 20 mm
Connected to fabric of building via thermally isolated concrete brackets
10 Balustrade (height 1370 mm):
20/10 mm parapet made of steel flats
Duplex coating of liquid paint and hot-dip galvanising, grey
11 Composite thermal insulation system:
18 mm reinforcement coat and mesh
140 mm mineral fibre laminated insulation panel, bonded
12 External sun protection:
Exterior aluminium blinds, cable guided, coated grey, 1280/3000/60 mm
13 Aluminium window, insulation glazing,
8 mm toughened glass + 18 mm argon-filled cavity + 6 mm toughened glass, low-E coated
14 Mullion/transom construction, in the centremost of the three adjacent fields with lift and slide door, otherwise fixed glazing:
180 mm steel supporting posts
Lift and slide element: 70 mm thermally insulated aluminium profile, coated with grey liquid paint, in some cases with electric drive
15 Mesh floor:
Pressure locked grating, hot-dip galvanised, 32 mm, mesh width 11/33 mm, screwed to fixed, height-adjustable feet, precast concrete slab, water impermeable, 5 % pitch, draining inwards
16 Mesh ceiling:
Pressure locked grating 30 mm, mesh width 33/33 mm, hot-dip galvanised, suspended with threaded rods, height-adjustable
17 Ceiling light fixture

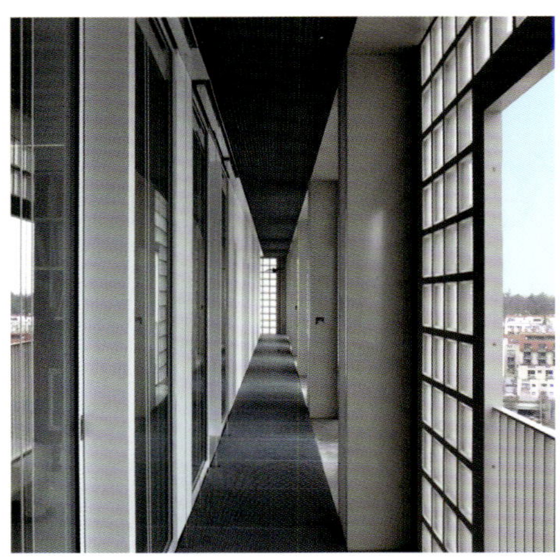

Advertising agency in Tokyo

Architects: Klein Dytham architecture, Tokyo

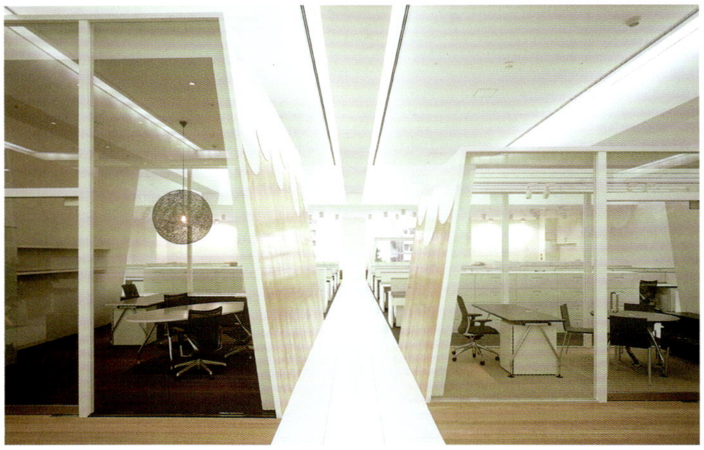

Parallel white walkways run in straight lines through the big open-plan office space with its park-like atmosphere.

This joint venture between one of the biggest international and one of the biggest Japanese advertising agencies has located its new business premises in an eight-storey leisure complex in downtown Tokyo. The space, formerly used as a bowling alley, extends over a whole floor which, following removal of the false ceiling, became a two-storey space. The lack of pillars made it possible to partition the office areas in any desired configuration, which inspired the architects to create a landscape reminiscent of a small village. The furnishings and the layout of the circulation zones are oriented on the rhythm of the dominant ceiling beams; parallel white "footpaths" run through the room, with the workplaces lined up along these. Small green spaces containing a mixture of real plants and lawn-like shaggy carpets are scattered between the "boxes" that are used as meeting rooms and individual offices, with accessible roofs on which the employees can relax, exchange ideas or have a picnic.

The office landscape is entered via the higher entrance level on which the reception area and conference rooms are located. From here, wide open stairs lead directly down to the small "village square" with a café, the communicative centre of the open-plan office. On the other side the staircase is designed in such a way that it can even serve as a seating option. It is also possible to use the stairs area for presentations and events.

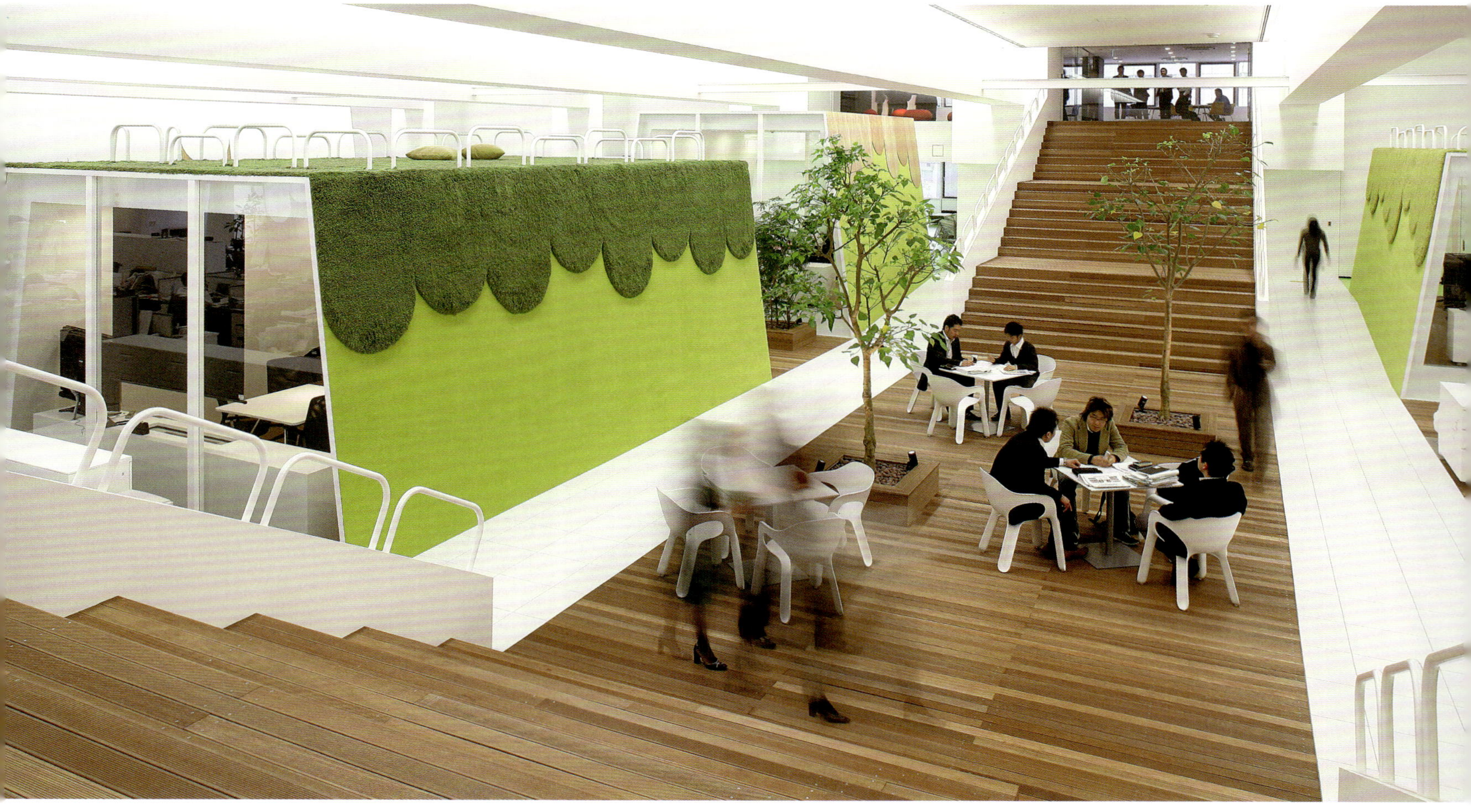

Sections • Floor plans
Scale 1:800

1 Administration
2 Reception
3 Open stairs
4 Café
5 Conference room
6 Individual office
7 Server room

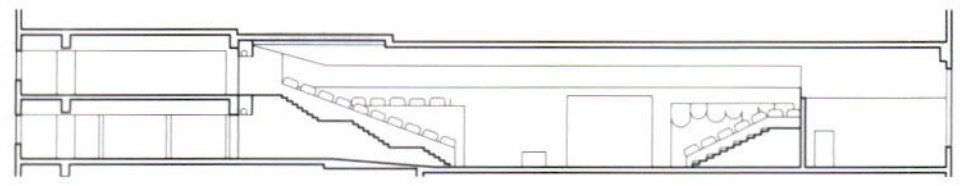

aa

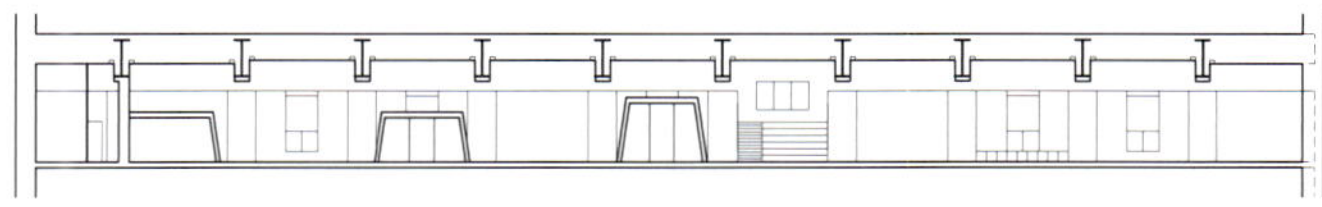

bb

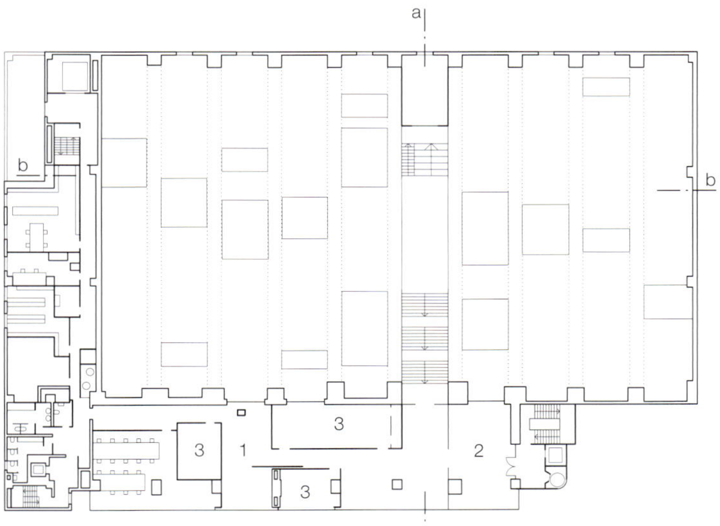

Entrance level (6th floor)

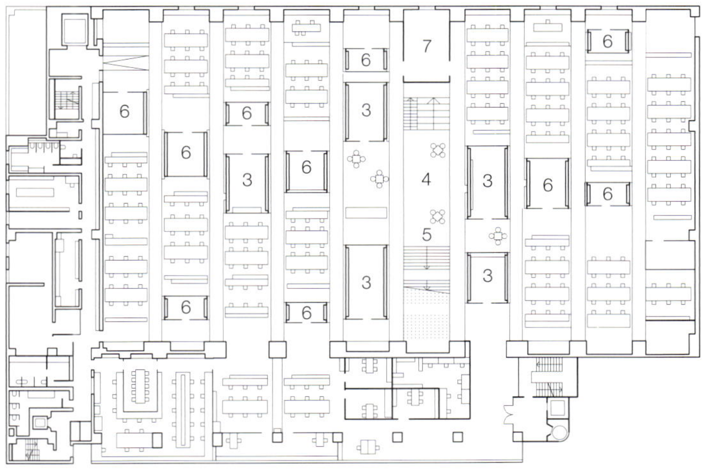

Office level (5th floor)

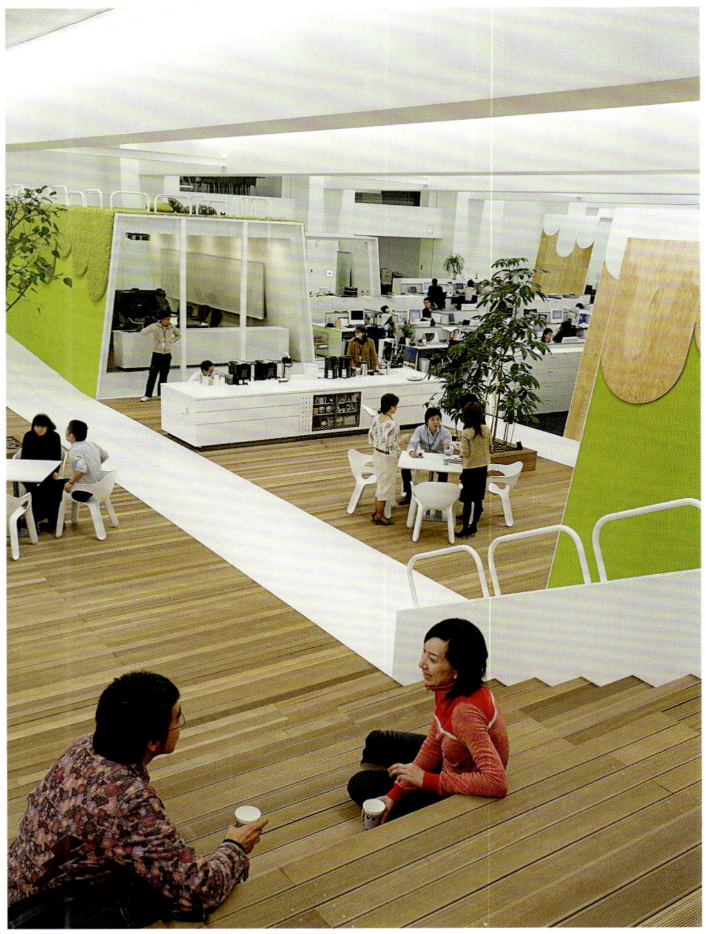

Project data:

Use:	Office
Access:	Staircase, "walkways" highlighted in white
Number of floors:	2
Clear room height:	5.50 m
Gross floor area:	4,215 m^2
Year of completion:	2007
Construction period:	4 months

Office building in Bolzano

Architect: Markus Scherer, Meran

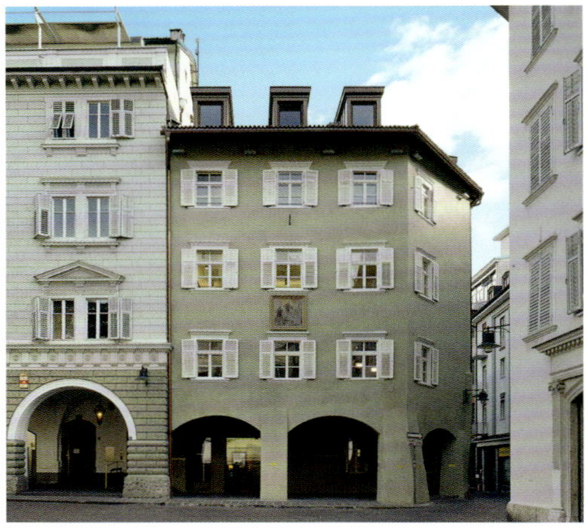

Following the conversion, a free-standing central staircase provides access to the offices located in this listed building, a former inn.

Instead of relocating their company headquarters to the edge of town, Fri-El Green Power, a company involved in the area of alternative energy, opted for a listed building in the centre of Bolzano which comes with a long history. However, little remained of the historic elements; apart from the basement levels, only the external facade and the middle wall on the upper floors were retained during the conversion. The original line of the stairs and thus the former means of access could no longer be traced. However, it transpired that the building had formerly had a large central light dome on the roof, a typical element used in Bolzano townhouses to illuminate the stairwell. Inspired by this, the architect developed his access concept in the form of a central, free-standing staircase with natural illumination from the reinstated light dome.

Separation of old and new

The new access point to the building is via the existing portico. From the two-storey entrance hall, the new staircase rises upwards, initially in three flights, then in two offset flights. They are flanked by wall sections which are designed either as transparent glass surfaces allowing views into the rooms beyond or, in places where the rooms are to be screened off, as backlit glass constructions with laminated photo motifs showing the company's wind farms.

In order to separate the new fixtures clearly from the historical fabric of the building, the staircase is in a free-standing position based on the presumed location of the original access stairs. It is designed as a lightweight metal structure, with glass balustrades creating a feeling of openness and transparency. The individual flights of stairs with their supporting construction of square hollow sections were all completely prefabricated in the workshop and then assembled in the building. The glass balustrades were also assembled on site. All the visible parts are finished in architectural bronze, the surface of which develops a certain patina over the course of time, with subtle colour nuances lending it a pleasant character. A lift is positioned to the side of the staircase which runs from the basement to the attic level, providing barrier-free access to the offices.

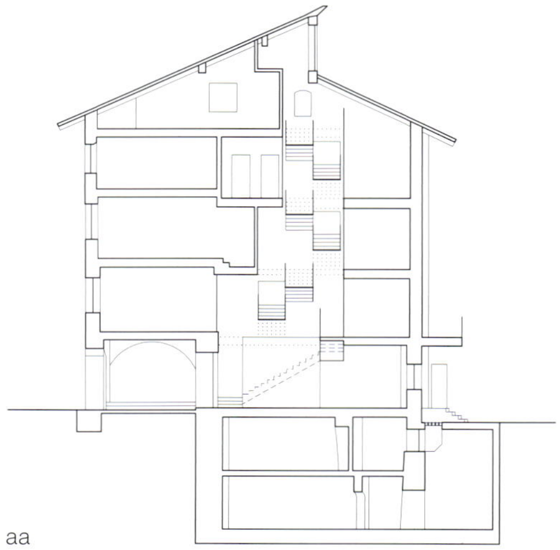

aa

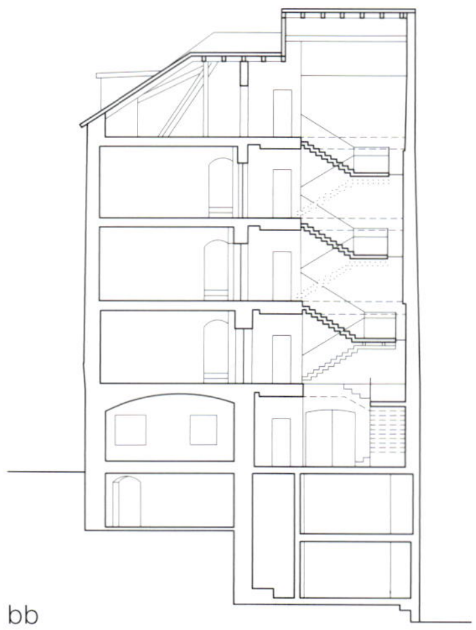

bb

Project data:

Use:	Office
Access:	Central steel stairs, lift
Number of floors:	5 + 2 basement levels
Clear room height:	2.53–3.30 m
Gross volume:	3,500 m^3
Gross floor area:	815 m^2
Year of completion:	2007
Construction period:	12 months

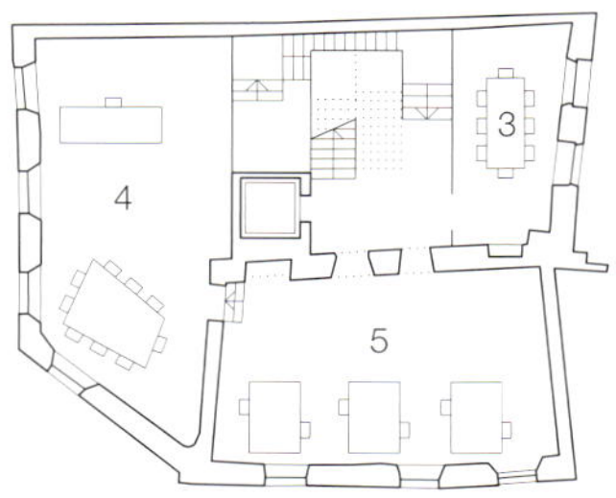

1st floor

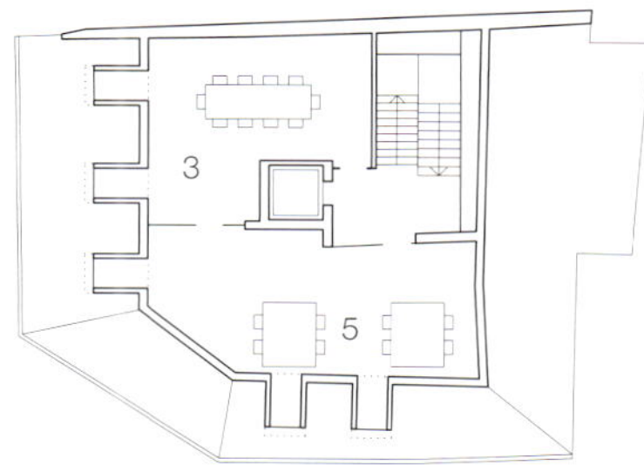

4th floor

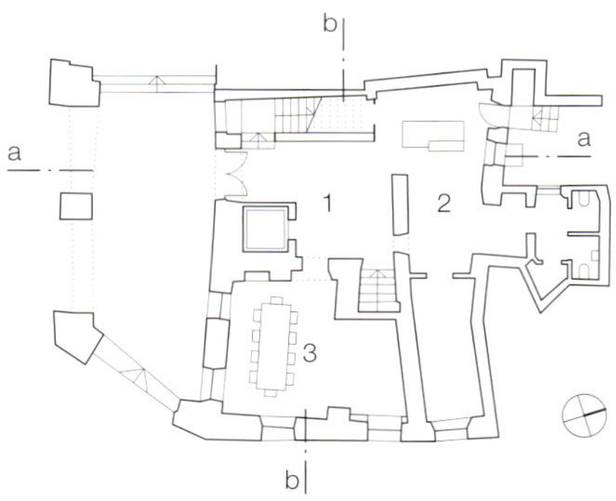

Ground floor

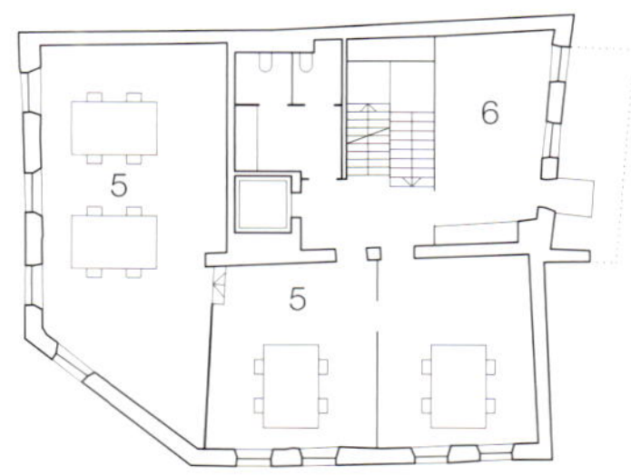

3rd floor

Sections • Floor plans
Scale 1:400

1 Entrance hall
2 Reception
3 Conference room
4 Executive office
5 Staff office
6 Amenity room

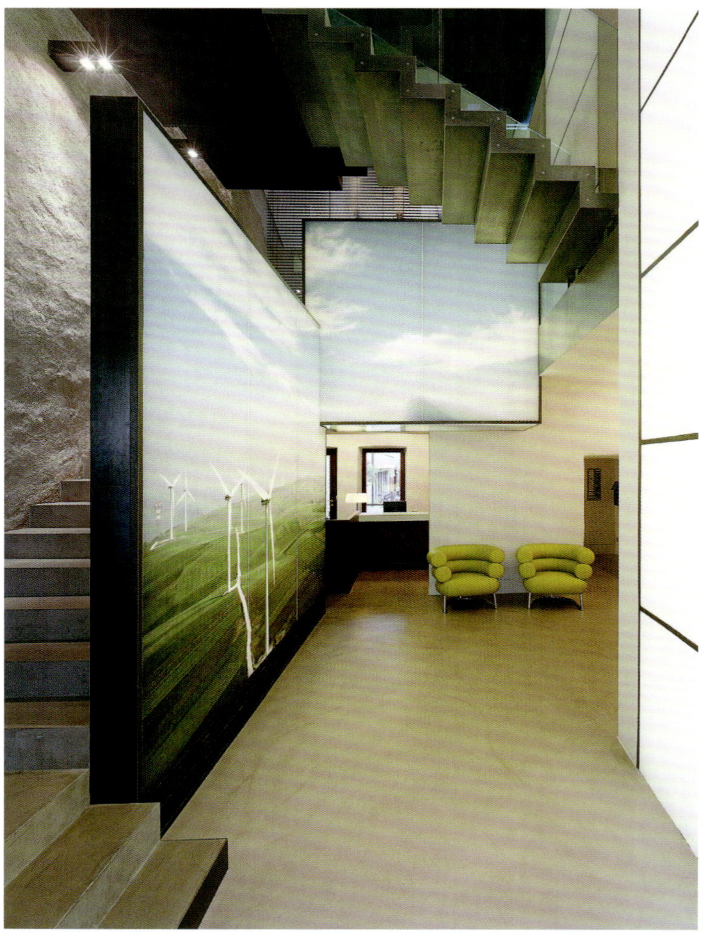

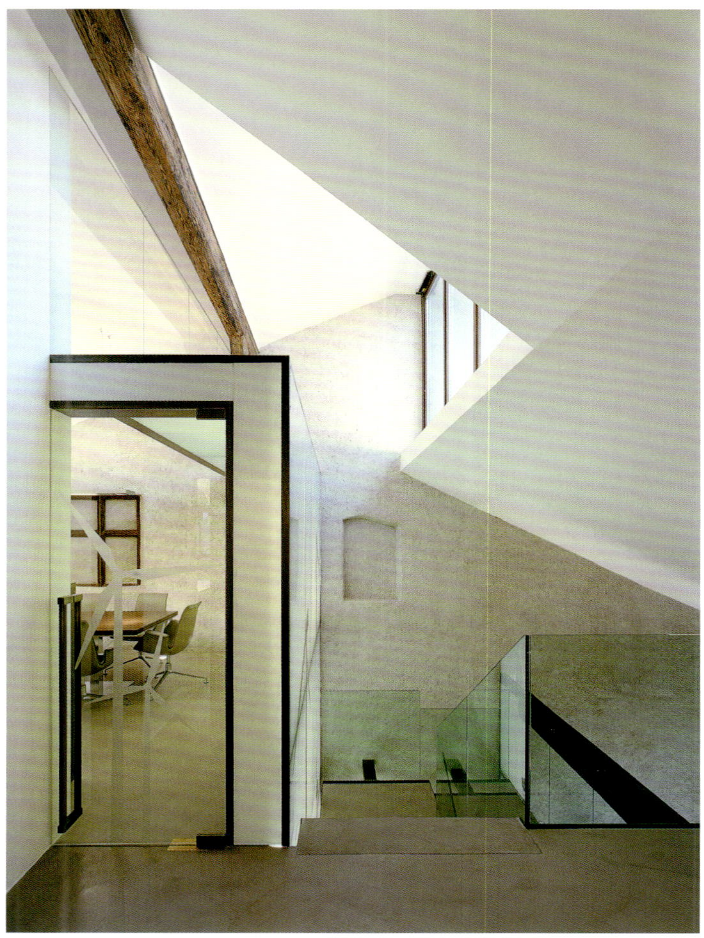

Sectional details of staircase
Scale 1:20

1 Laminated safety glass balustrades consisting of 2× 10 mm toughened glass with 1.5 mm PVB foil sandwiched between
2 Outer clamping plates in Tombak 8 mm architectural bronze, screwed to steel profile, with glass balustrades clamped between these
3 5–7 mm smoothing cement
20 mm calcium sulphate panels
1–2 mm cork mat; 4 mm sheet steel; 60/60 mm square hollow steel sections
Tombak 2 mm architectural bronze sheeting
4 6 mm cement screed
32 mm calcium sulphate panel
12 mm impact sound insulation
80 mm concrete screed
250 mm reinforced concrete
12.5 mm plasterboard suspended ceiling
5 15 mm steel flat, welded, with three stiffening ribs
6 Facing: Tombak 8 mm, concealed fixing from behind
7 Concealed lighting strip
8 15 mm steel angle, welded, continuous
9 Tombak 8 mm architectural bronze
10 80/80 mm steel profile stair supports

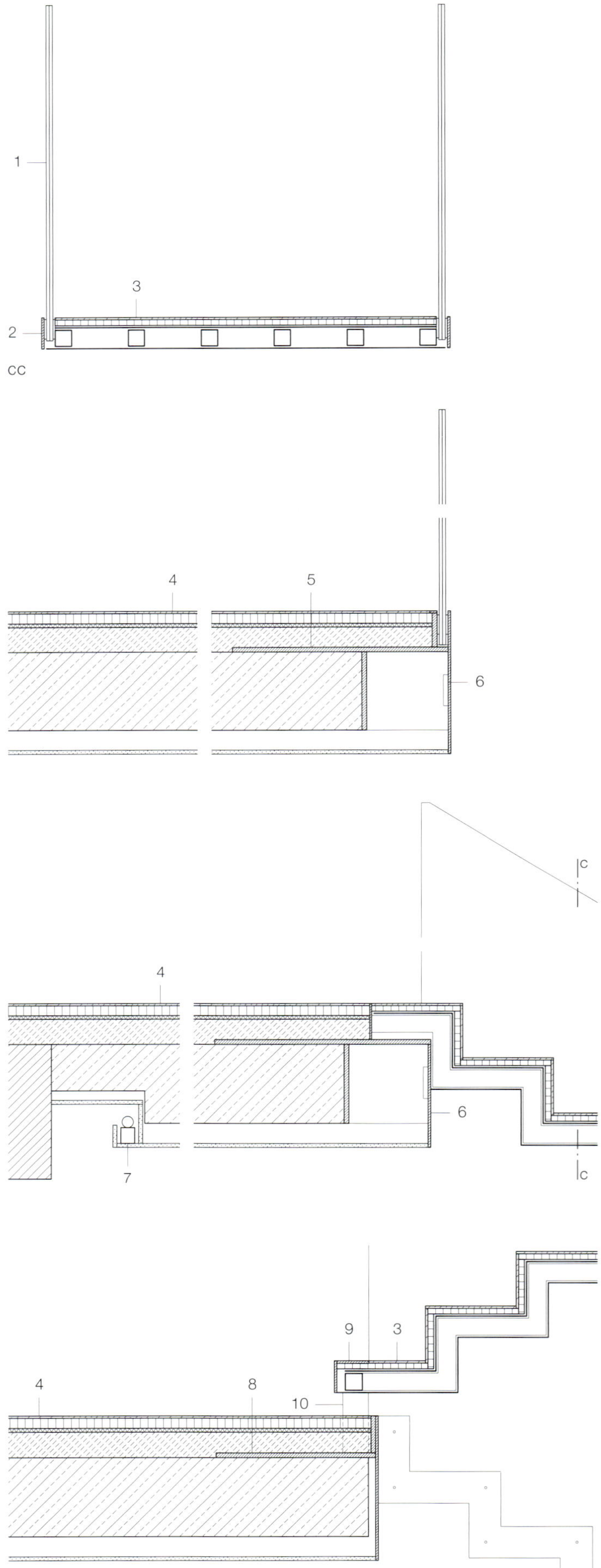

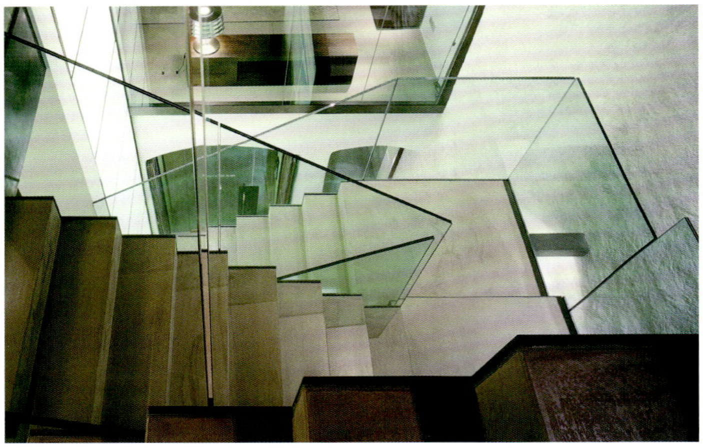

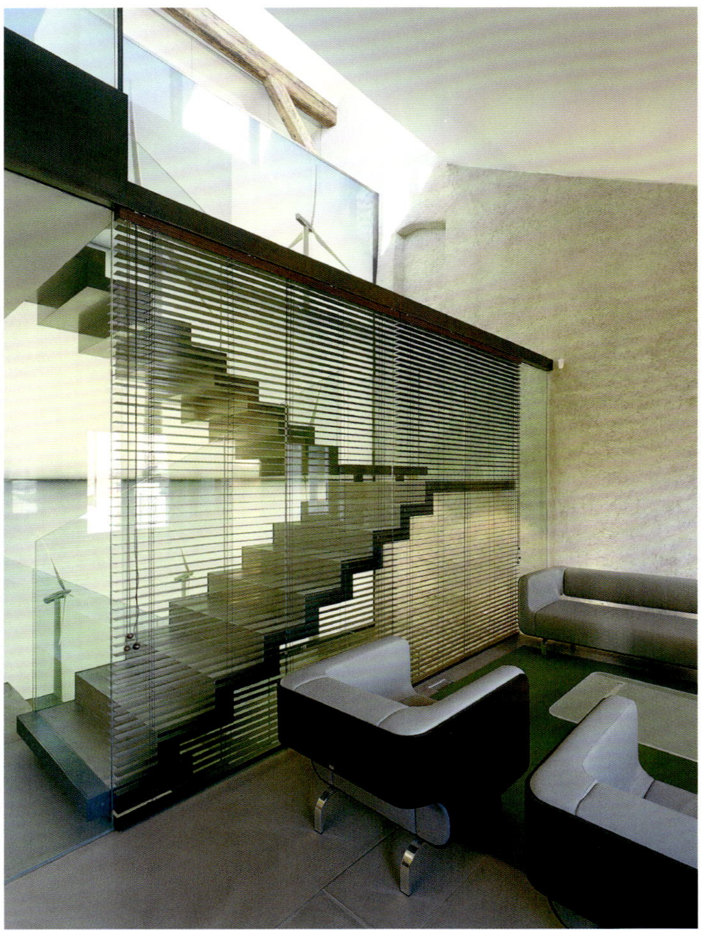

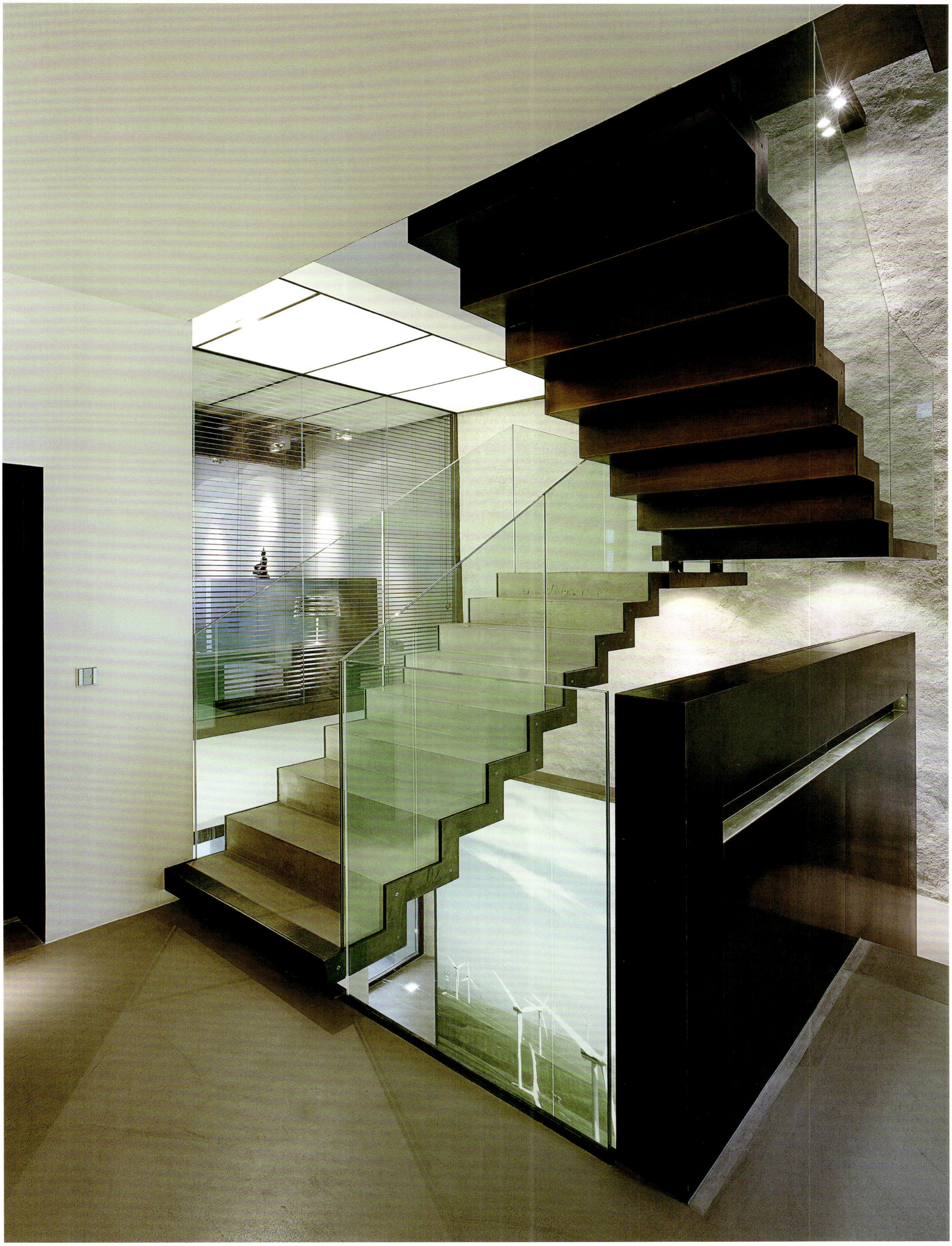

AachenMünchener headquarters in Aachen

Architects: kadawittfeldarchitektur, Aachen

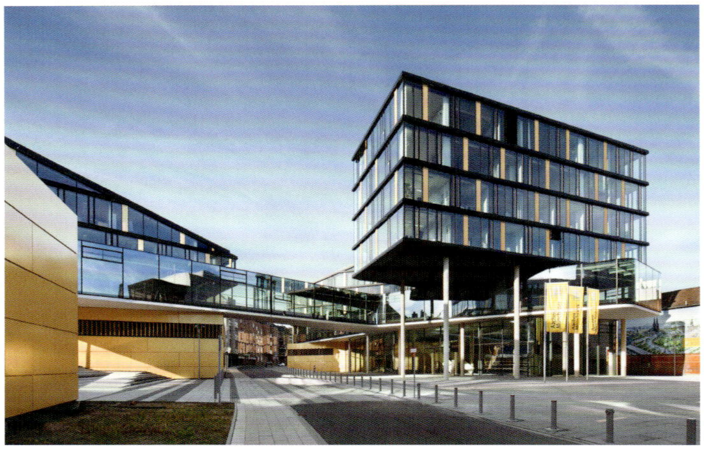

A central boulevard connects an existing structure and new buildings to form a single complex which is integrated into the surrounding urban environment.

The insurance group wanted to concentrate its various business divisions, which had previously been spread over several different locations around the city, into a single building complex. The challenge was to integrate an office landscape providing 30,000 m^2 of floor space into an existing urban setting, incorporating existing buildings, without closing the site off from the public. The architects succeeded in both meeting the client's needs for a prestigious central corporate headquarters and accommodating the interests of the city and its residents. The municipal authorities wanted a link between the redeveloped railway station area and Aachen's historical city centre, which for a long time had been blocked by the insurance company's two slab-like 10-storey office blocks dating from the 1970s and various ancillary structures. In order to achieve this, public pedestrian zones were created crossing the essentially private site at crucial points to form the "Via Culturalis" pedestrian link between the central station and the city centre. Differentiated spaces on the site, including small squares and a broad flight of steps, now serve as a public amenity for residents.

Boulevard as a space for communication

In total, four buildings were created to accommodate the insurance group's offices and a fifth building which is leased out, providing everyday facilities such as a post office, shops and restaurants that enliven the area. In order to satisfy the need for a communicative architectural ensemble characterised by transparency and permeability, the required floor space was distributed over building structures set at different angles which are linked together by a glassed-in walkway, the so-called Boulevard. In the interior of the buildings, this connection, with its widened and narrowed sections, serves the users as a space for chance encounters and communication. In addition, all commonly-used areas such as the staff restaurant, cafeteria and conference, training and seminar rooms are located here. All the necessary technical installations such as sprinklers and smoke alarms are concealed behind the suspended ceiling made of white painted expanded metal. The ceiling-high glazing of the walkway, which is continued over the edges of the roof, provides spacious views of the surrounding urban environment.

Spacious flight of steps

The newly created, open AachenMünchener-Platz provides the company headquarters with a prestigious address on Borngasse in the city centre. Here, the Boulevard is linked to street level via a 20-metre-wide flight of stairs and opens up to the urban space with an imposing two-storey foyer. The decision to locate the main entrance in the centre of the ensemble added a further valuable open urban space to the sequence of squares.
A serpentine ramp winds down through the generously dimensioned steps with their intermediate levels, allowing unproblematic bicycle and pushchair access.

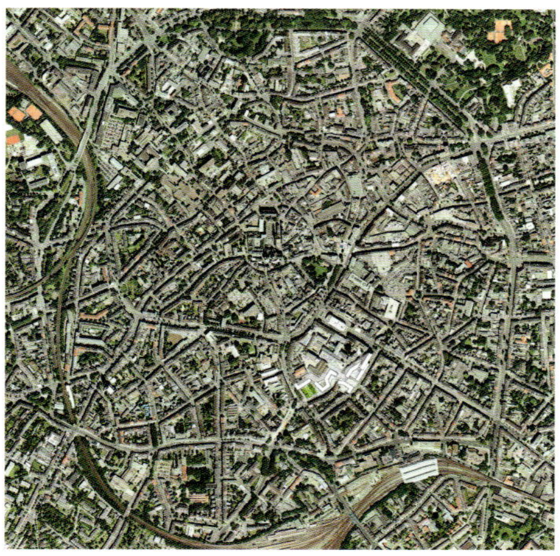

"Via Culturalis"

1 Cathedral
2 Münsterplatz
3 Elisen Fountain
4 Theatre
5 Alexianergraben
6 Kapuzinerkarree
7 AachenMünchener-Platz
8 Steps
9 Pocket park
10 St Marien
11 Main station

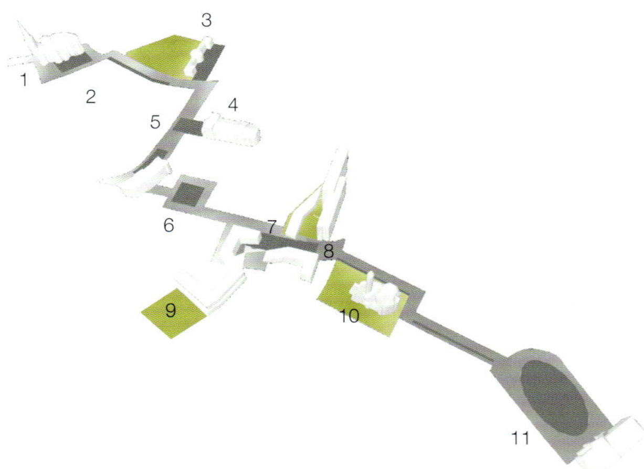

Project data:

Use:	Office
Access:	Main access in the centre of the complex, foyer with sculptural staircase, internal connection via "Boulevard"
Number of floors:	7 + 1 basement level
Clear room height:	2.75 m (office), 2.90 m (Boulevard)
Gross volume:	122,100 m³
Gross floor area:	34,900 m²
Year of completion:	2010
Construction period:	35 months

Part of ground floor
Scale 1:1250
Site plan
Scale 1:4000

12 AachenMünchener-Platz
13 Main entrance
14 Foyer
15 Access to basement garage
16 Existing building
17–19 New buildings
20 New building, leased out

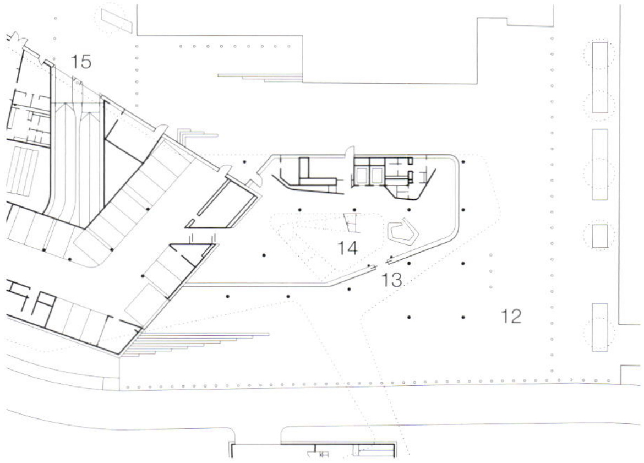

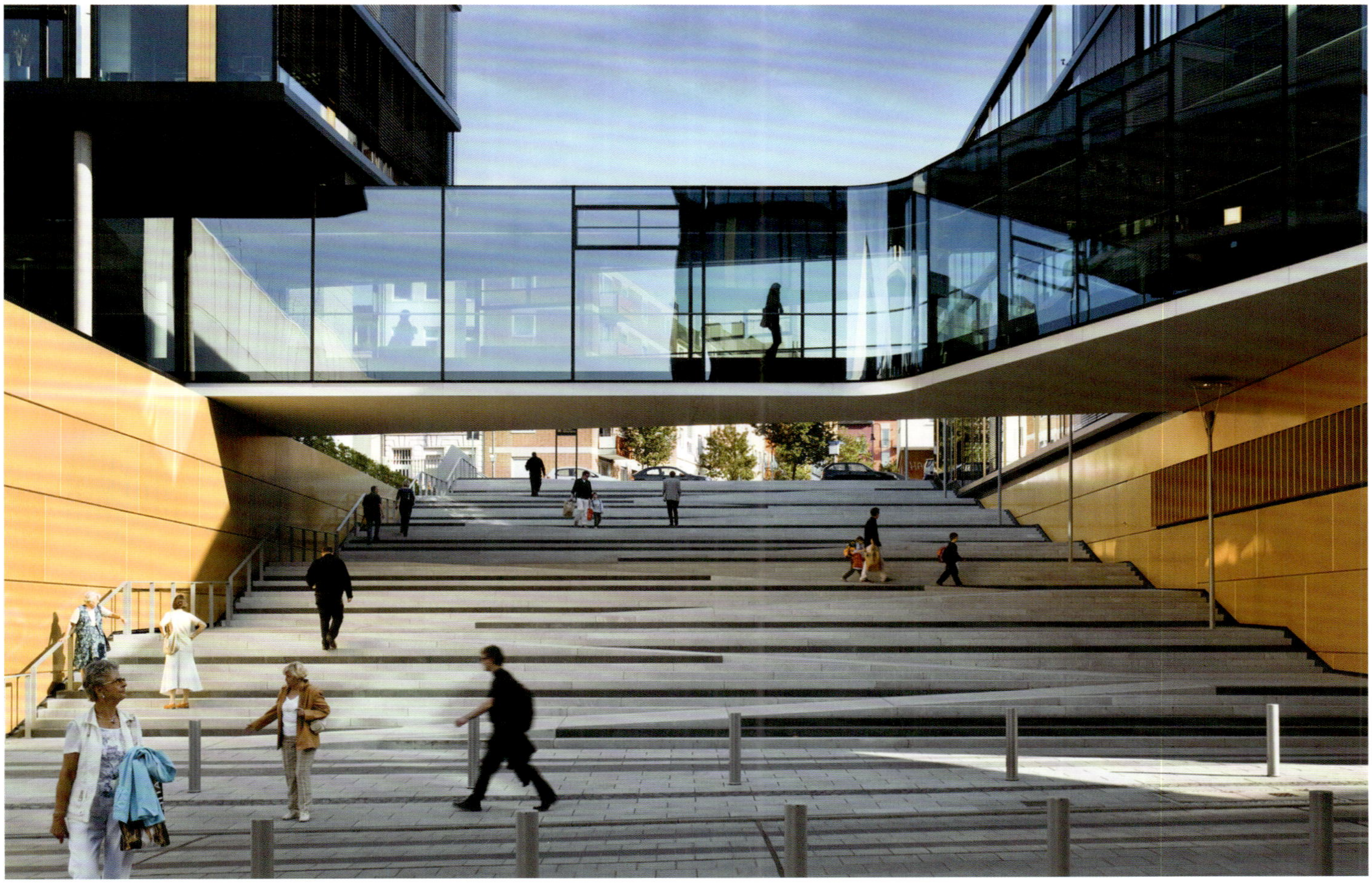

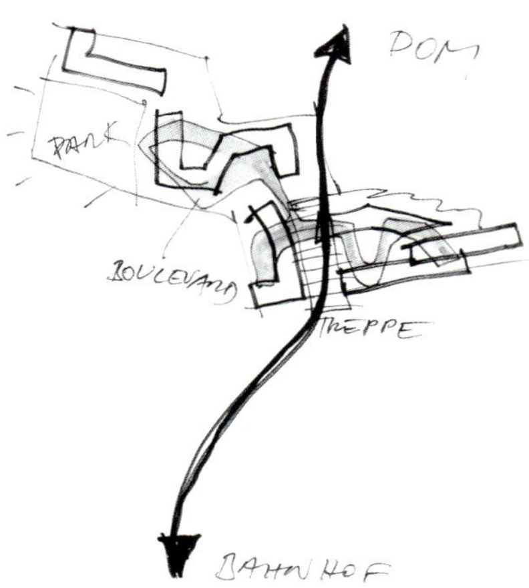

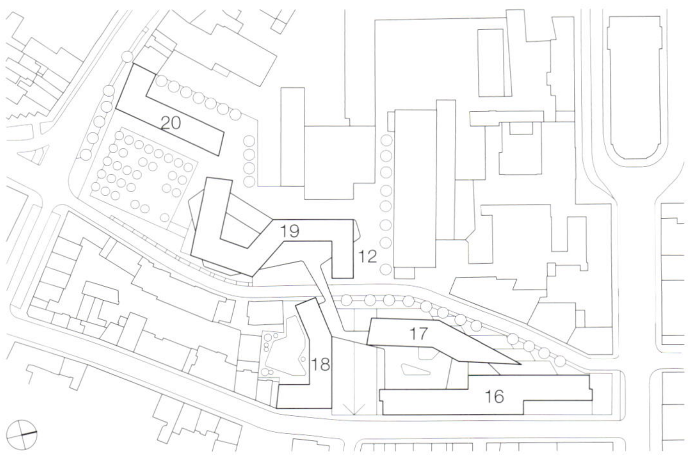

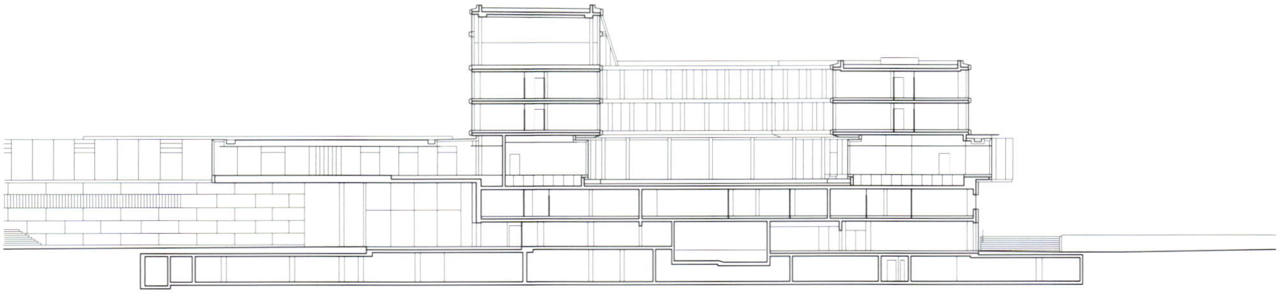

aa

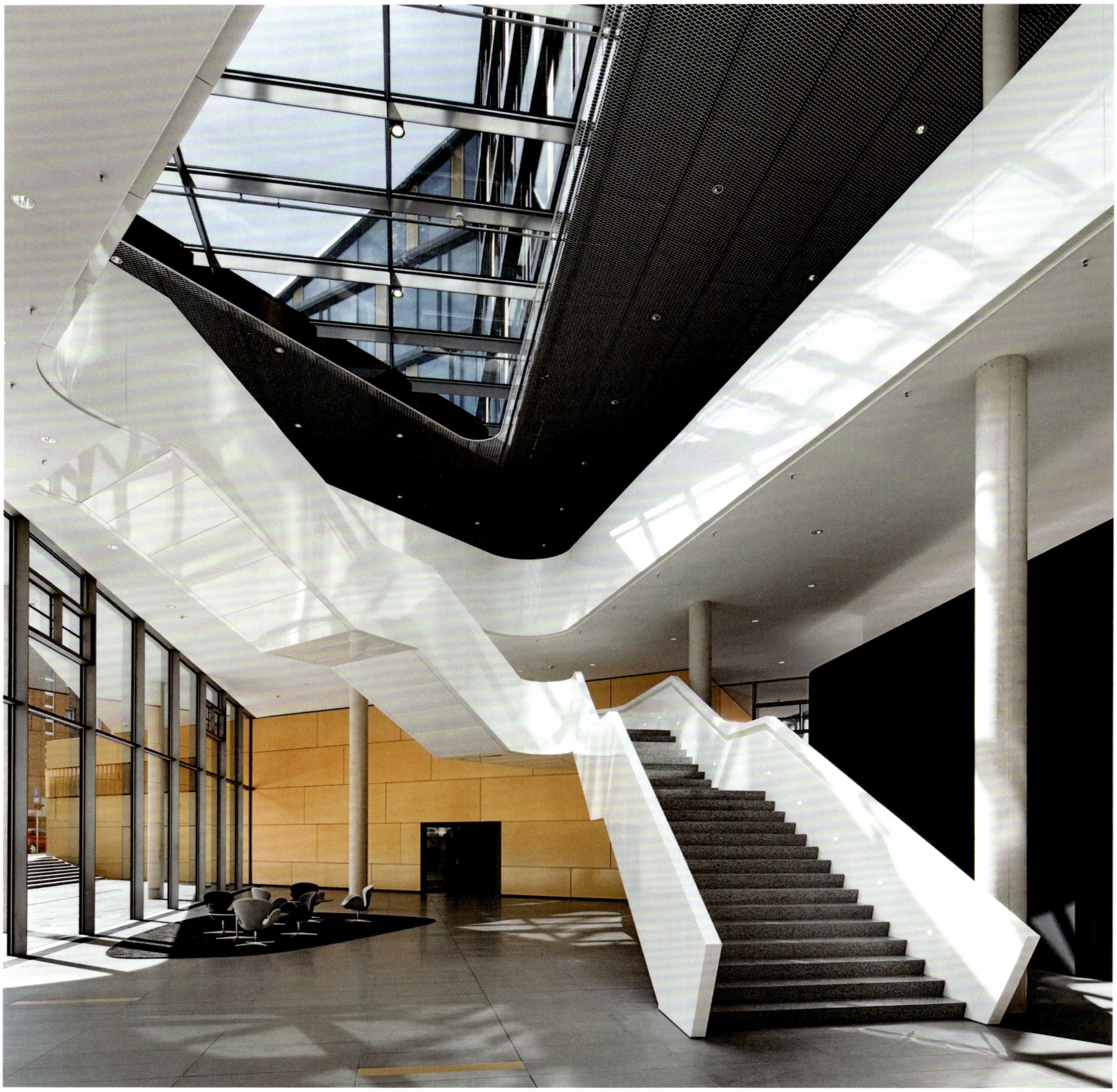

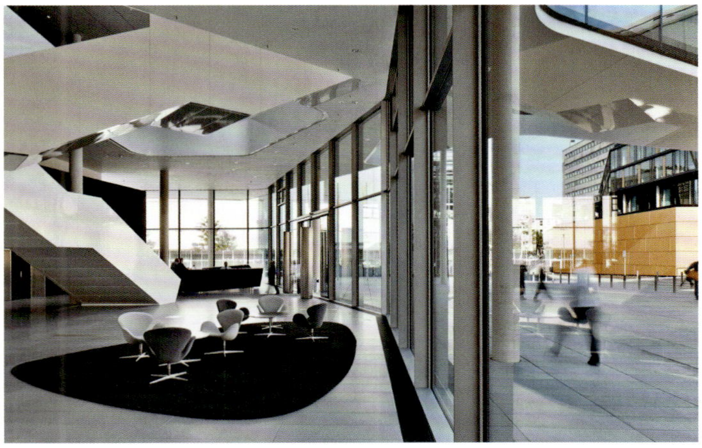

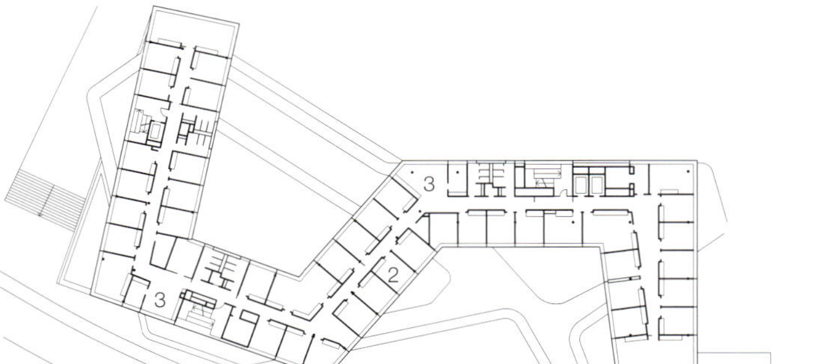

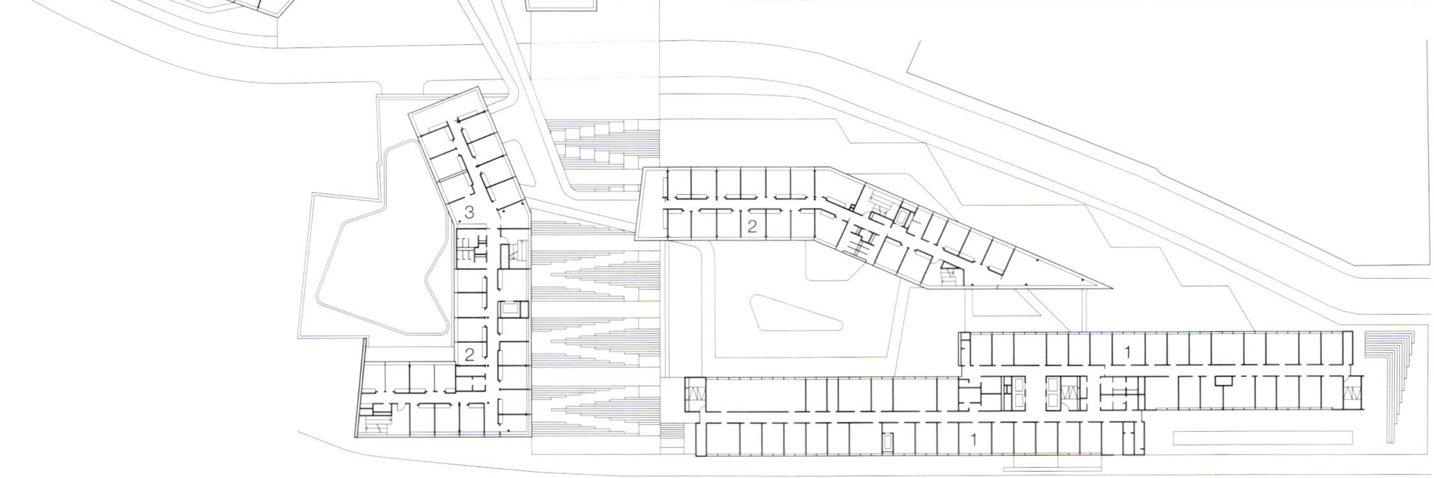

3rd floor

Section Scale 1:750
Floor plans Scale 1:1250

1 Single office in existing building
2 Single office in new building
3 Kitchenette
4 AachenMünchener-Platz
5 Boulevard
6 Meeting room
7 Conference room
8 Training/seminar room
9 Planted courtyard
10 Stairs from foyer
11 Cafeteria
12 Staff restaurant
13 Kitchen/serving counter

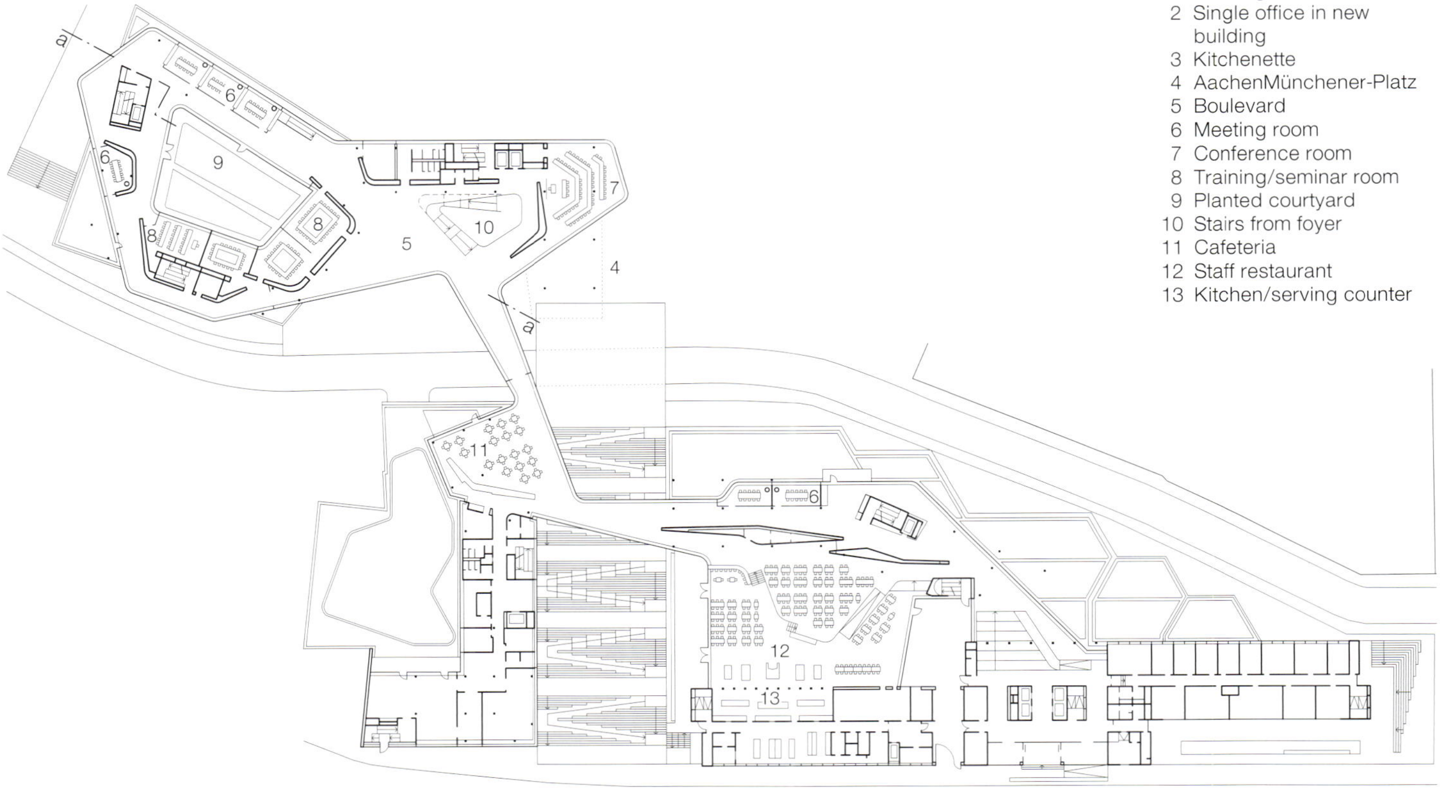

2nd floor

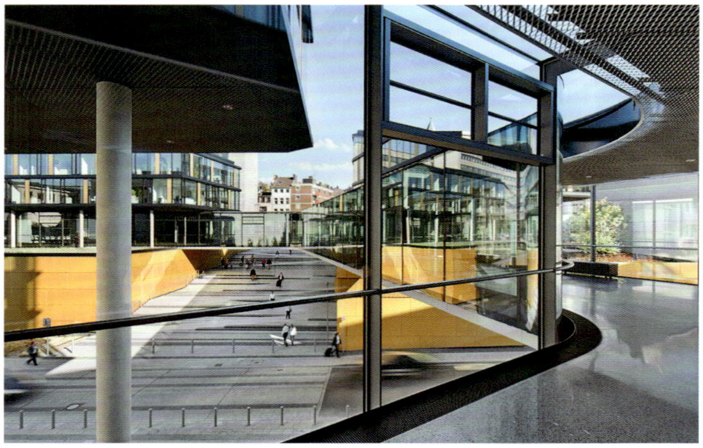

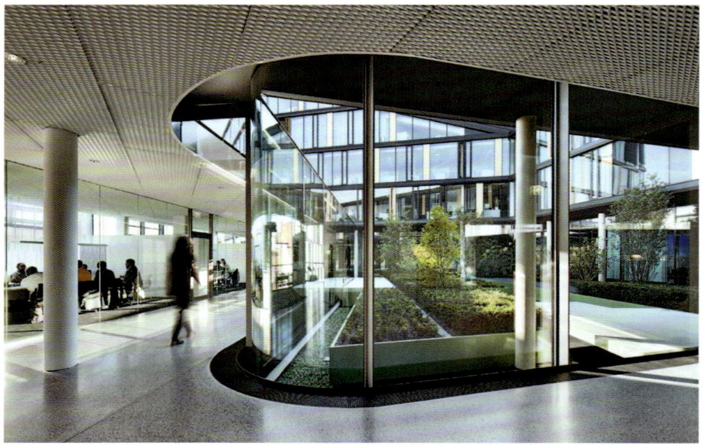

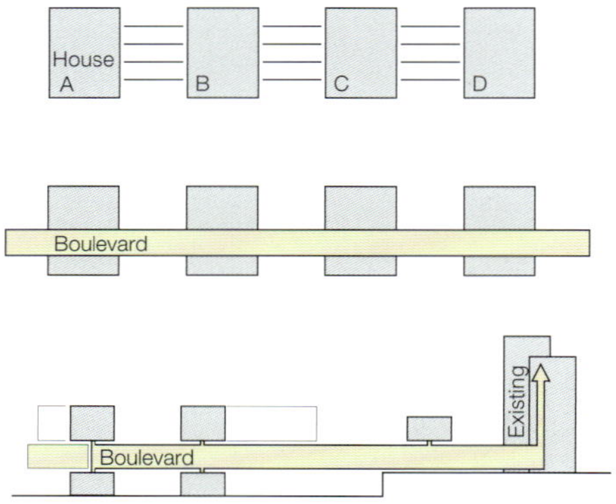

A

1 Meeting room
2 Conference room
3 Training/seminar room
4 Planted courtyard
5 Stairs from foyer
6 Cafeteria
7 Staff restaurant

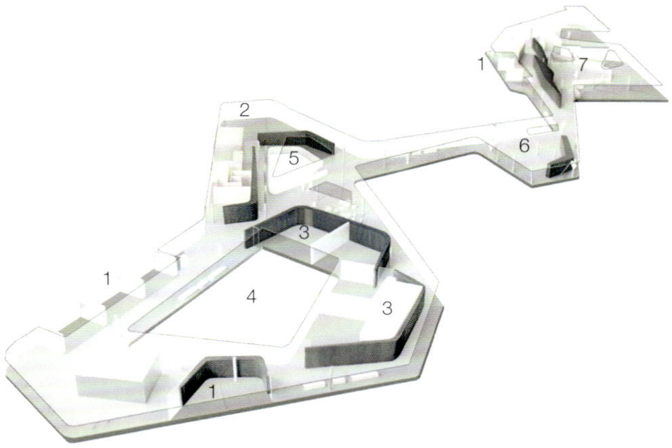

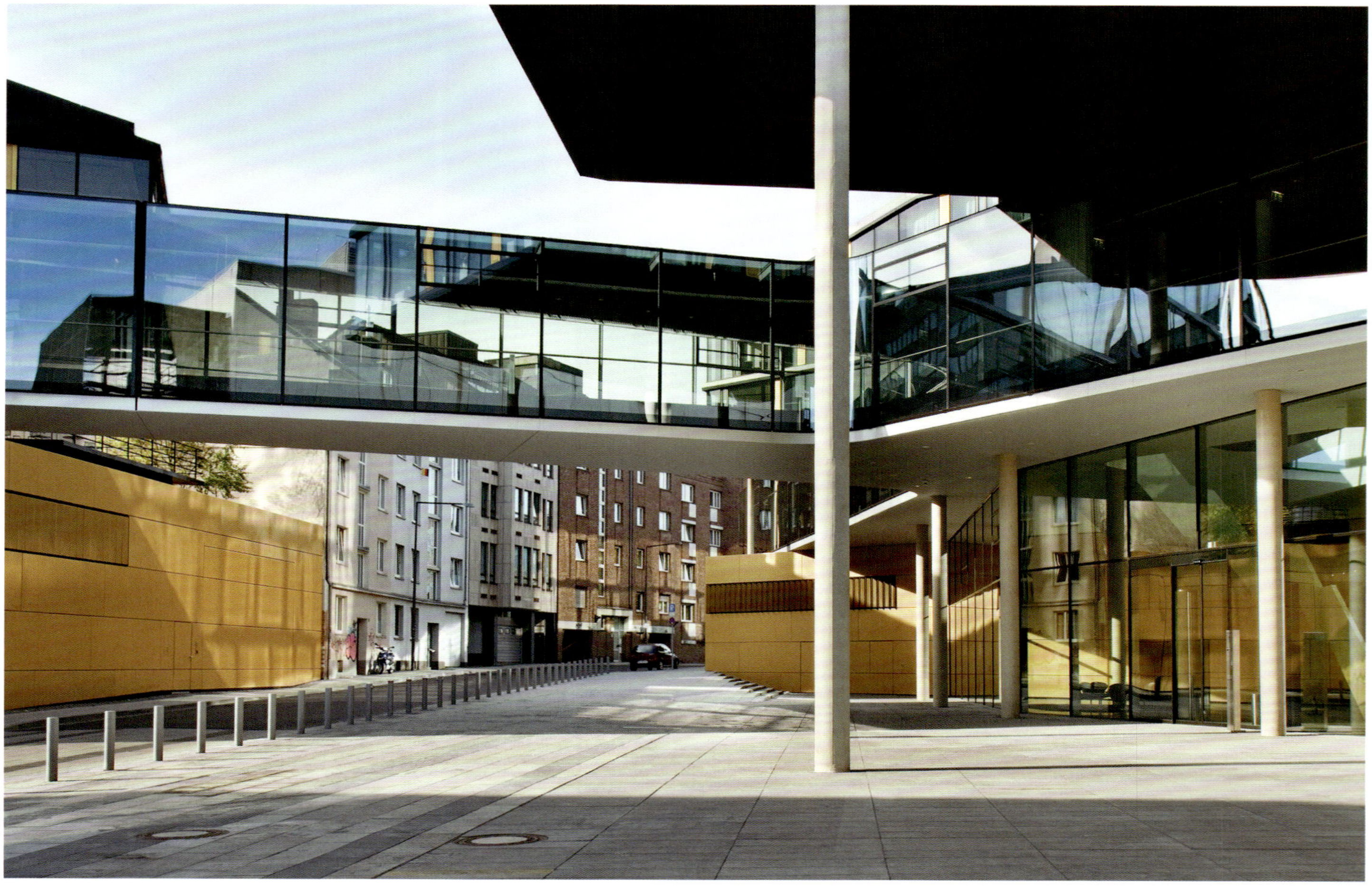

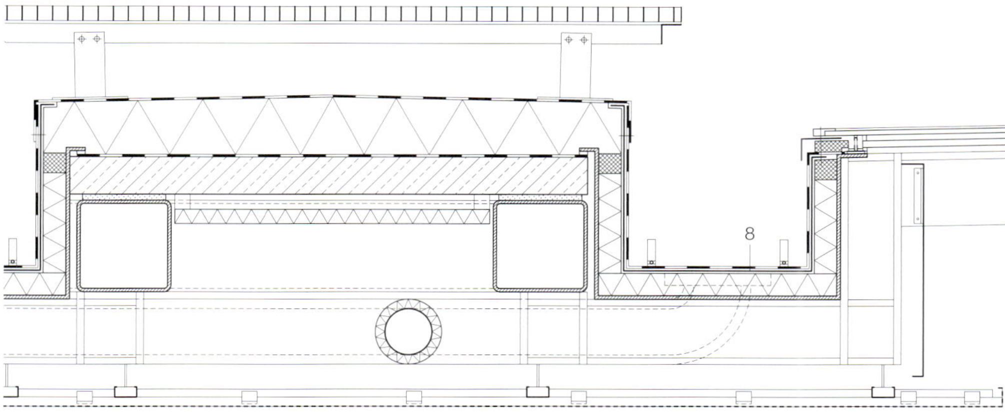

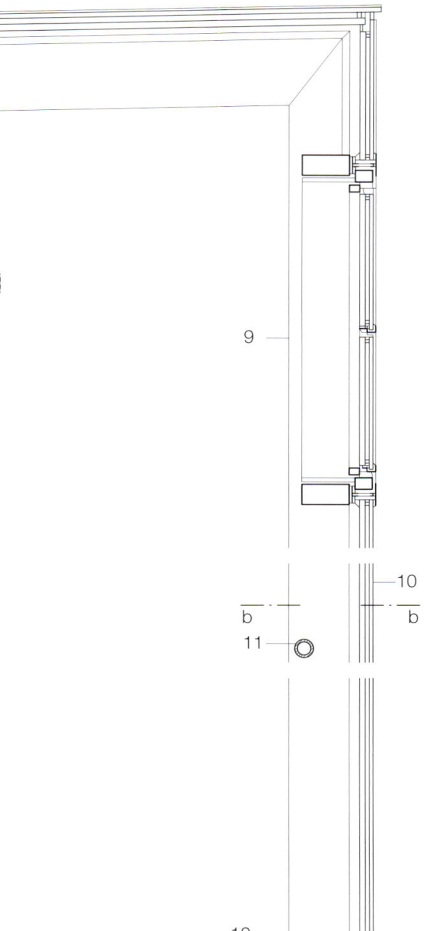

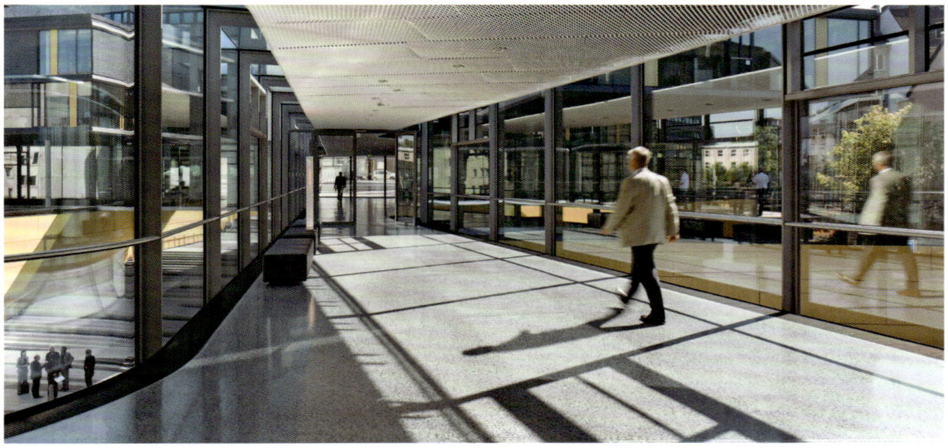

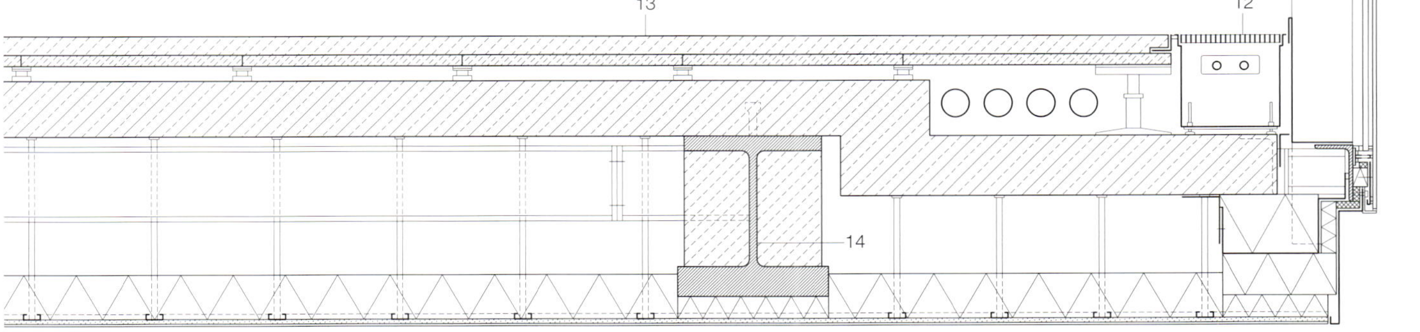

Vertical section
Horizontal section
Boulevard
Scale 1:20

A Typological development of boulevard: instead of corridors laid out on each storey with linking routes, the boulevard unites these on a single level. All jointly used facilities like the cafeteria and meeting areas are laid out along this main access route.

8 Liquid plastic seal on heet steel
2 mm felt inlay; 60 mm insulation
5 mm sheet steel
9 200/70/20 mm steel T-section
10 Sunscreen glazing: 10 mm toughened glass + 12 mm cavity + 15 mm laminated safety glass, U = 1.1 W/m²K
11 ∅ 57 mm satin-anodized stainless-steel tube
12 Floor convector
13 50 mm terrazzo paving
20 mm cement-bonded bearing layer
36 mm cavity; 160 mm reinforced concrete floor
120 mm mineral-wool insulation
12.5 mm impregnated plasterboard soffit, stopped and painted
14 Steel I-section 400/430/80/40/20 mm

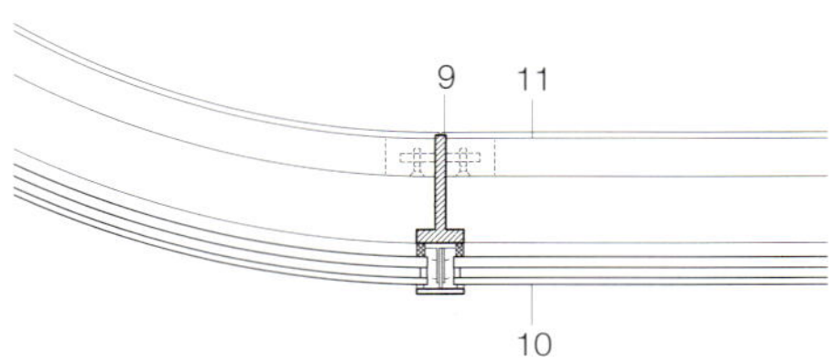

bb

Restaurant and bar in Zurich

Architects: Burkhalter Sumi Architekten, Zurich

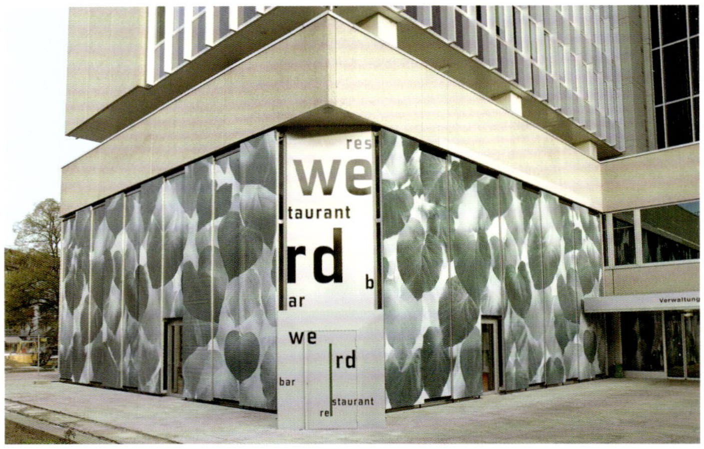

The curving staircase leading to the gallery level represents a key element in the vividly-coloured dynamic interior design.

During the course of the refurbishment of this high-rise ensemble dating from the early 1970s, the formerly open ground floor was enclosed. It now houses a restaurant with bar. A low, diagonally arranged windbreak was added to form an entrance area. A striking feature of the ground floor facade are the sunblinds with oversize printed leaf motifs which screen out excessively bright sunlight.
On entering the restaurant, the visitor is immediately struck by the powerful spatial dynamics and unconventional colour scheme, which extends throughout the various spaces and differentiates the different levels. As well as the floor, the circular staircase winding round a central column and the gallery level are integrated in the colour concept and are painted in intense shades of red and green. With their black colouring, the rectangular columns, dating from the original building, tend to recede into the background. A bar in dark stained oak curves around the staircase.
The spiral staircase curves upwards as a central element which dominates the room and provides access to the second level. With its enclosed balustrade it divides the space into a high-ceilinged entrance, lounge and bar area at the front and a low-ceilinged dining area to the rear below, and on, the gallery, which provides additional restaurant seating. Up on the gallery, the curving balustrade of the staircase develops into ergonomically shaped banquette seating which runs around the whole level.

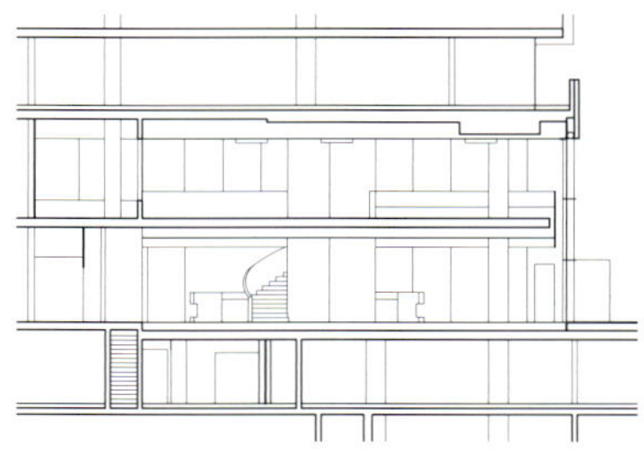

aa

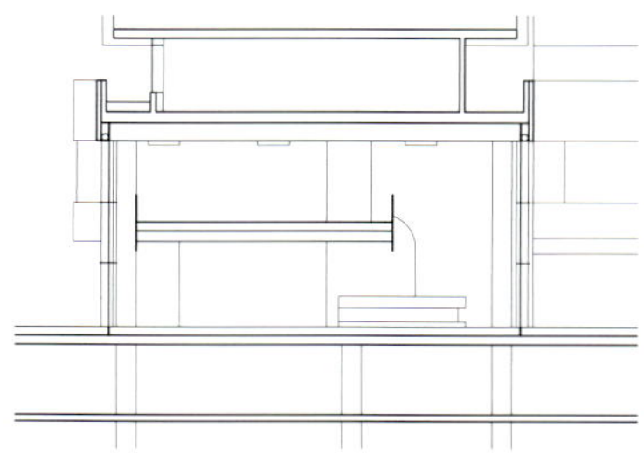

bb

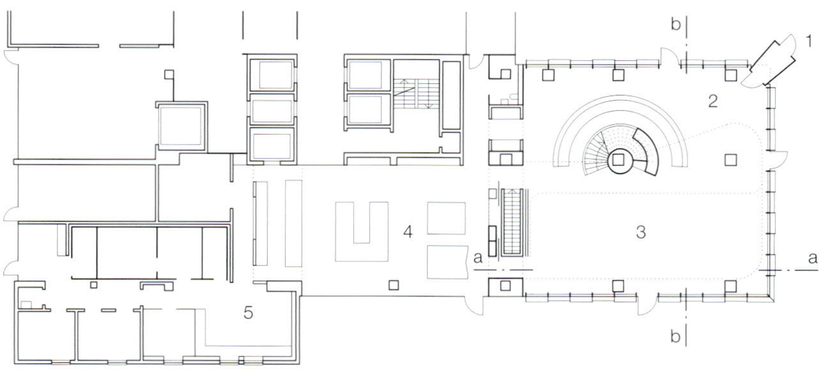

Ground floor

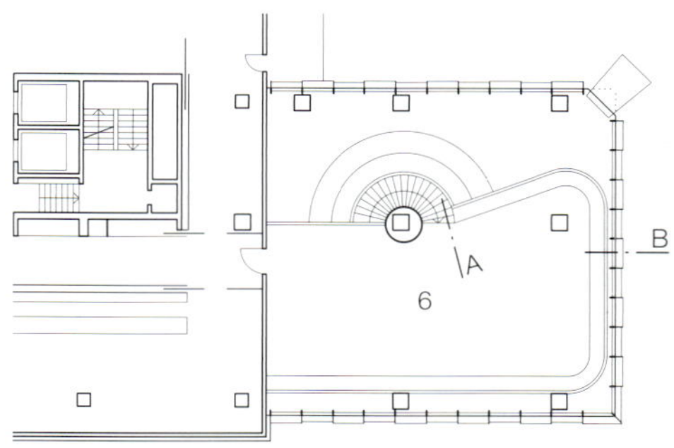

Gallery level

Site plan
Scale 1:2000
Sections • Floor plans
Scale 1:500

1 Entrance
2 Bar
3 Restaurant
4 Self-service area
5 Kitchen
6 Gallery

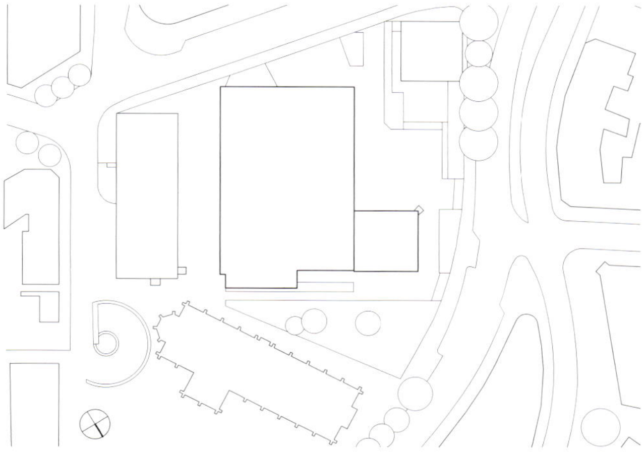

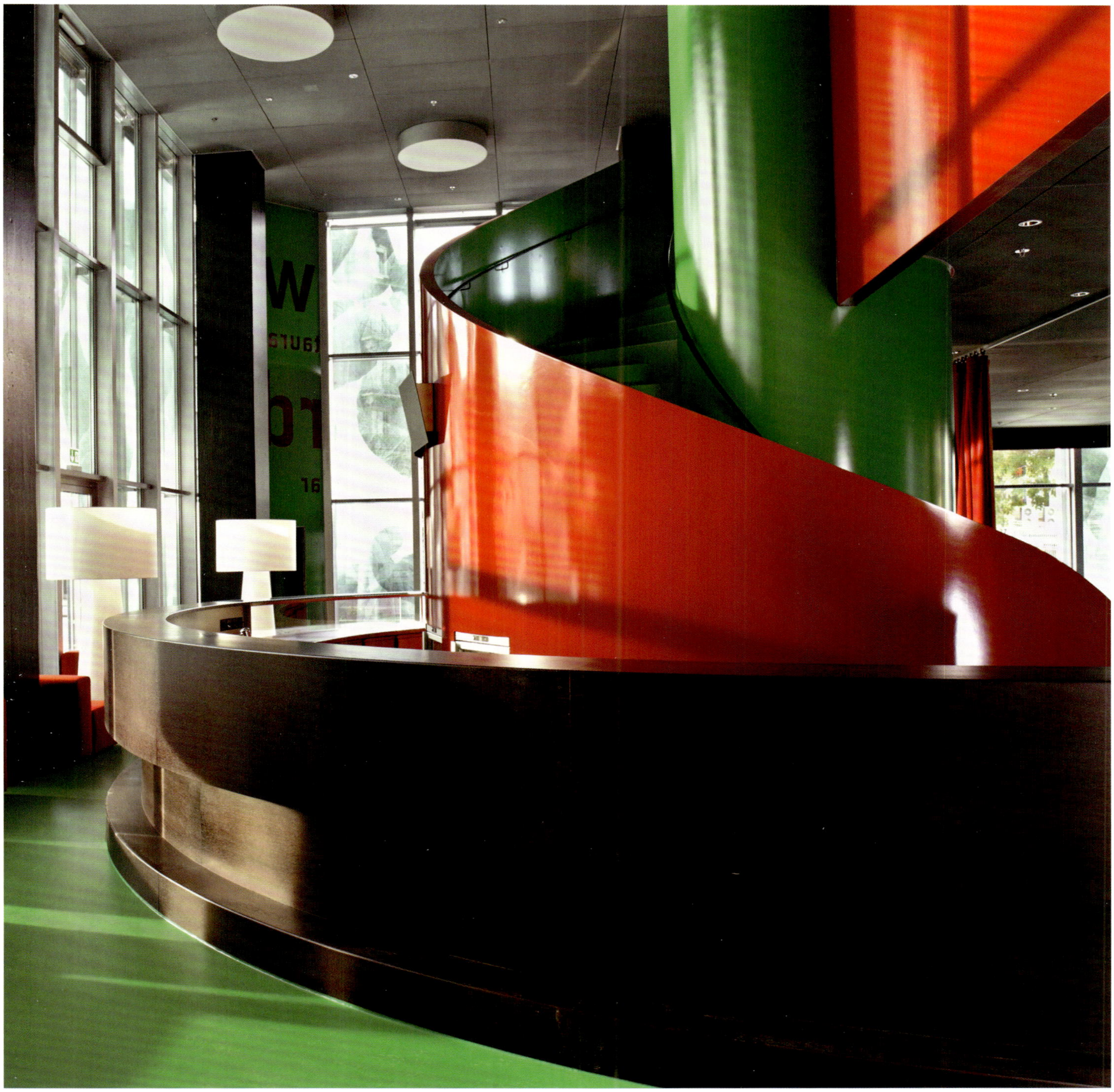

Project data:

Use:	Gastronomy
Access:	Windbreak, spiral staircase
Number of floors:	2
Clear room height:	3 m (restaurant, gallery), 7.50 m (bar)
Gross volume:	3,800 m³
Gross floor area:	880 m²
Year of completion:	2006
Construction period:	7 months

Sectional details
Scale 1:20

1 Existing facade
2 178 mm insulated panel
3 140 mm foam-glass thermal insulation
4 Powder-coated sheet-zinc soffit with Ø 40 mm openings 20 mm acoustic inlay
5 Printed fabric sunblind
6 Triple glazing (4 + 16 + 4 + 16 + 4 mm) in post-and-rail construction
7 19 mm chipboard bench, painted
8 40/5 mm steel-flat handrail
9 3 mm poured polyurethane-resin flooring
4 mm impact-sound insulation
43 mm anhydrite screed
320 mm reinforced concrete
10 Air-supply outlet
12 25/12 mm floor duct
13 Air-supply duct
14 Inbuilt unit: 30 mm MDF sheeting

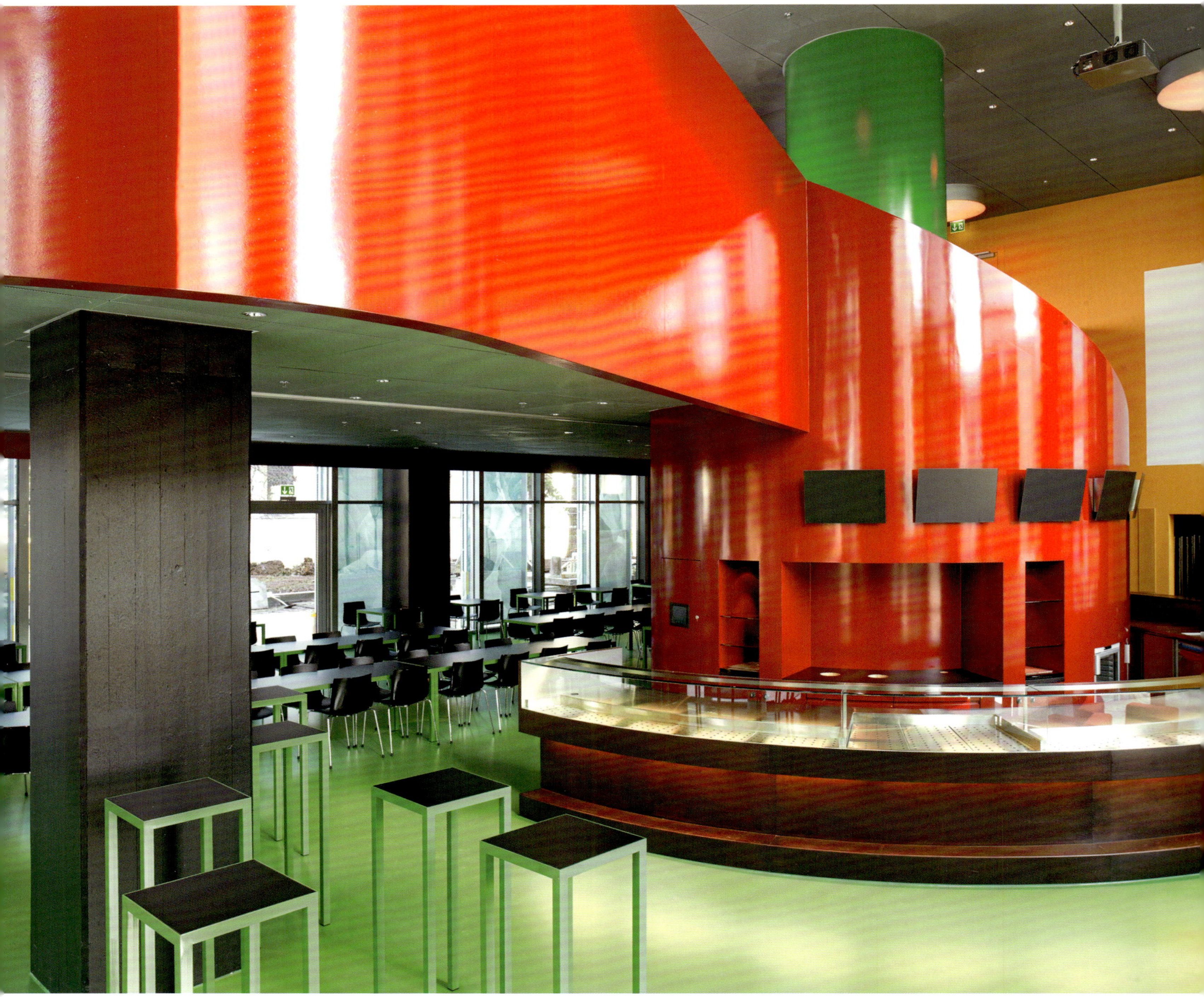

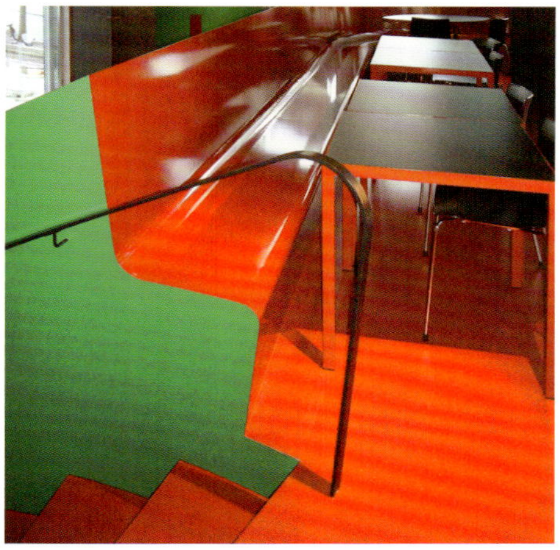

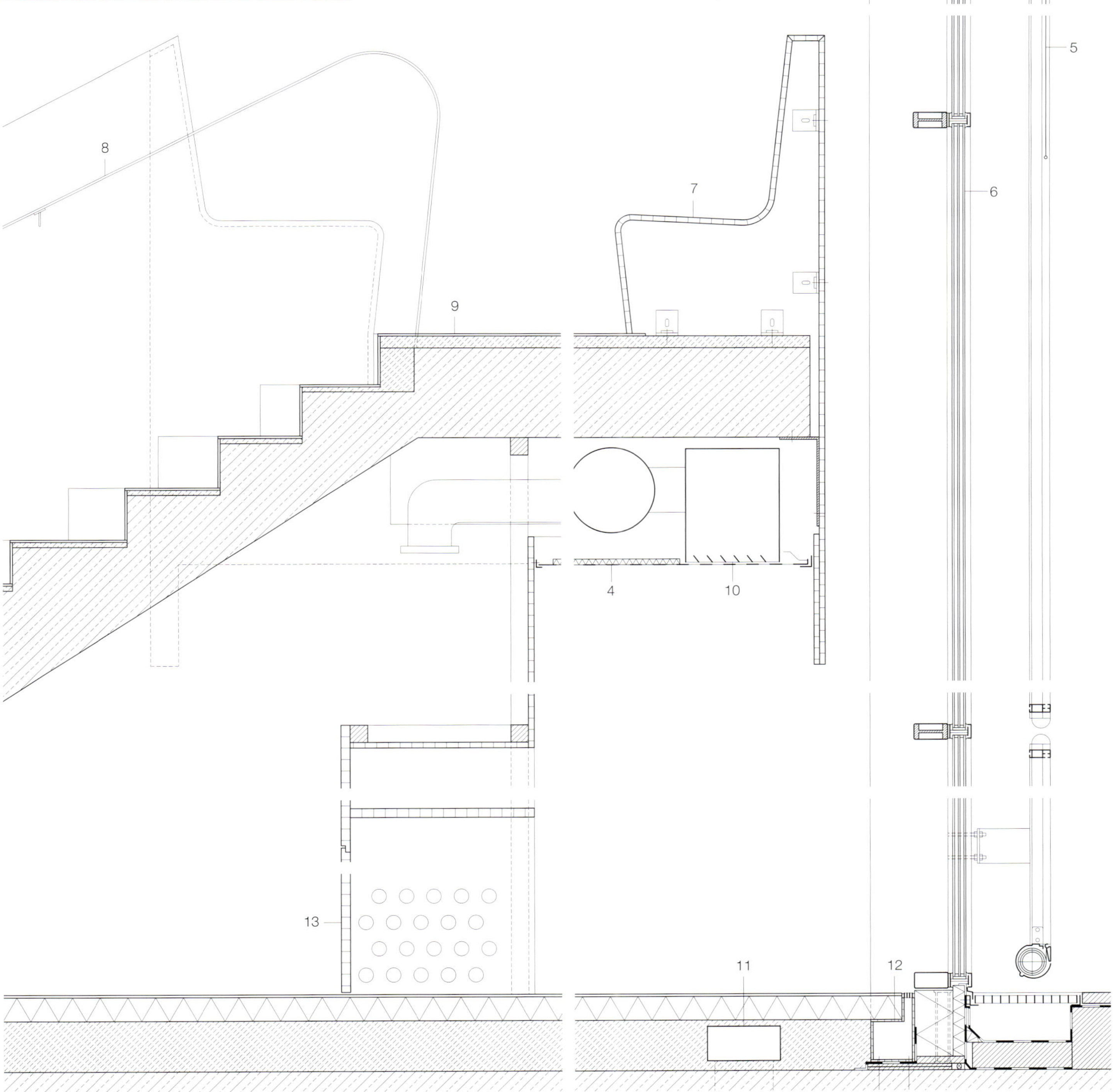
1
2
3
4
5
6
7
8
9
4
10
11
12
13
A
B

Pharmacy in Athens

Architects: KLab architecture, Athens/London

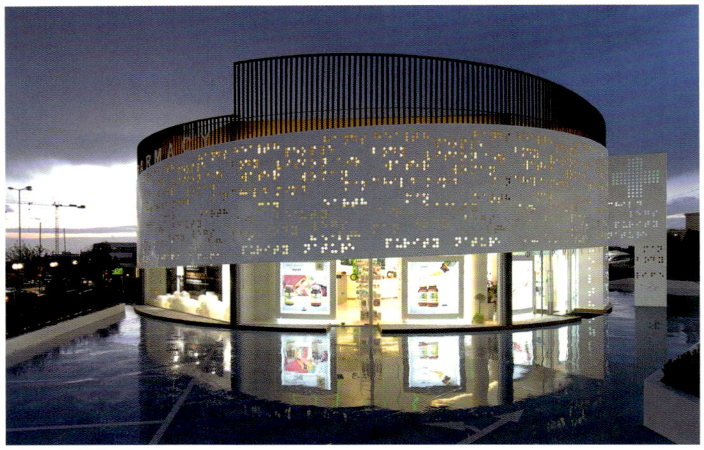

Light effects on the facade and a free-standing ramp in the interior transform a visit to the pharmacy into a shopping experience.

The Placebo Pharmacy is situated on one of the busiest streets in the district of Glyfada, an upmarket suburb of Athens. A white metal band, with perforations resembling super-sized Braille – a reference to its use on pharmaceutical packaging – is wrapped spirally around the free-standing building, which converted from an octagonal car showroom. Above this, the building is topped by bamboo cladding.
The interior is predominantly white, with the plants arranged between the glass facade and the perforated band providing a green accent. In addition, the different structures create interesting lighting effects. A circular counter stands in the centre of the pharmacy, with the shelves displaying over-the-counter products and the dispensary for prescription medicines arranged concentrically around this. These circular and elliptical product displays, between which the customers can wander freely, divide the open, very spacious room.
A curving free-standing ramp that follows the curvature of the facade leads to the mezzanine floor which stretches like a bridge over the central part of the building. Here, glass partition walls provide transparency, with their green colouration enlivening the colour scheme of the gallery. With their wave-formed design, the balustrade and walls of the upper spaces echo the dynamic character of the facade and the ramp.

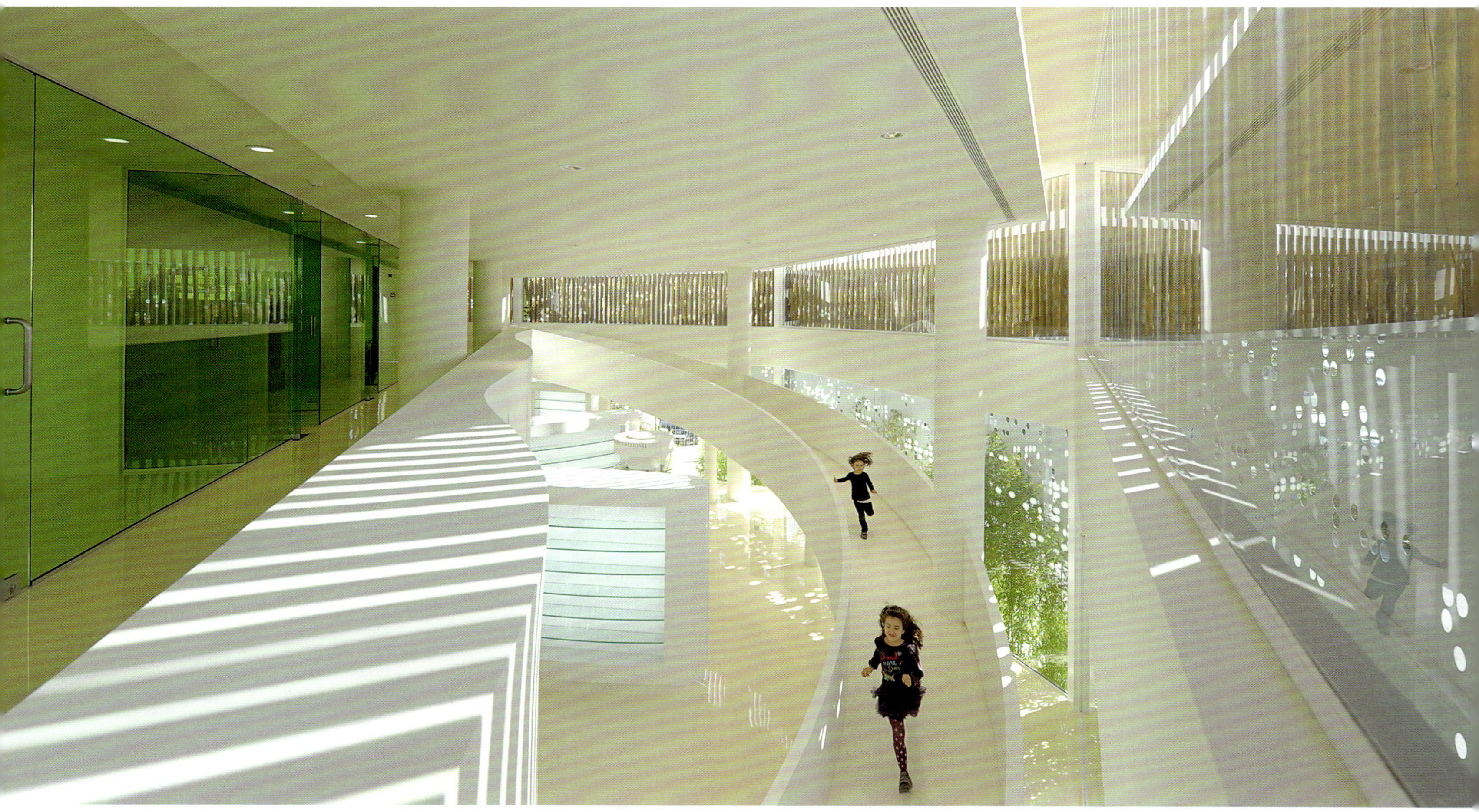

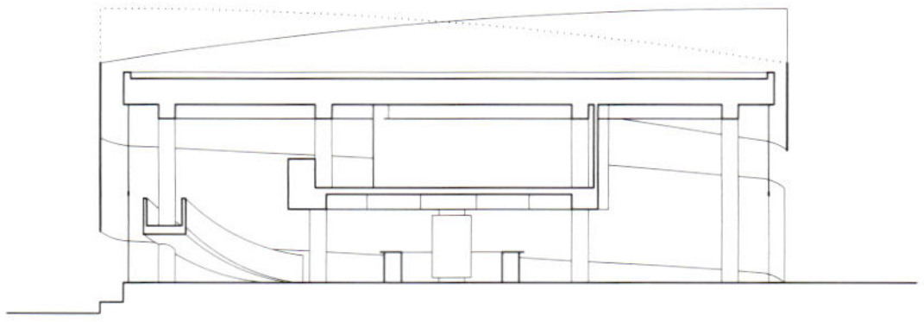

aa

Section • Floor plans
Scale 1:400

1 Central sales counter
2 Dispensary
3 Office
4 Stockroom
5 Shop-in-Shop area
6 Events room

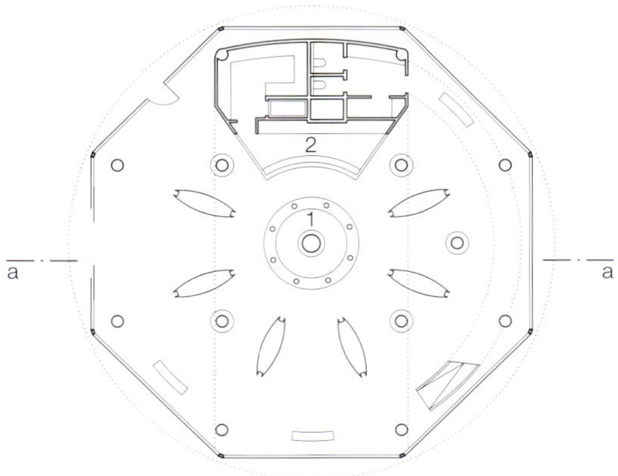

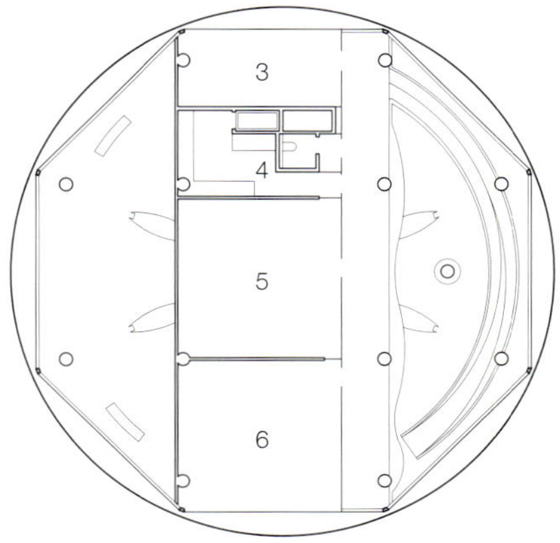

Project data:

Use:	Retail
Access :	Ramp
Number of floors:	2
Clear room height:	2.30–5.60 m
Gross volume:	1,685 m³
Gross floor area:	600 m²
Year of completion:	2010
Construction period:	4 months

Section through ramp
Scale 1:20

8 3 mm epoxy resin
30 mm concrete screed
170 mm reinforced concrete
9 12.5 mm plasterboard
65/65 mm rectangular steel profile
2× 12.5 mm plasterboard
10 2 mm edged aluminium sheeting

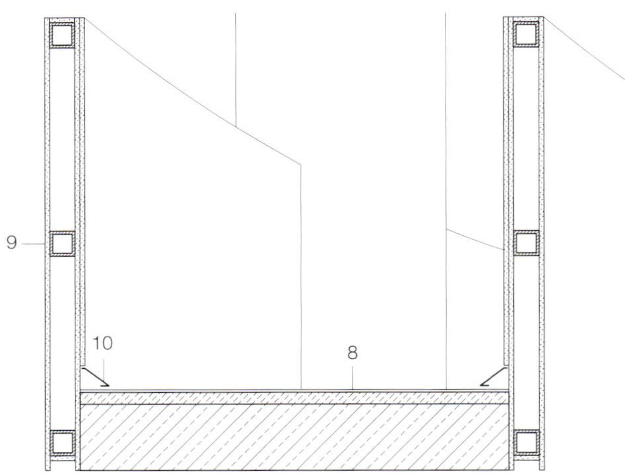

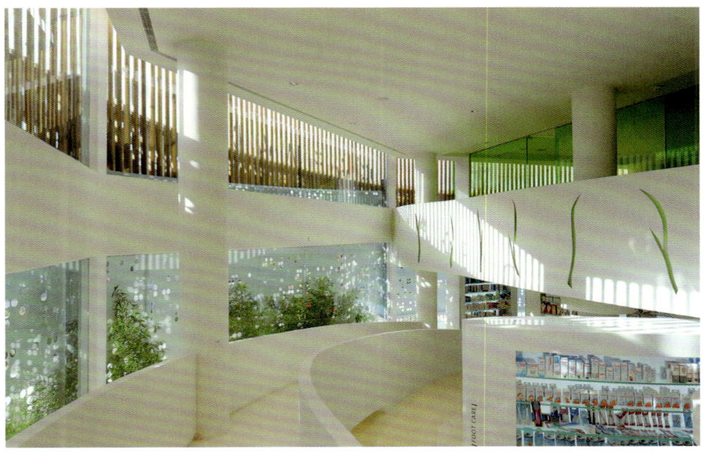

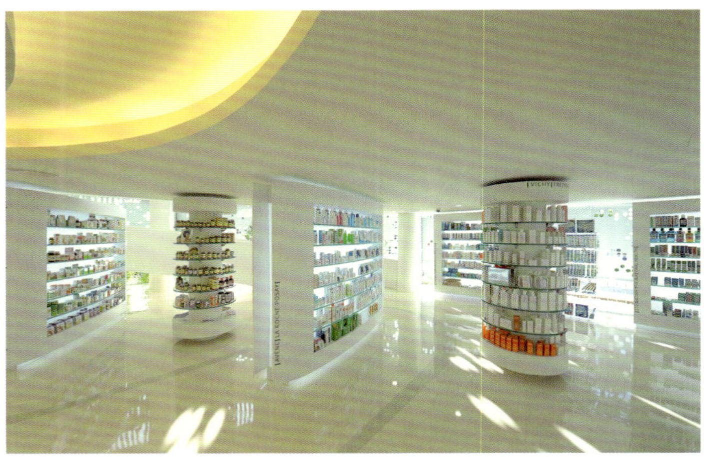

Town hall in Bad Aibling

Architects: Behnisch Architekten, Munich

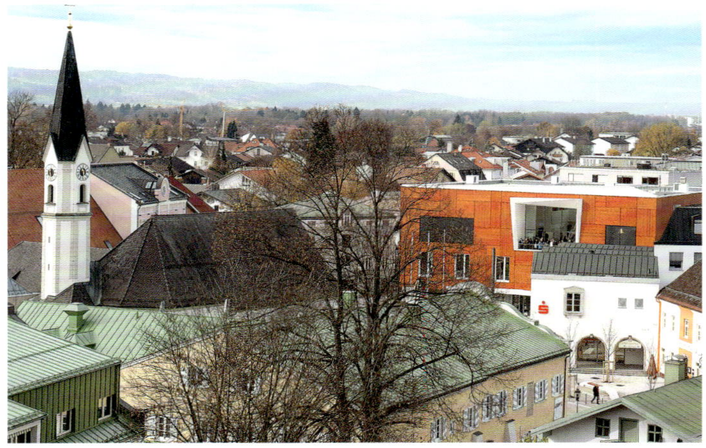

The light coloured granite in the square in front of the town hall is continued into the building as a public thoroughfare lined by shops, council offices and a café.

During the course of extensive redevelopment work, the town of Bad Aibling in Upper Bavaria was given a new town hall. The old building dating from the 1970s was demolished down to ground level and replaced with a new building which is intended to contribute to life in the town by encouraging communication. Because the original basements and foundations had been retained, the architects opted for a lightweight concrete frame construction with wooden walls. The brief also included the redesign of the adjacent Marienplatz together with the adjoining streets. The town hall itself is designed as an open building: three publicly accessible terraces on three different sides of the facade provide views of the town and surrounding area. In addition, a passage flanked by shops, council offices and a café leads from the square through the building to the Mühlbach river which flows behind it. The passage and entrance area lead into a multi-storey atrium which, as a relaxed meeting place, provides access to the town library, register office, offices and meeting rooms. The north wall of the atrium reflects the light entering through skylight strips by the almost sculptural form of its cladding of plastered wood composite panels. The wall elements continue into the atrium space as balustrades, benches and counters. Rolled steel profiles form the basic structure of the central stairs, with balustrades of veneered plywood and steps faced with artificial stone.

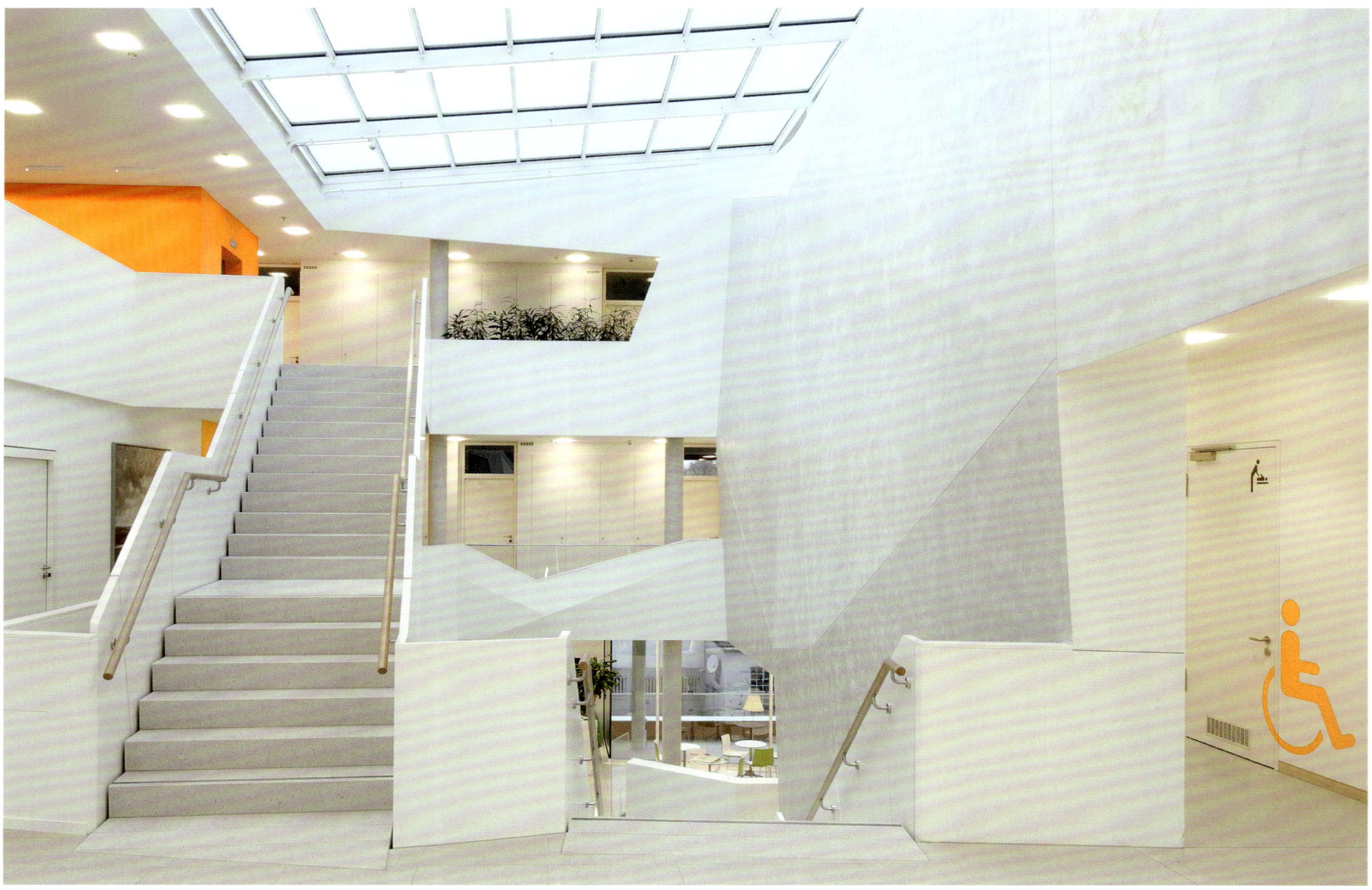

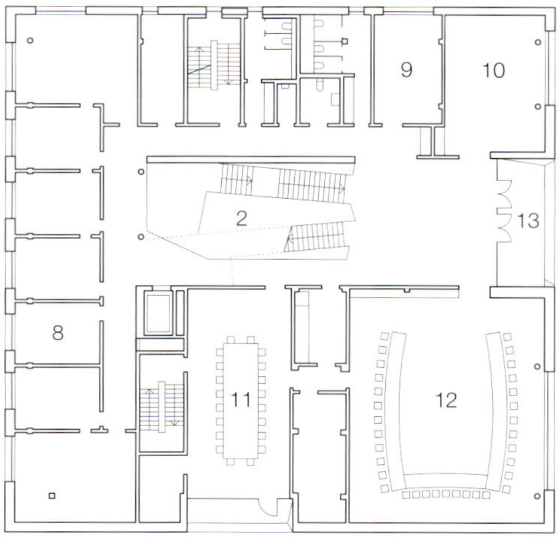

2nd floor

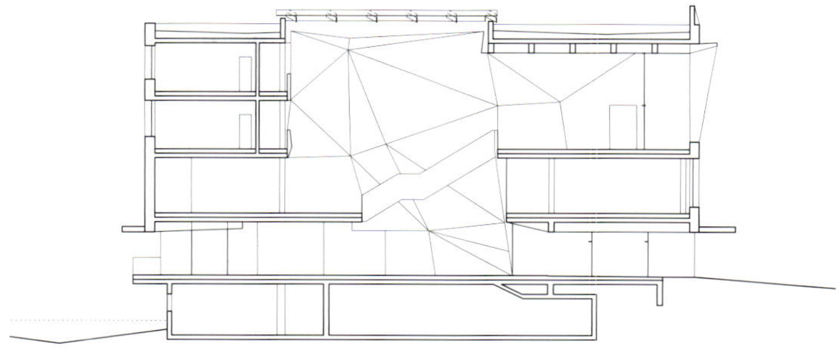

aa

Floor plans • Section
Scale 1:500

1 Windbreak
2 Entrance area/ atrium
3 Shop
4 Council office
5 Library
6 Reading café
7 Reading terrace
8 Office
9 Mayor's office
10 Register office
11 Small meeting room
12 Large meeting room
13 Terrace

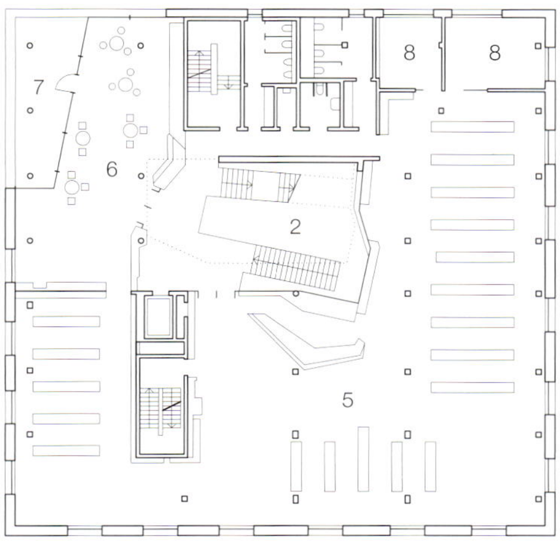

1st floor

Project data:

Use:	Administration, shops, café, library
Access :	Stairs, atrium, lift
Number of floors:	4 + 1 basement level
Clear room height:	3.30 m (library), 2.80 m (offices)
Gross volume:	15,000 m³
Gross floor area:	4,300 m²
Year of completion:	2012
Construction period:	18 months

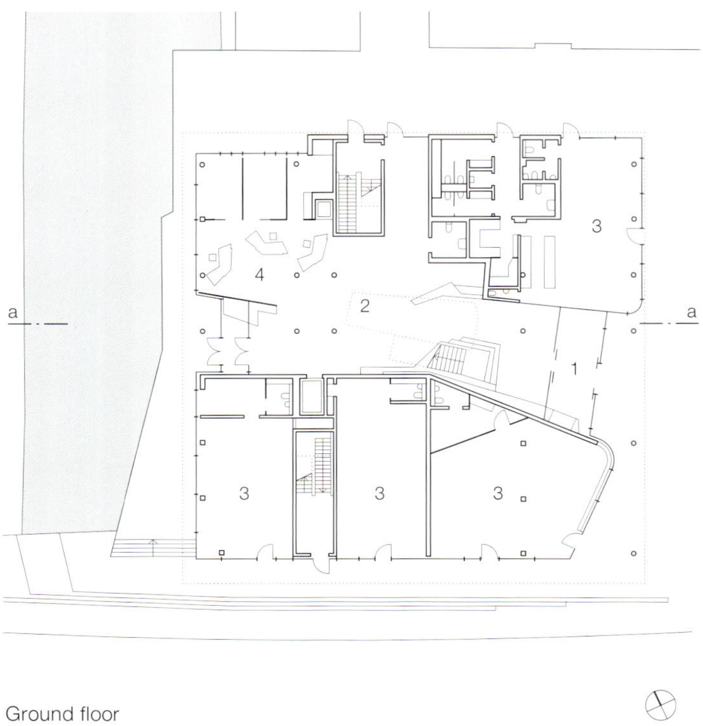

Ground floor

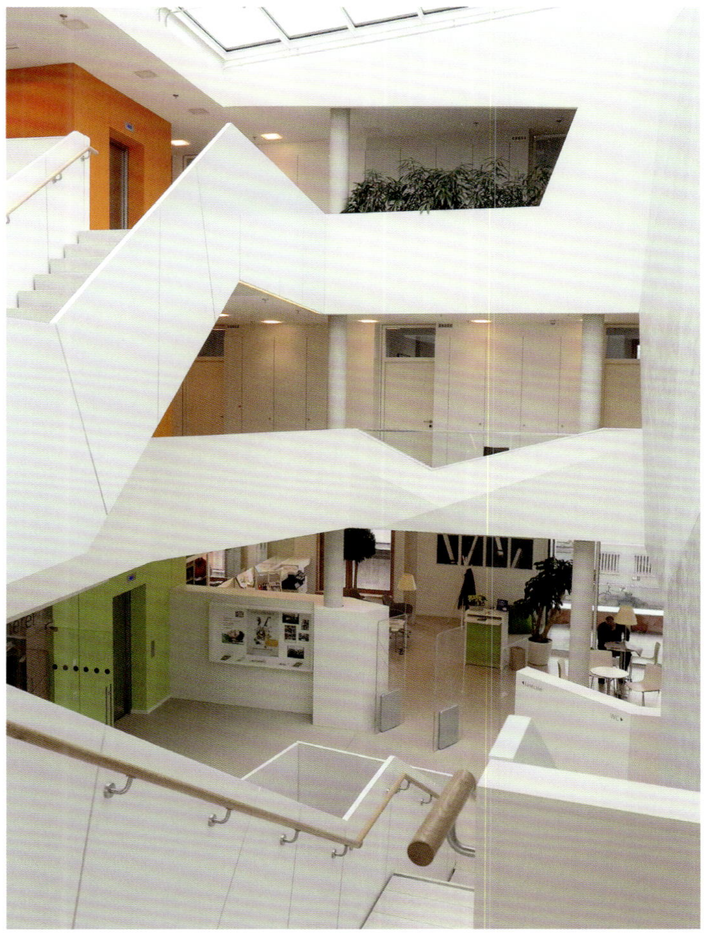

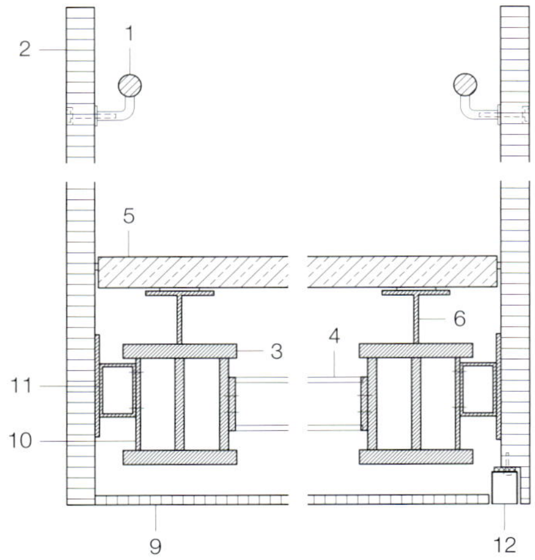

Sectional details
Stairs in foyer/atrium
Scale 1:20

1 Handrail Ø 50 mm solid oak, clear varnished fixed to Ø 20 mm round steel bar
2 63 mm wood composite panel
3 Main beam 220 mm HEM steel profile, with T-profiles welded on
4 Secondary beams 120 mm HEB steel profile with welded 10 mm head plates
5 70 mm reinforced artificial stone step
Epoxy resin bonding on 5 mm neoprene underlay
6 Step support 150/120/10 mm welded steel T-profile, 2× per step
7 70 mm artificial stone
10 mm steel plate on 10 mm steel webs
Footfall insulation: 10 mm elastomer compression layer, trapezoidally profiled on both sides
8 Floor structure:
30 mm reinforced artificial stone
15 mm mortar bed
70 mm heated screed
Separating polyethylene foil layer
70 mm insulation
250 mm reinforced concrete floor, surface plastered and painted
9 19 mm wood composite panel, surface plastered and painted
10 10 mm flat steel profile
11 120/80/5 mm hollow rectangular section
12 LED lighting strip

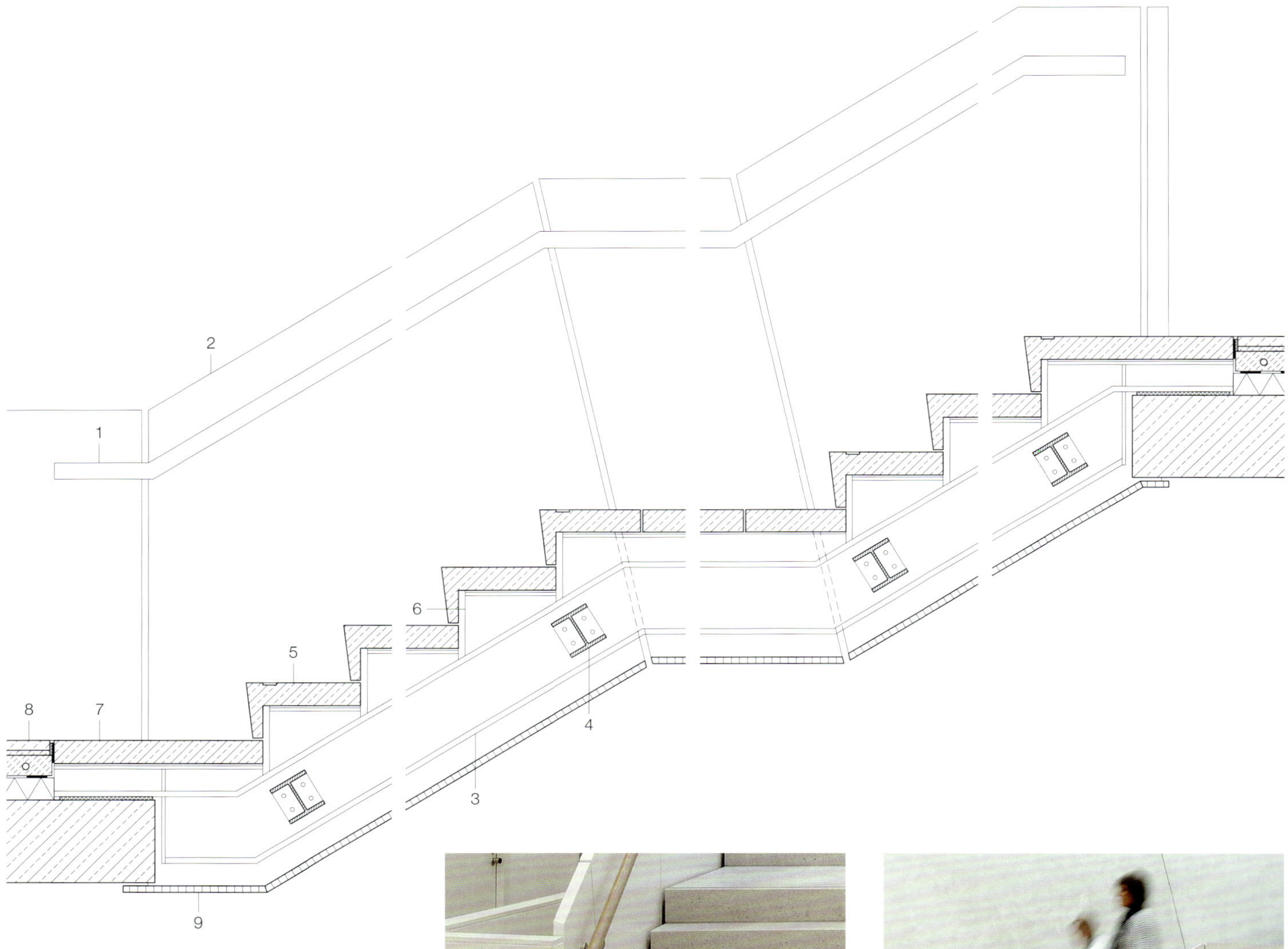

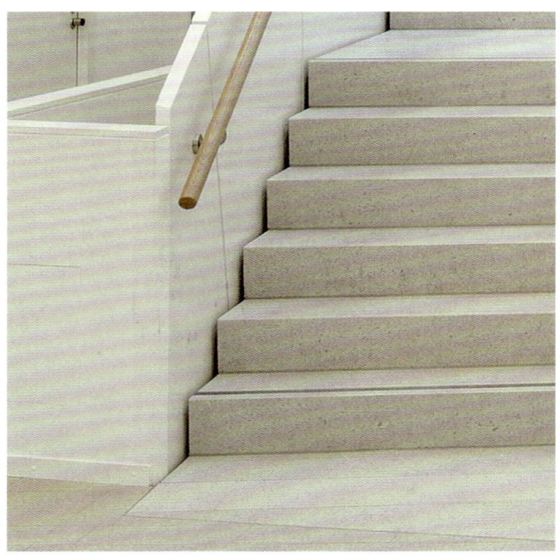

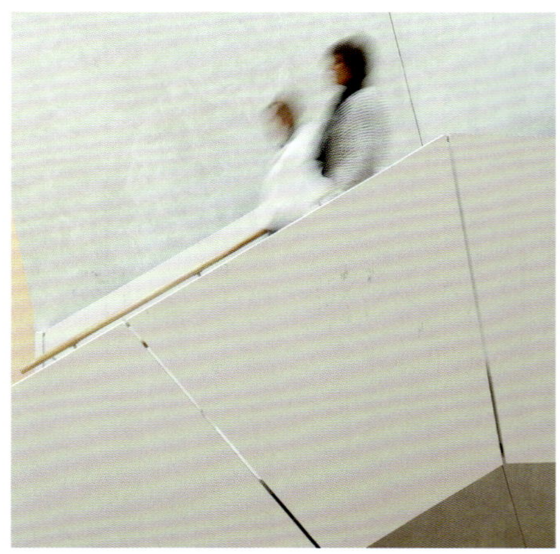

Café and exhibition spaces in Bragança

Architect: Giulia De Appolonia, Brescia

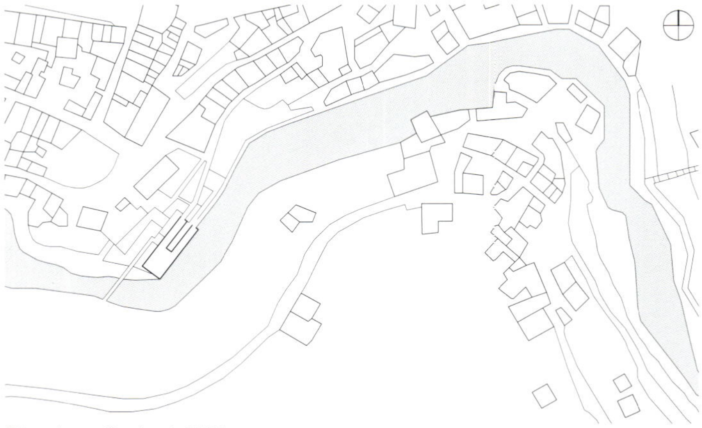

Site plan Scale 1:6000

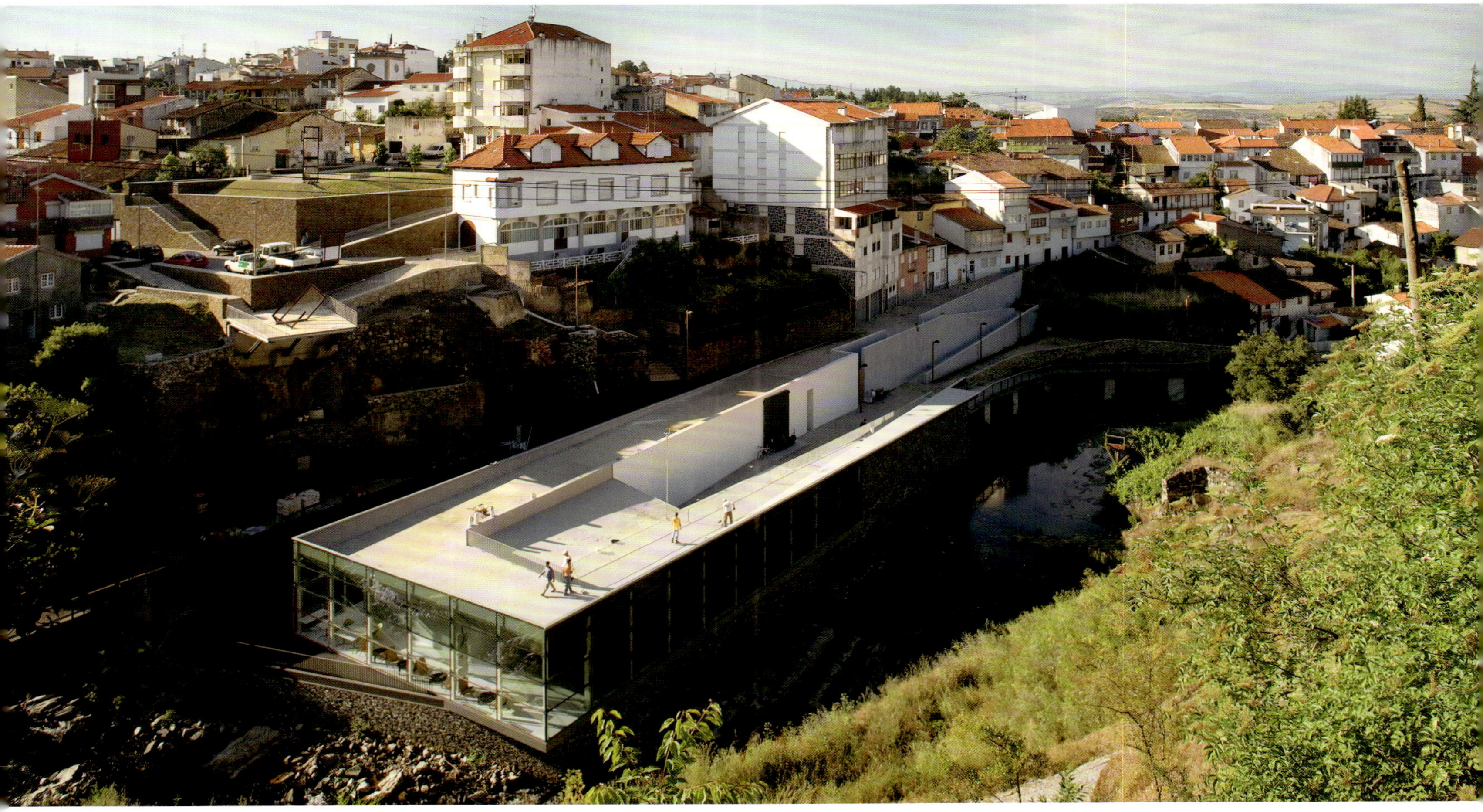

A sequence of ramps on the roof of the building connects the historical town centre with the river-side below.

The accessible roof of the exhibition pavilion serves primarily as a public footpath. Gently sloping ramps and more direct flights of steps connect the historical centre of the northern Portuguese city of Bragança with the riverside below. The three-dimensional roof landscape, including the parapets, is coated throughout with synthetic resin textured with a finely grained quartz sand and, with its prominent location, invites visitors to stop for a while and enjoy the views of the surrounding area. Coming from above, and having walked along the facade parapet which rises dynamically in opposite directions, just as the visitor has almost left the roof they can either continue in the direction of the riverside or take a U-turn and head for the entrance to the building that is located at the end of a ramp walkway framed by concrete side walls. This is because, almost incidentally, the roof also forms the upper termination of a building which, in its interior, presents objects and video installations on the theme of the environment and renewable energies. The interior of the building is dominated by two main rooms and a connecting service zone including a café that also offers seating on the front terrace. Room-height pre-oxidised steel panels arranged behind the south-facing glazing absorb the solar radiation passing through the glass. Motor-operated ventilation flaps can be used to divert the solar-heated air in the intervening cavity into the interior or expel it to the outside as required.

Project data:

Use:	Exhibition space, café
Access:	Ramps and stairs on the roof
Number of floors:	1 + accessible roof
Clear room height:	2.50–4.55 m
Gross volume:	2,026 m^3
Gross floor area:	579 m^2
Year of completion:	2007
Construction period:	21 months

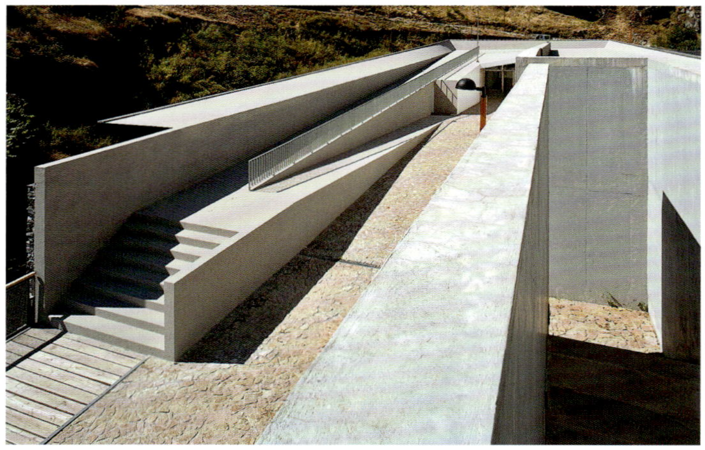

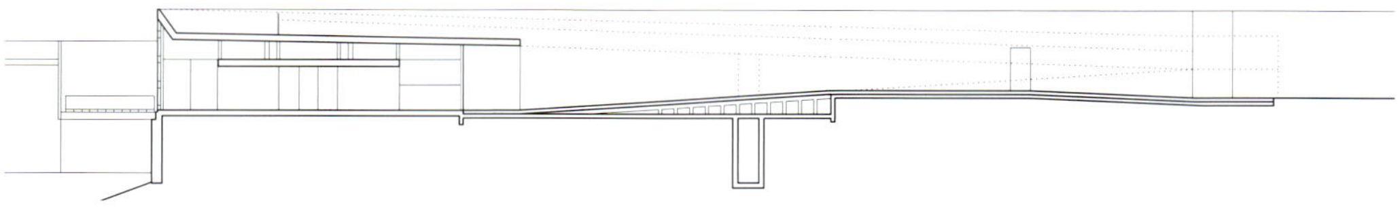

aa

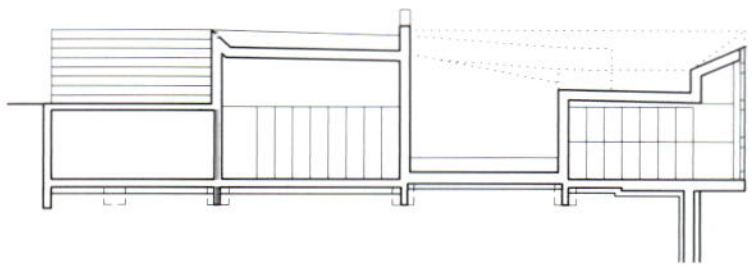

bb

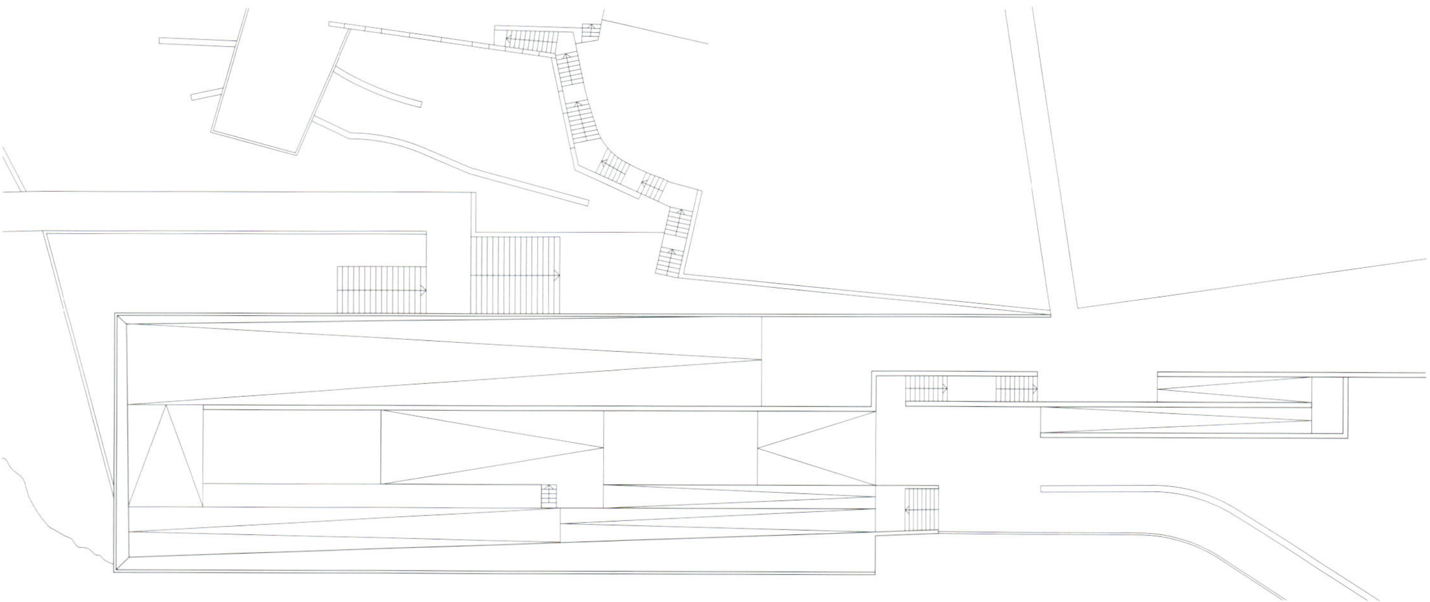

Roof plan

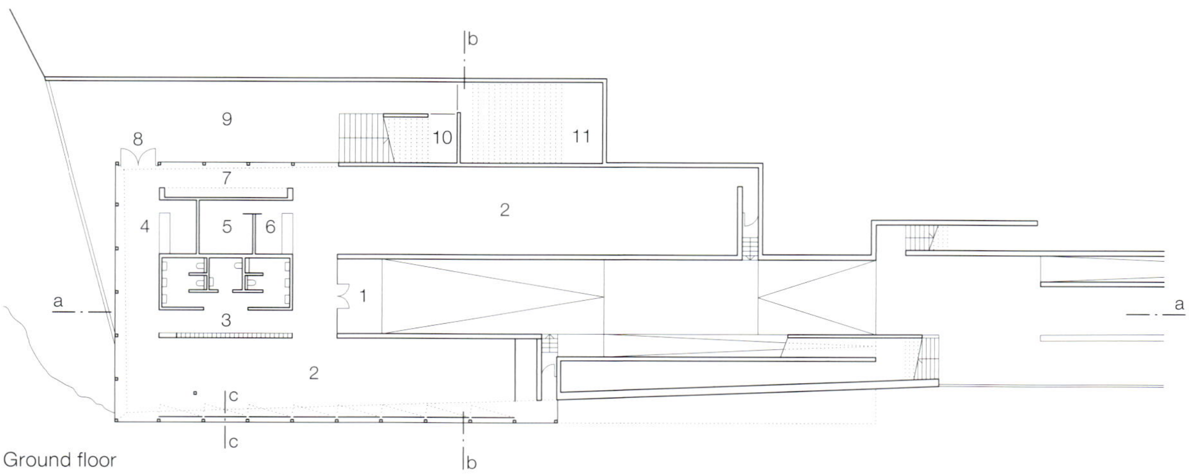

Ground floor

Sections
Roof plan • Floor plan
Scale 1:500

1 Entrance
2 Exhibition hall
3 Lockers
4 Bar/café
5 Office
6 Ticket sales
7 Computers
8 North entrance
9 Terrace
10 Services
11 Turbines

Vertical section roof/facade
Scale 1:10

12 Sun protection glazing:
6 mm float glass + 12 mm cavity +
6 mm toughened glass
13 Ventilation opening, motorised
14 140/140/7 mm structural steel
15 Stainless steel connection element
16 2 mm stainless-steel glass clamp
17 3 mm pre-oxidized steel sheet
absorber
18 12 mm wood fibreboard,
with cement binding agent,
on 12 mm MDF board
19 2 mm synthetic resin coating
textured with quartz sand
3 mm two-component cement
sealing compound
50 mm cement screed, reinforced
60 mm EPS rigid foam board
reinforced concrete to falls
20 2 mm synthetic resin coating
textured with quartz sand
3 mm two-component cement
sealing compound, reinforced
with glass-fibre mesh
7 mm calcium silicate board,
bonded
60 mm EPS rigid foam board,
bonded
250 mm reinforced concrete
21 LED lamp
22 2 mm synthetic resin coating
on 3 mm steel flat

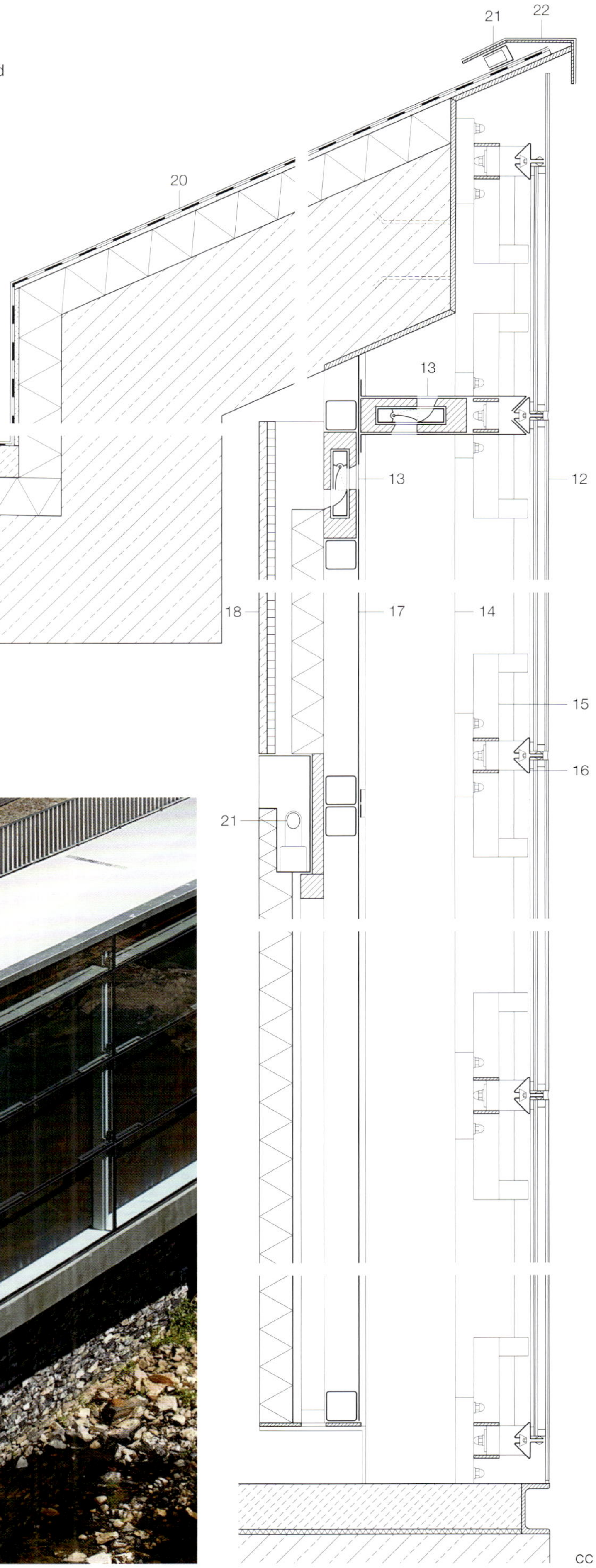

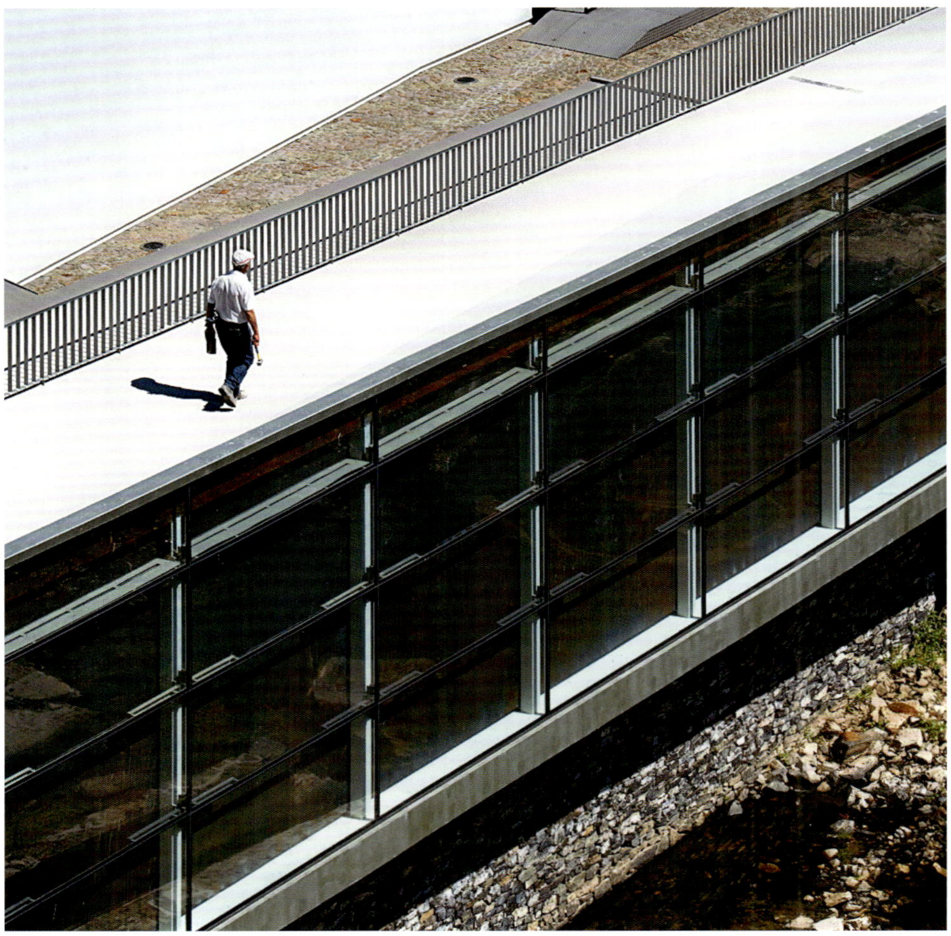

cc

Beachfront promenade in Benidorm

Architects: OAB – Office of Architecture in Barcelona

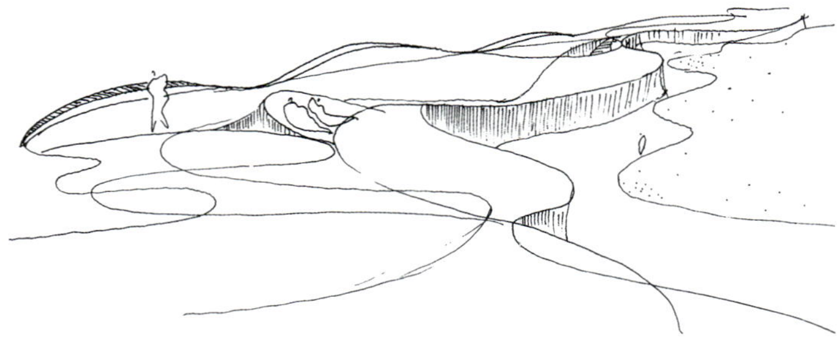

With stairs, ramps and wave-shaped meanders, the beachfront promenade is a public urban space which provides access to the beach.

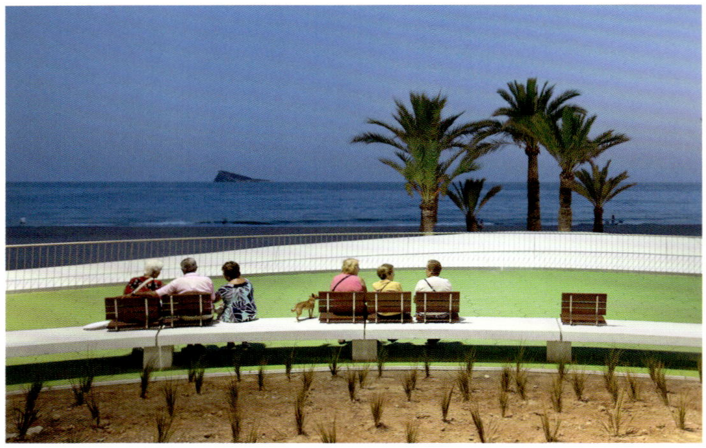

Benidorm, situated 45 km north-east of Alicante on Spain's Costa Blanca, was given a newly designed beachfront promenade in 2009. The approval of a revolutionary development plan in the 1950s and the permission to wear bikinis rapidly transformed the fishing village into a busy tourist resort with a high-rise silhouette of densely clustered hotels. At times during the peak season, the population rises to over 1.5 million people. With its organic lines, the 1.5-kilometre-long promenade snakes along the beach, advancing and retreating in wave-like curves, its vibrant colour scheme enlivening the transition between the hard urban edge and the water. The width of the promenade varies between 8 and 30 m. The "Paseo de Poniente" rests on a base zone consisting of a white concrete wall which is sometimes concave, sometimes convex in profile. The inwardly bulging curves allowed a total of 3,000 m^2 of additional beach space to be created. These niches provide shelter from the sun or rain or privacy for changing. As a hinge between city and sea, the promenade provides improved access to the beach while at the same time offering holidaymakers a certain "protection" against the masses of high-rise buildings.

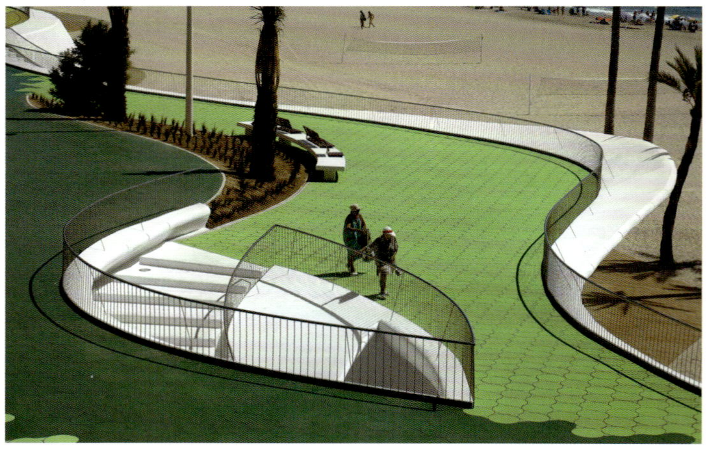

Glazed ceramic tiles

Curving stairs and ramps bridge a height difference of 4.50 m. At street level, a new public space was created offering viewing platforms with bench seating and planted islands as well as an attractive promenade for strollers, joggers or roller skaters. The former four-lane highway was scaled back during the course of the building works and now only provides access for residents. The Paseo is surfaced in glazed ceramic tiles coloured in 22 different shades – from dark blue and purple through yellow, orange and red to shades of green. These serve as an aid to orientation, helping holidaymakers to locate their hotel among the uniform hotel blocks. The round, extruded tiles with a diameter of 43 cm were laid closely together, with the intervening spaces filled with small rounded triangular tiles and LED spotlights, creating a homogenous honeycomb-pattern surface. Edge strips, stairs and ramps are made of white painted concrete. A 4-metre-wide strip of impregnated wooden decking runs in front of the base; here too, lights are embedded in the floor in order to facilitate orientation in the dark.

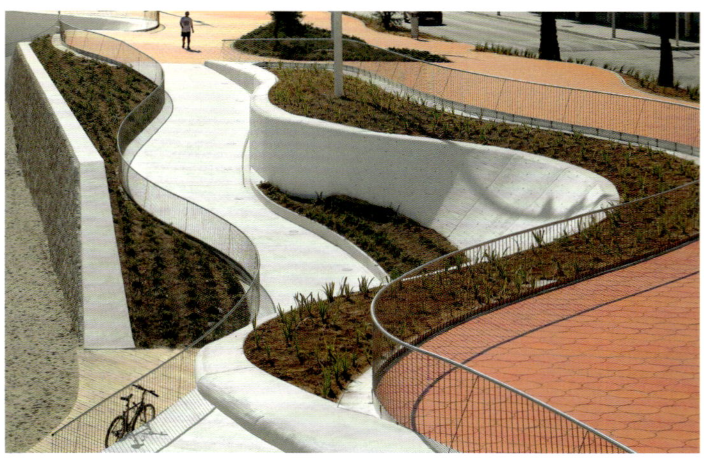

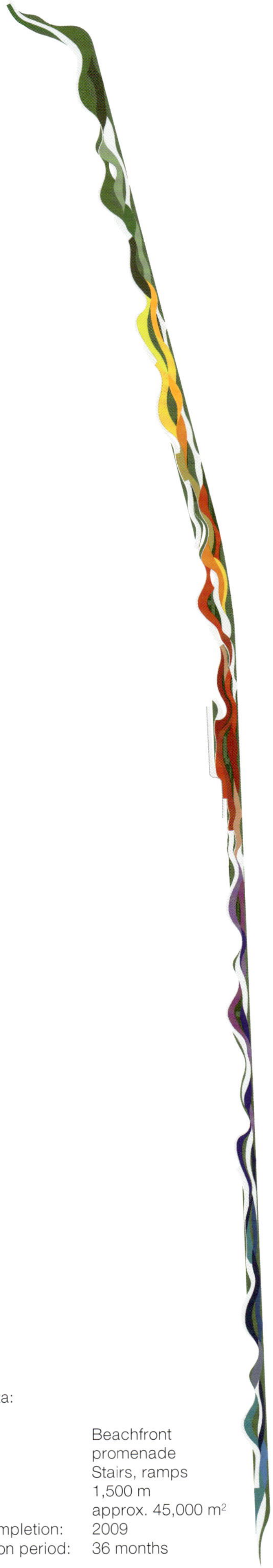

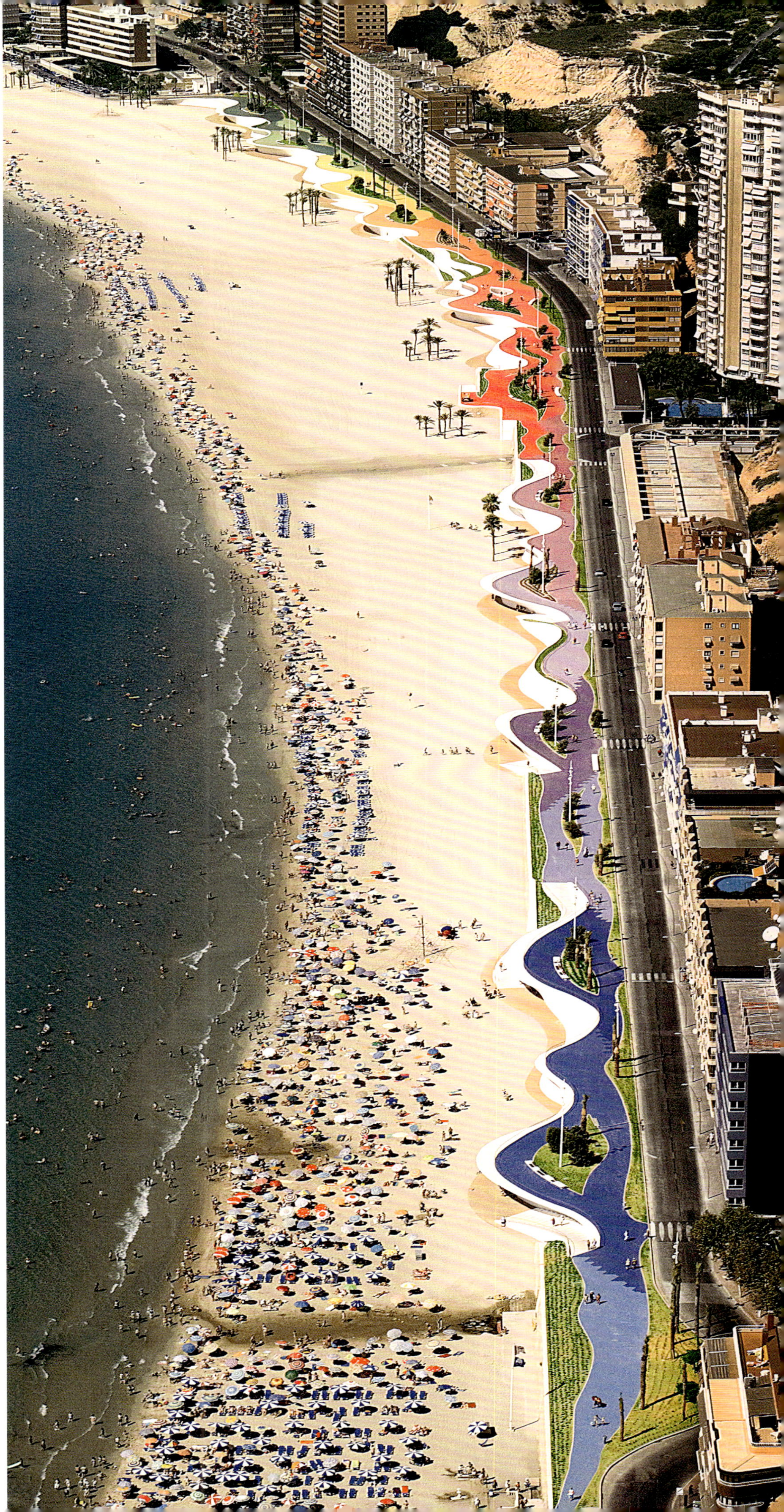

Project data:

Use:	Beachfront promenade
Access:	Stairs, ramps
Length:	1,500 m
Area:	approx. 45,000 m²
Year of completion:	2009
Construction period:	36 months

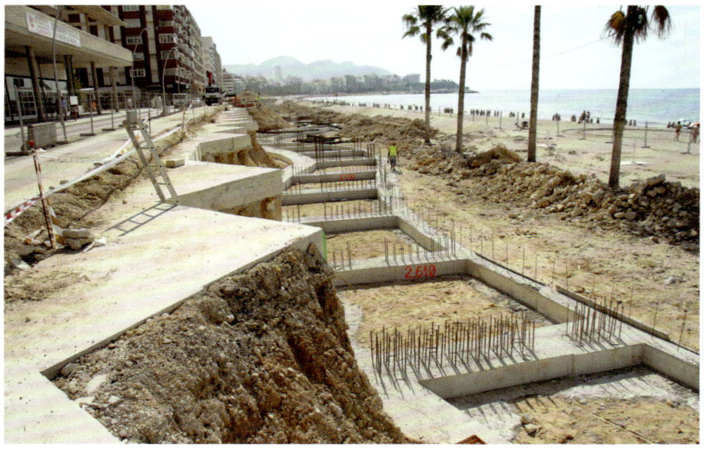

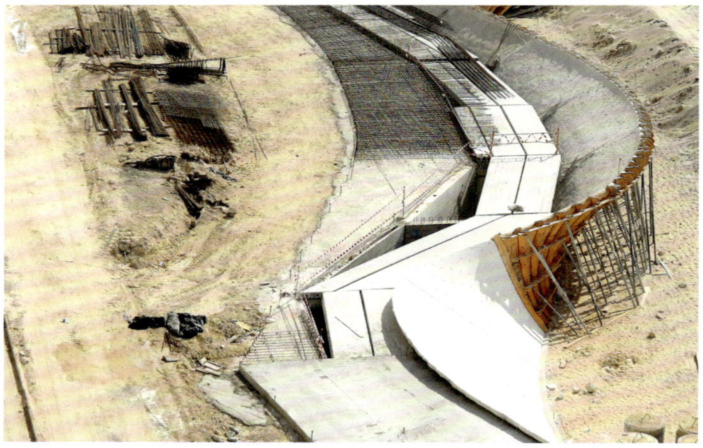

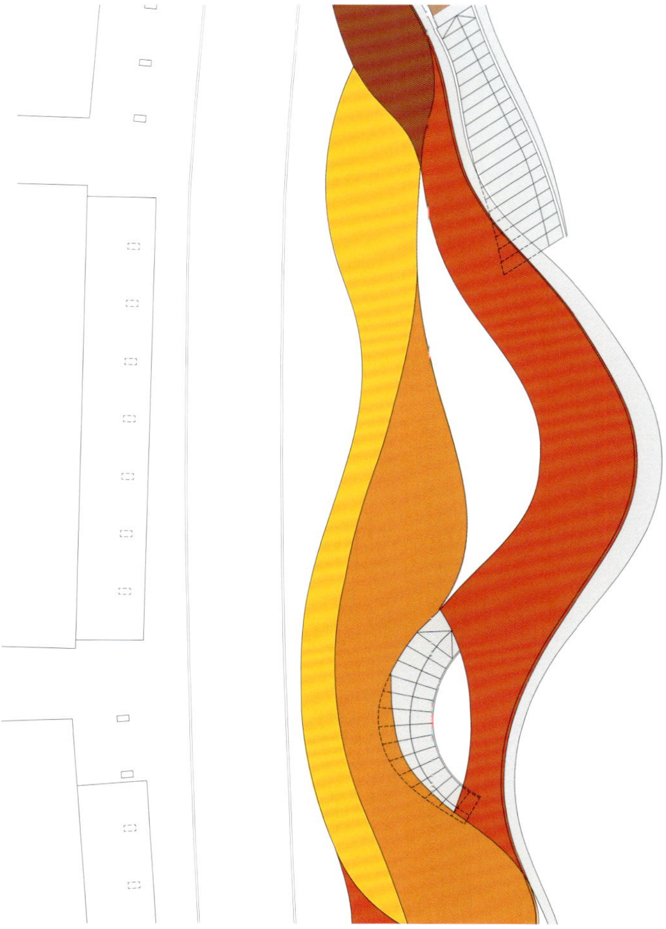

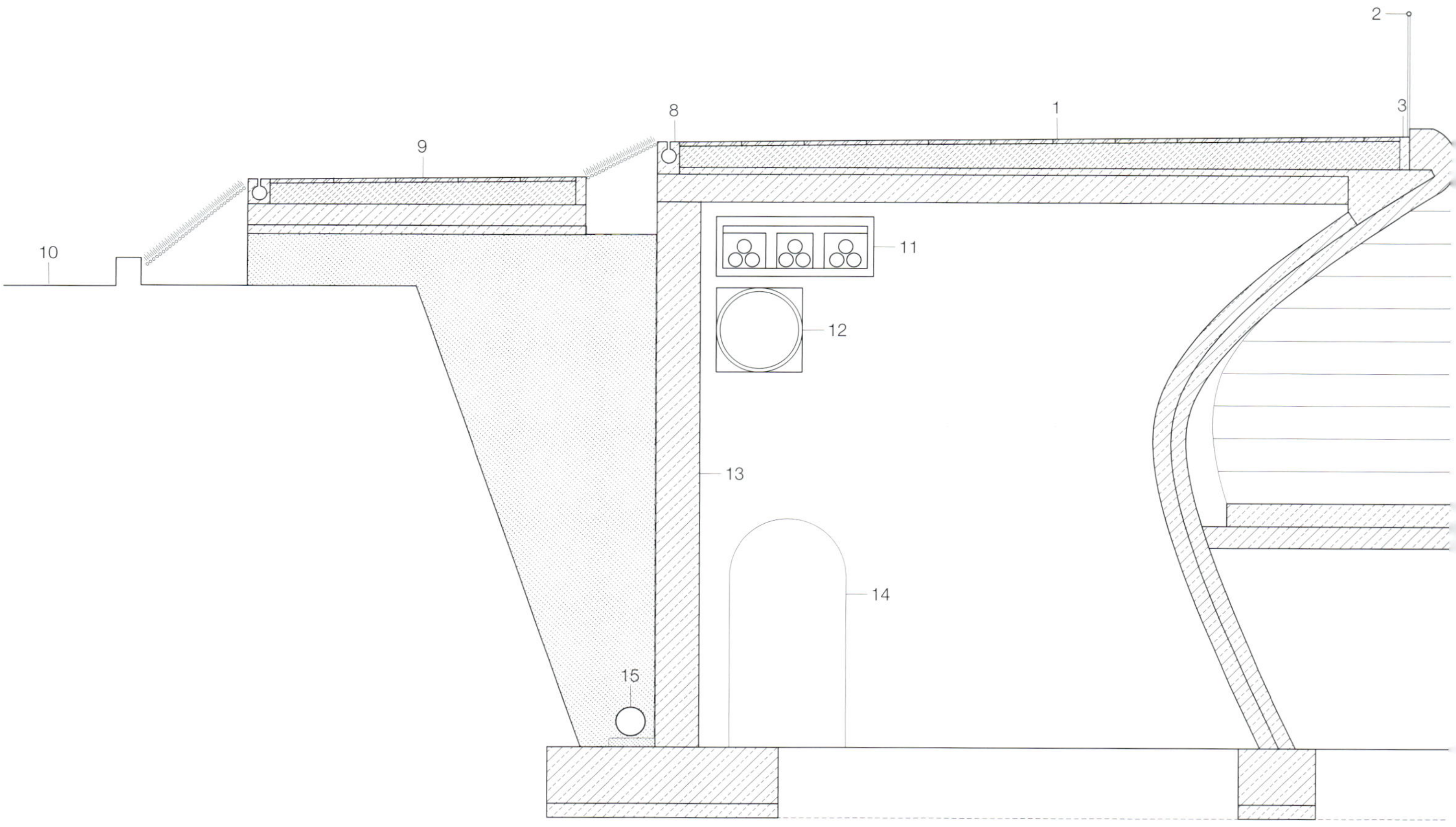
2
8
1
3
9
10
11
12
13
14
15

Floor plan Scale 1:500
Section through promenade/steps down to beach Scale 1:50

1 25 mm glazed ceramic flooring
150 mm levelling layer, 50 mm concrete
200 mm cavity concrete slab
2 Handrail: Ø 10 mm stainless steel
7 mm balustrade post
3 Base edging element:
70/230 mm precast concrete
4 Hollow slab: lightweight concrete
5 120 mm reinforced shotcrete, white painted with silicate-based solution
6 Unreinforced concrete stair
reinforced concrete ramp, both white painted with silicate-based solution
7 Flooring (beach level):
30 mm pine decking, treated with copper salt, 30 mm counterbattens resting on sand compacted with mortar
8 Prefabricated element for drainage:
Ø 100 mm drainage pipe
9 25 mm glazed ceramic flooring
150 mm levelling layer
150 mm concrete slab
75 mm concrete levelling layer
200/300 mm gravel underlayer
10 Roadway
11 Electrical installations
12 600 mm drainage pipe
13 Rear wall:
300 mm cavity concrete element
Sealing, fine gravel
14 Maintenance passage
15 Ø 200 mm porous drainage pipe
Mortar bed for pipe

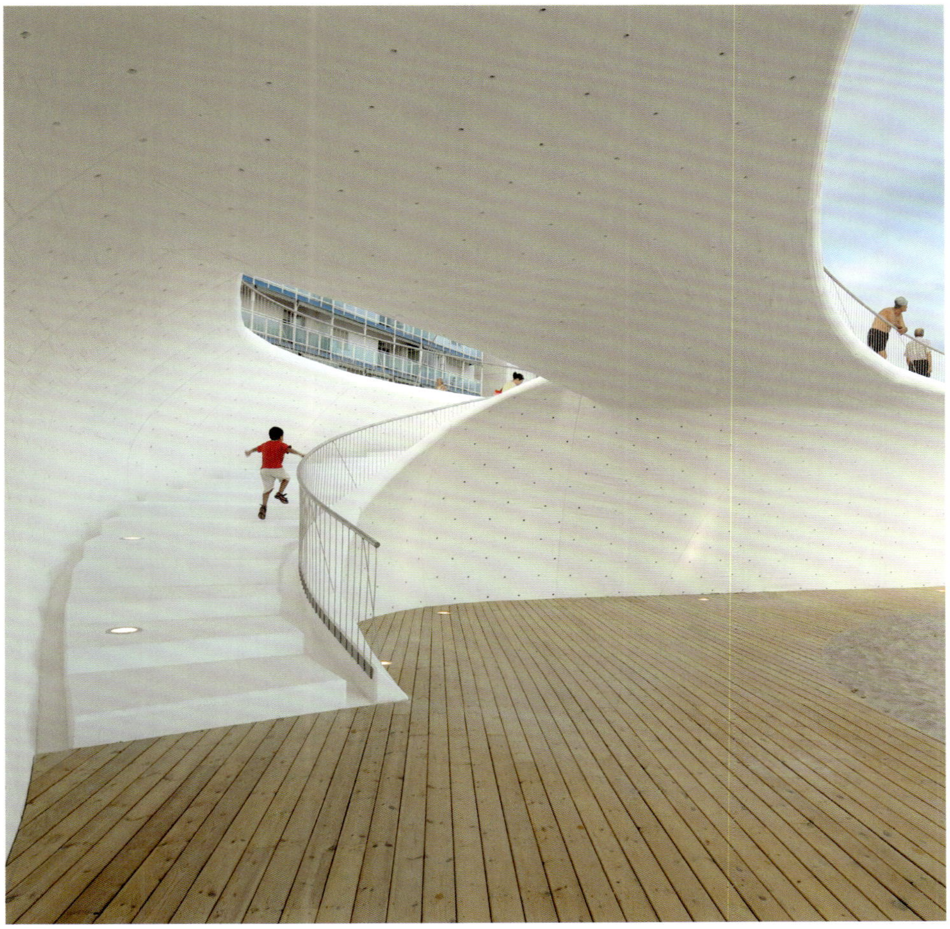

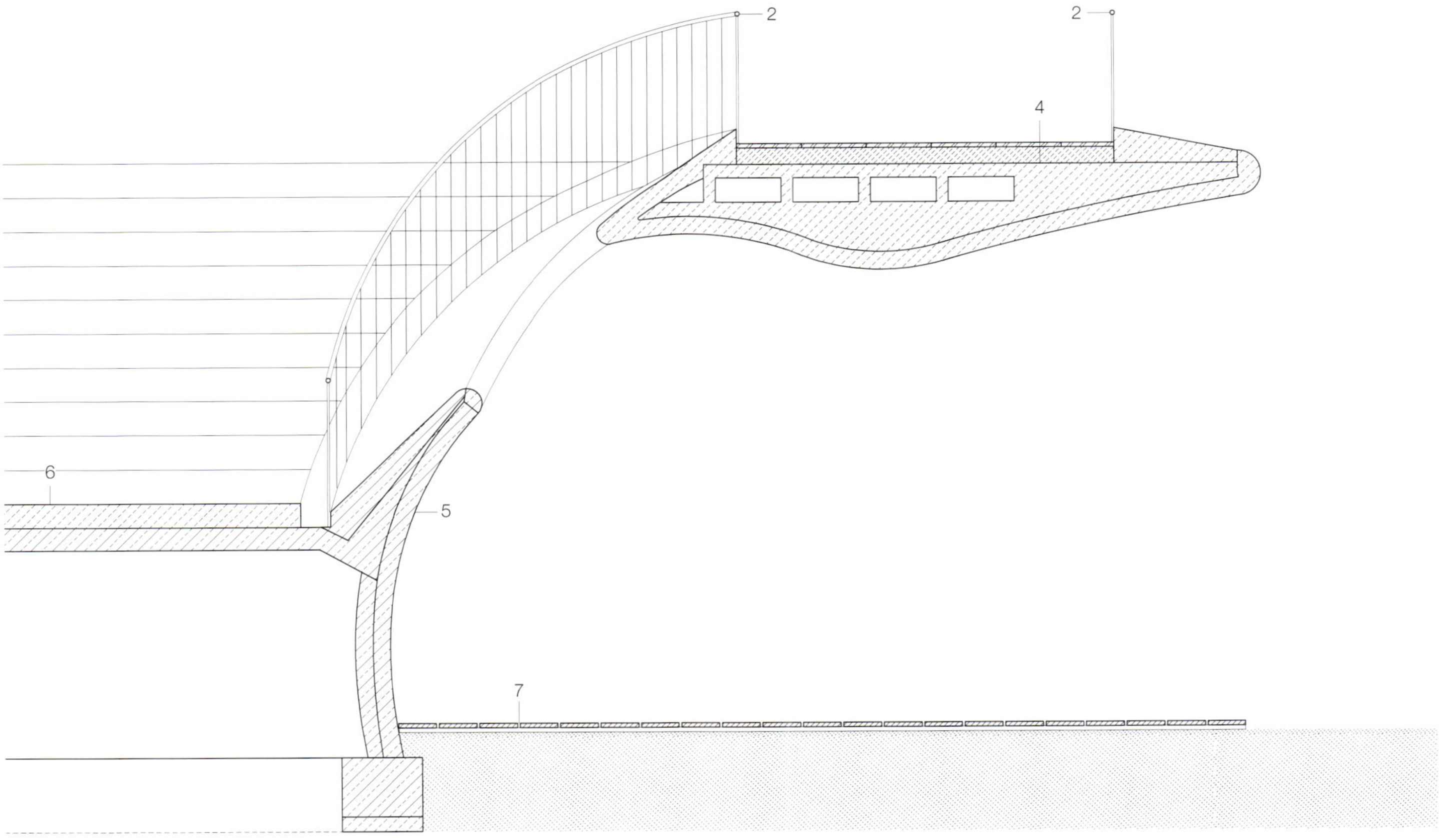

Festival arena in Roman quarry in St. Margarethen

Architects: AllesWirdGut Architektur, Vienna

Theatre architecture, access elements and the natural geology of the setting create a dramatic spatial experience.

The Roman quarry near St. Margarethen on Neusiedler See is one of the oldest quarries in Europe. 2,000 years ago, the calcareous sandstone quarried here was used to build Roman settlements. Later, it was used to build St Stephen's Cathedral and a number of buildings on Vienna's Ringstraße. Since the 1960s, an important symposium of European sculptors has been held here, as well as passion plays which are staged every five years. With the opera festivals which have also been held here annually since 1996 attracting increasing audience numbers – up to 6,000 each evening – a renovation and expansion of the infrastructure became necessary. A clear formal language was employed in order to integrate the access infrastructure and various utility areas, as well as the surrounding geology, into all phases of a visit to the theatre.

Barrier-free ramp

The access route leads the visitor from the car park directly down to the entrance structure and ticket control. After passing through a short connecting tunnel, the visitor is greeted by a first breathtaking view of the arena. From here, the centrepiece of the design, a 330-metre-long wheelchair-friendly ramp, leads the visitor on in a zigzag course over rocks, ravines and the roof of the service building, finally leading down to the festival site, which is 19 m lower. Visitors can pause to admire the sculptural quality of the project along this route, especially at the turning points. The ramp and the facade of the entrance building are clad with 20 mm thick sheets of weatherproof structural steel. Not only do these serve as a reminder of the quarry's industrial past, with their pre-corroded surface they provide protection against the effects of weathering during the winter months. In areas like the box office, toilet entrances or the catering counters, where visitors are likely to come into direct contact with the steel surfaces, these are covered with white fibre-cement sheeting. This creates a bright contrast with the rough sandstone and rusty steel. Materials sourced from the quarry itself and related materials are primarily used for the surfaces in the open areas. In the foyer park, different types of gravel are used to form a water-bound covering which guarantees the necessary seepage and prevents the formation of dust during the dry summer months. Where circumstances permit, the rock itself becomes a building material: the VIP and backstage areas use the surrounding rock as a space-defining element. The latter is concealed behind the rocks of the natural stage; a separate access route for performers and staff leads from the car park via an outside stairway directly into a forecourt-like vestibule. So as not to impair the clear view of the starry night sky or interfere with the action on stage, no lanterns or spotlights are used. Different lighting techniques are used for the basic illumination of the arena which deliberately accentuate the individual areas and assist orientation: strip lights on the floor edging and integrated into ramp balustrades or benches in the picnic area, broad-beamed side-lights on the floor and emergency spotlighting in the aisles make it possible to move around during the performance without tripping over.

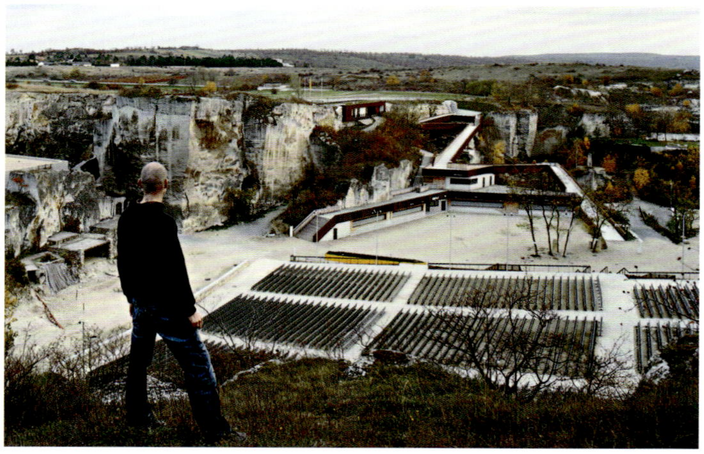

Project data:

Use:	Festival site
Access:	Barrier-free ramps, stairs
Number of floors:	1 (entrance structure)
Gross floor area:	5,580 m²
Year of completion:	2009
Construction period:	30 months

Site plan
Scale 1:2000

1 Visitors' entrance
2 Artists' entrance
3 Entrance structure
4 Tunnel
5 Access ramp
6 Turning point with view
7 Catering
8 Foyer park
9 Backstage structure
10 Existing natural stage
11 Existing main seating area
12 WCs
13 Children's opera
14 Parking area

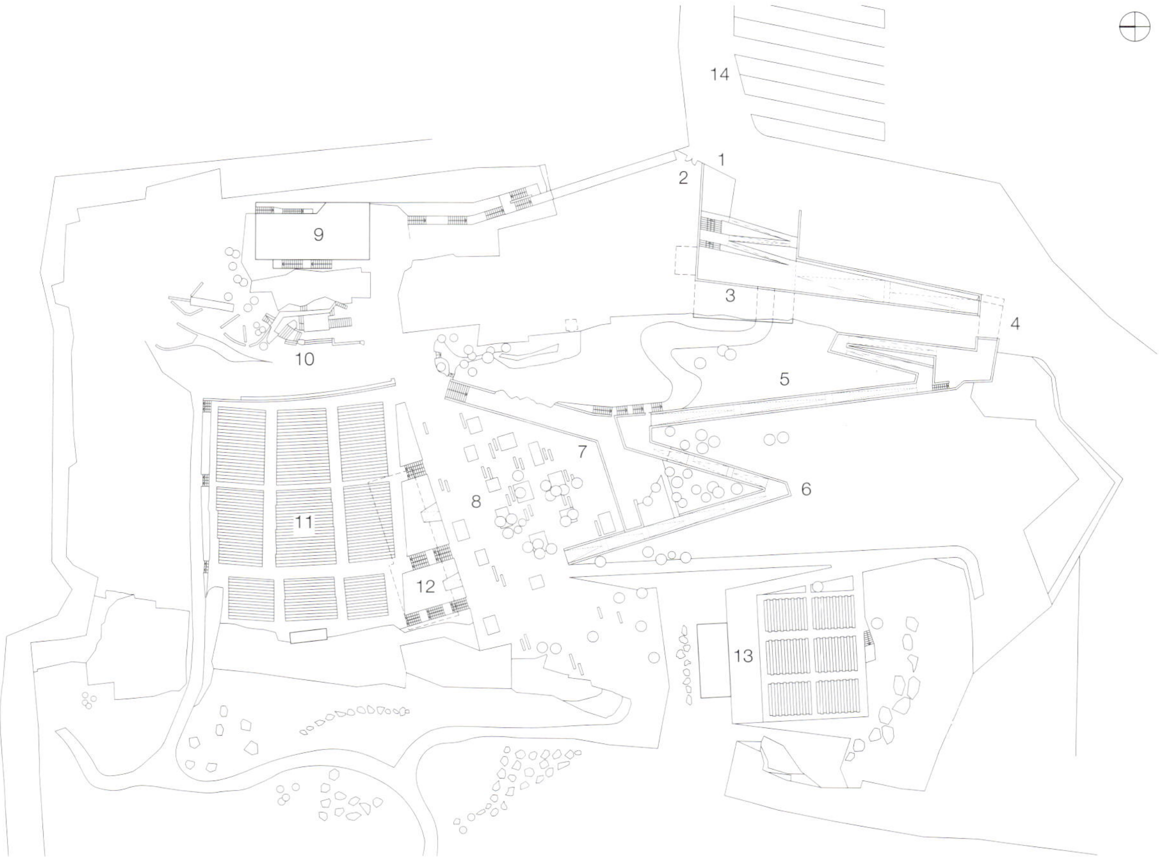

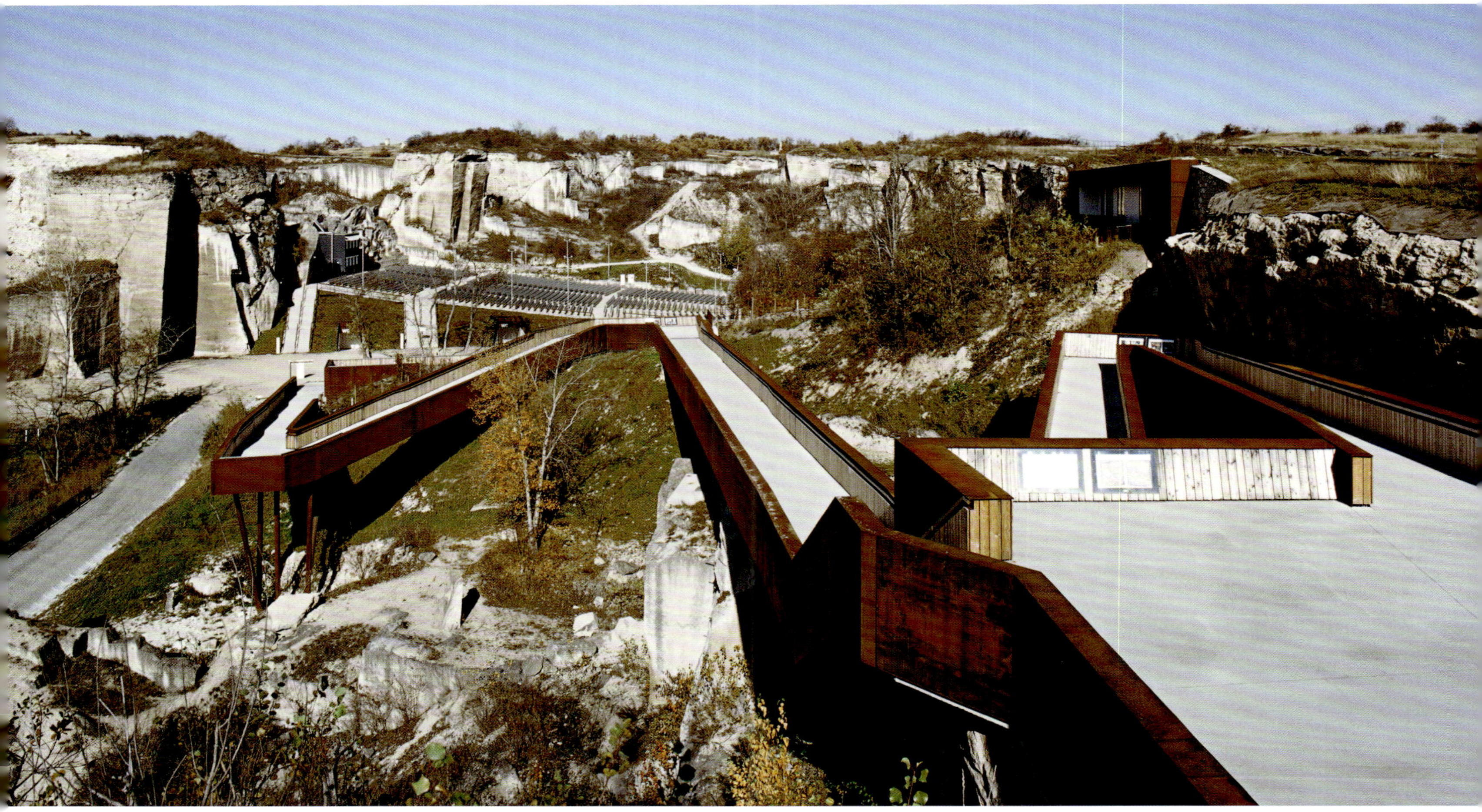

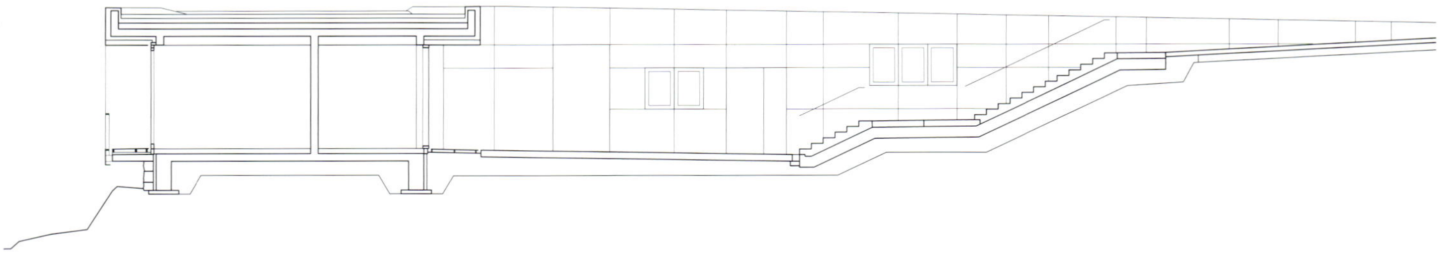

aa

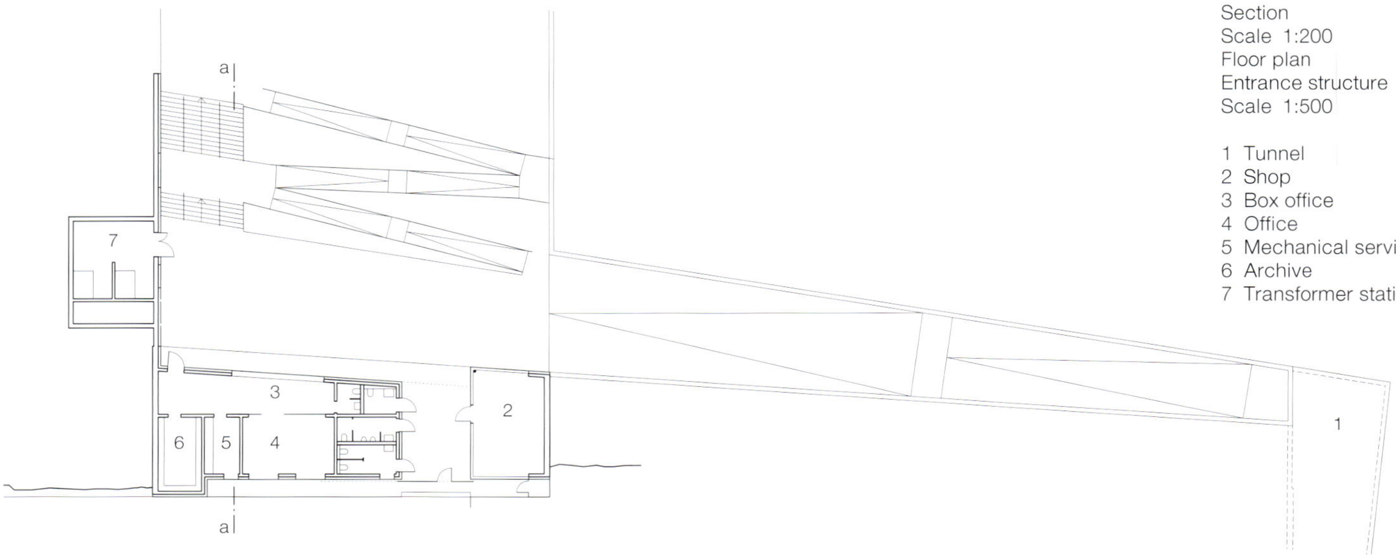

Section
Scale 1:200
Floor plan
Entrance structure
Scale 1:500

1 Tunnel
2 Shop
3 Box office
4 Office
5 Mechanical services
6 Archive
7 Transformer station

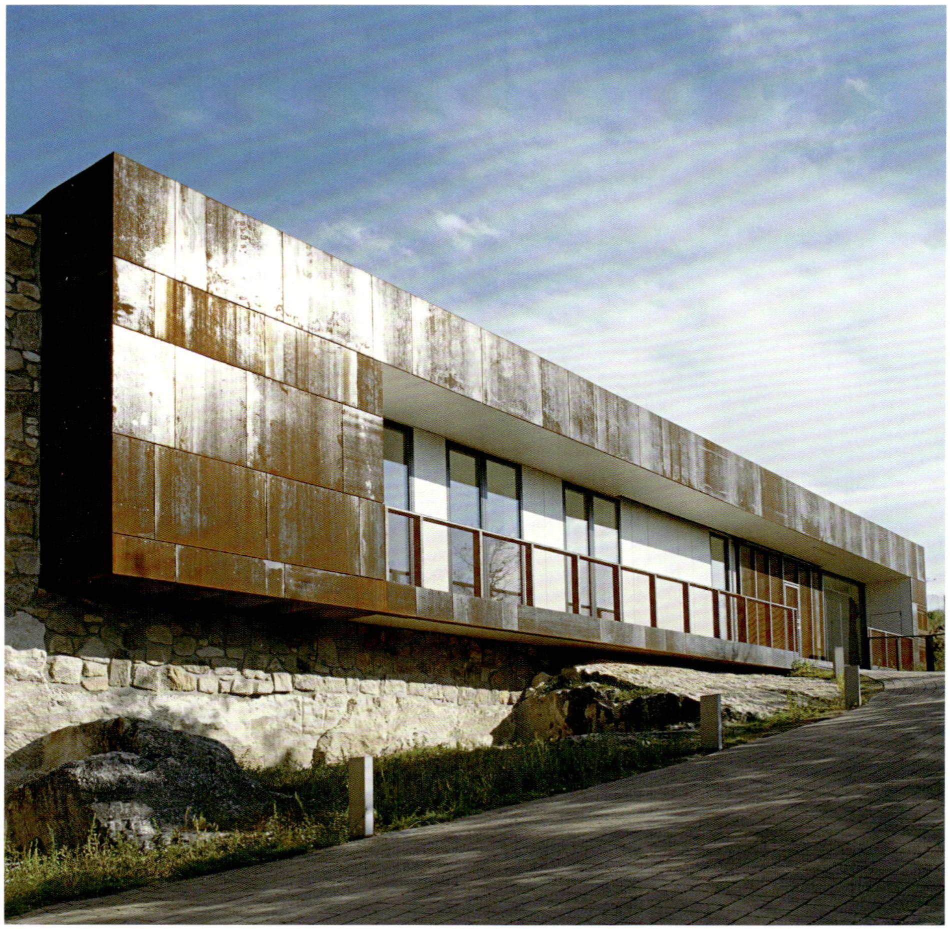

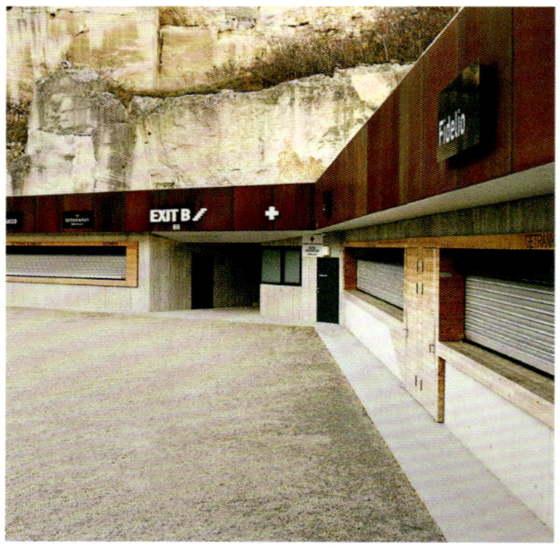

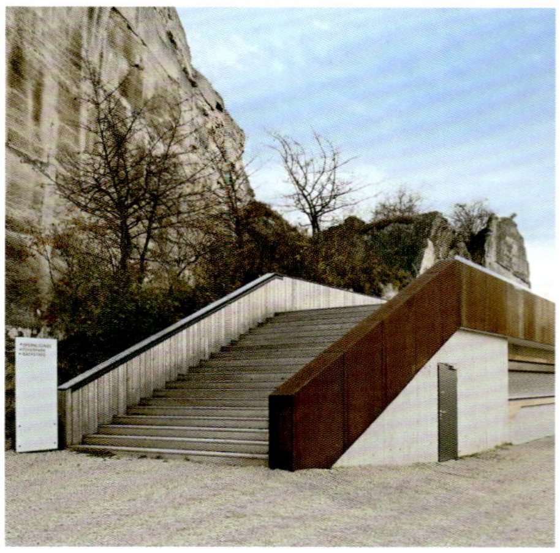

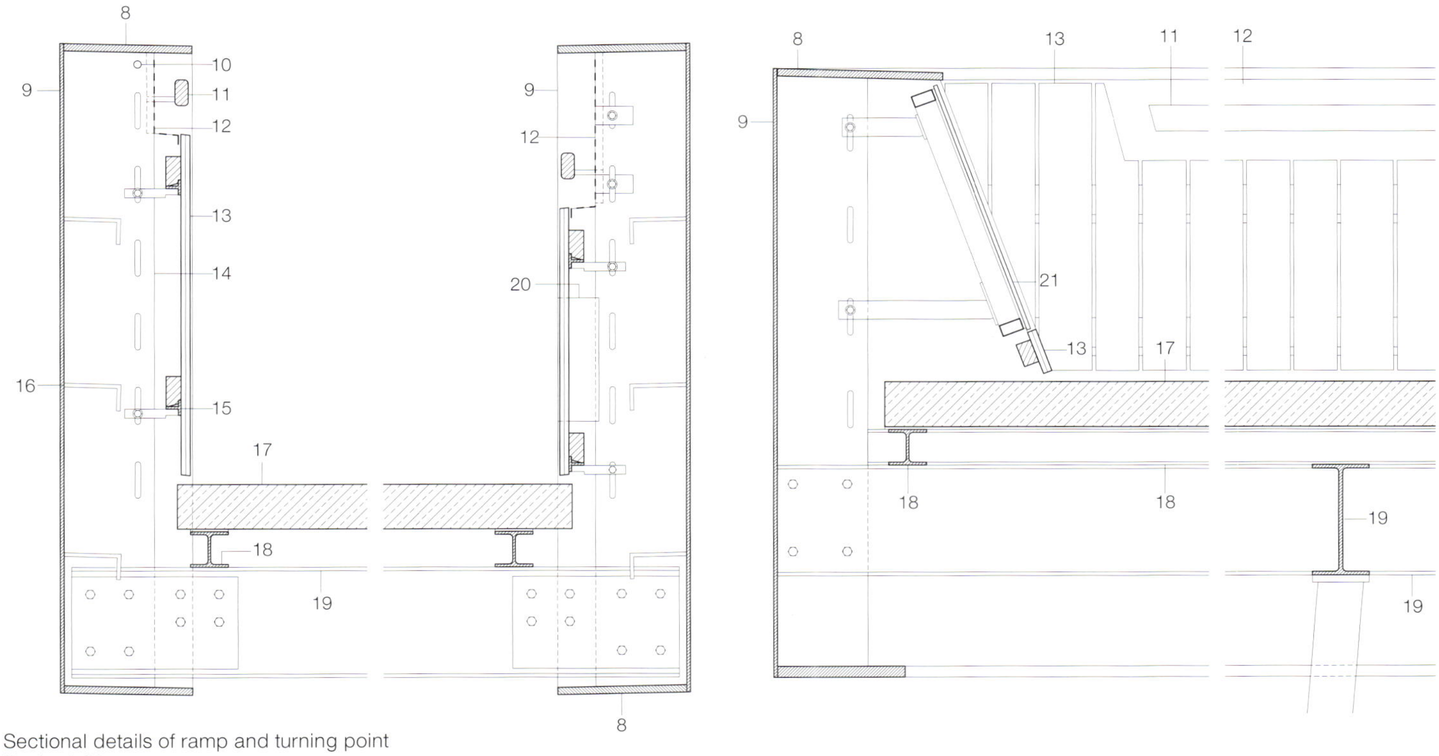

Sectional details of ramp and turning point
Scale 1:20

8 20 mm preoxidized steel plate
9 10 mm preoxidized steel sheeting
10 Cold-cathode fluorescent tube on side towards hill
11 35/70 mm larch handrail
12 Perforated metal sheeting, powder-coated
13 24 mm larch boarding
14 10 mm preoxidized steel bracing
15 40/40/5 mm steel T-section fixed with lugs to 14
16 150/70/10 mm steel angle longitudinal bracing
17 120 mm precast concrete slab
18 HEA 100 steel beam
19 IPE 300 steel beam
20 Recessed light fitting on valley side
21 775/1,120/76 mm aluminium showcase with frameless glazing on top

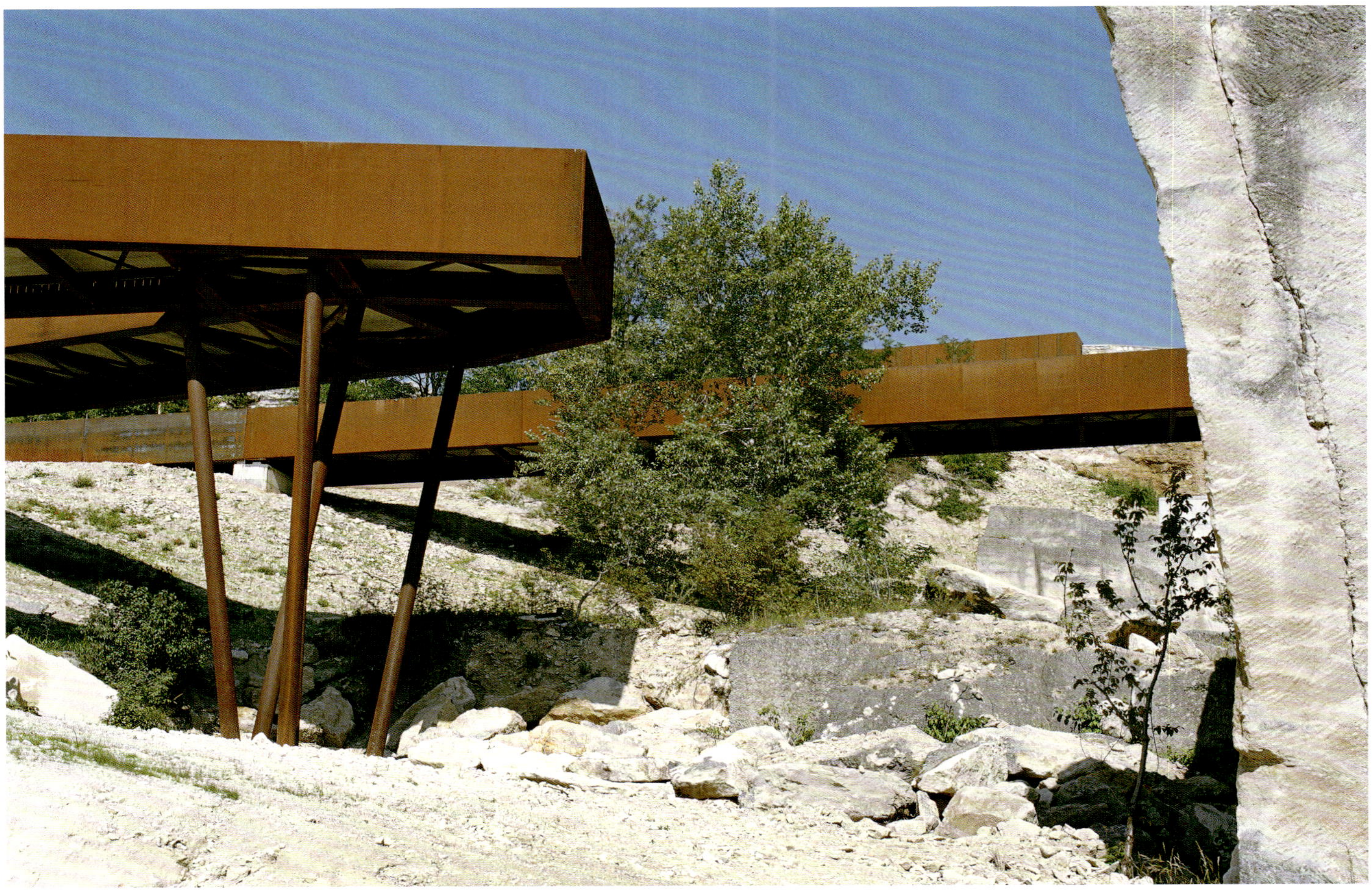

Opera house in Oslo

Architects: Snøhetta, Oslo

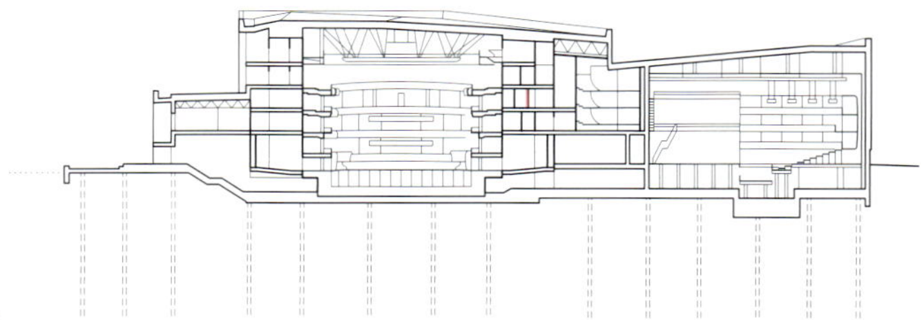

aa

The accessible roof landscape is establishing itself as a new public urban space, offering views over the city and fjord.

With its sloping roof surfaces rising from the waters of the fjord, the Oslo Opera House conjures up any number of associations, evoking comparisons with rafted ice floes, the snout of a glacier or mounds of snow. The opera was the first project in the planned revitalisation of the port area in the Bjørvika inlet, a district of Oslo which had long been neglected and that has been scheduled for regeneration. A new tunnel under the fjord has already largely freed the area of through-traffic. As a solitary structure in the inner harbour, the building forms a connection between the fjord and the city centre.

Exterior and roof

As well as creating a new home for the Norwegian State Opera, at the same time the architects created a new public urban space with the publicly accessible roof landscape. Since its opening, this has become a popular promenade for pedestrians and joggers, offering views over the city and fjord. The accessible architectural sculpture rises like a spiral in broad ramps leading from the water level up to the highest point of the roof. In the winter, narrow paths are formed on the snow-covered roof when the underfloor heating thaws a number of pedestrian routes and the terrace in front of the main entrance, since de-icing salt would damage the marble. Making use of the entire site, the huge volume of the building with its two stages, foyer and an extensive back-stage and workshop area is cleverly concealed beneath the roof slabs. Not just reserved for opera-goers, the café and restaurant in the foyer are much used by the public. This gesture also reflects the Scandinavian ideal of large public spaces which, like the natural landscape, are freely accessible to all.

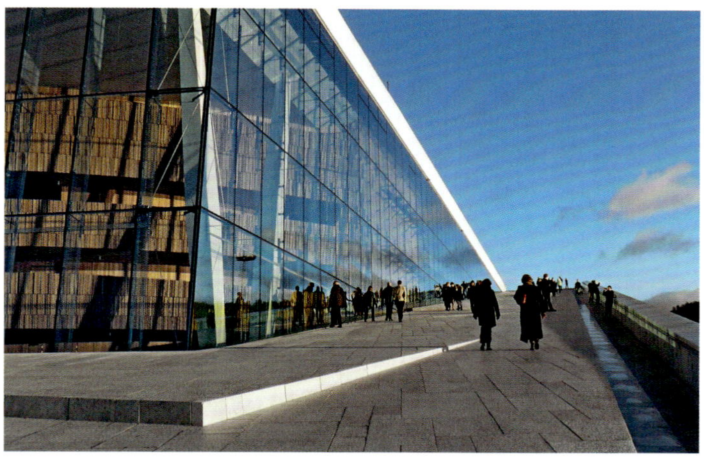

Materials and interior

Carrara marble was chosen for the floor covering, both inside and out. This stone retains its brilliance and the desired bright white colour even when wet, and proved the most weather-resistant. The 18,000 m^2 accessible roof area consists of a total of 36,000 ind vidually cut 10–20 cm thick slabs. In collaboration with artists, a non-repeating pattern featuring offsets in the surface level and polished, rough and rippled surface structures was developed which reflects the light in different ways but is also uneven in places – a public footpath of this kind would probably not comply with regulations in other countries. In contrast to the bright marble and the reflective glass surfaces of the envelope of the building, the key material used in the interior is oak. The light-flooded foyer is dominated by the curving wall clad with CNC-cut wooden slats. It is composed of small elements in order, on the one hand, to cope with the complex geometry consisting of connected conical segments, and on the other hand in order to achieve an acoustically damping effect in the foyer. Ramps and stairs in this so-called wave wall lead to the balcony and upper circles of the large auditorium. As they ascend, visitors can enjoy different views over the city and water framed by inclined columns. Atmospheric lighting is provided by recessed lighting fixtures in the lower area of the parapet as well as in the shadow gap between ceiling and wall. The horseshoe-shaped auditorium is also panelled with oak, since this is pleasant to touch, easy to shape and yet robust. Also, it can reflect sound waves without starting to vibrate itself. The wood was steamed with ammonia, which makes it darker and even stronger. The colours and illumination become increasingly warm and intimate the farther one penetrates into the building. After the light-flooded foyer, the large auditorium feels like a cosy dark cavern. The fluidly curving wooden cladding, designed for optimal acoustics, was installed by a Norwegian boatbuilding firm. The relatively small auditorium has very good spatial acoustics, which were developed in close collaboration with acoustic designers. Even the LED chandelier functions as an acoustic reflector.

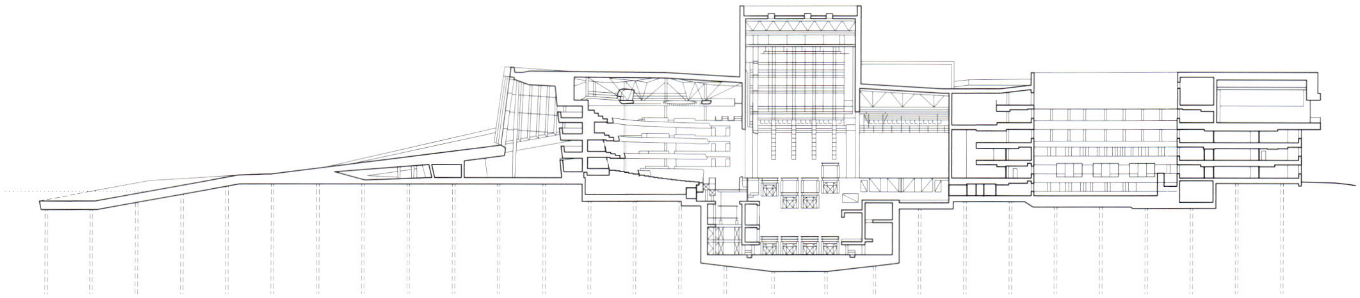

bb

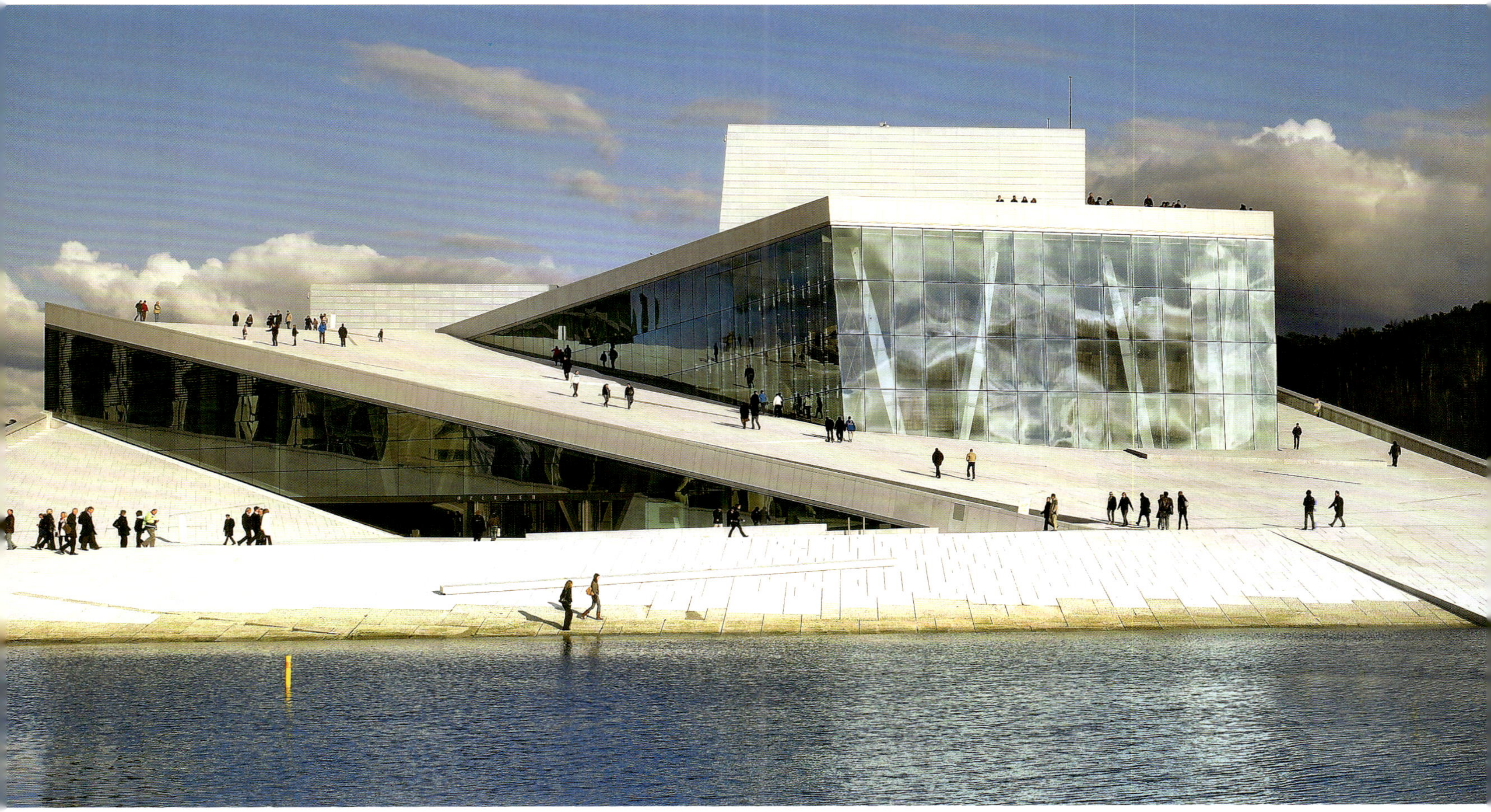

Project data:

Use:	Opera house
Access:	Accessible roof, foyer, ramps, stairs
Number of floors:	5 + 3 basement levels
Gross floor area:	38,500 m²
Year of completion:	2008
Construction period:	74 months

Sections
Scale 1:1500
Site plan
Scale 1:13000

1 Opera house
2 Central station
3 New development with offices and apartments
4 Urban freeway tunnel

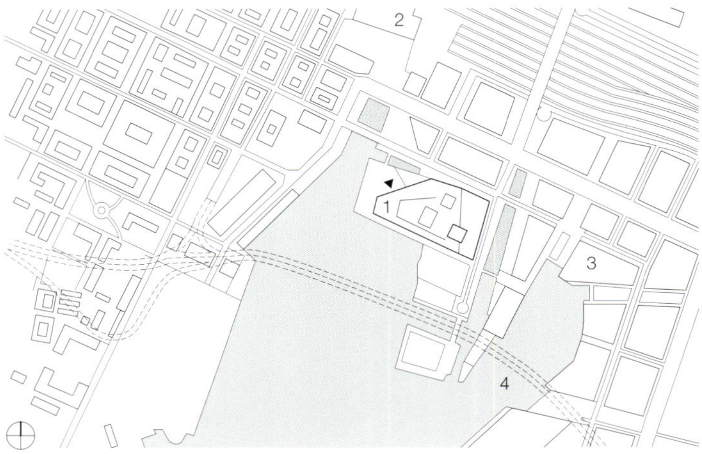

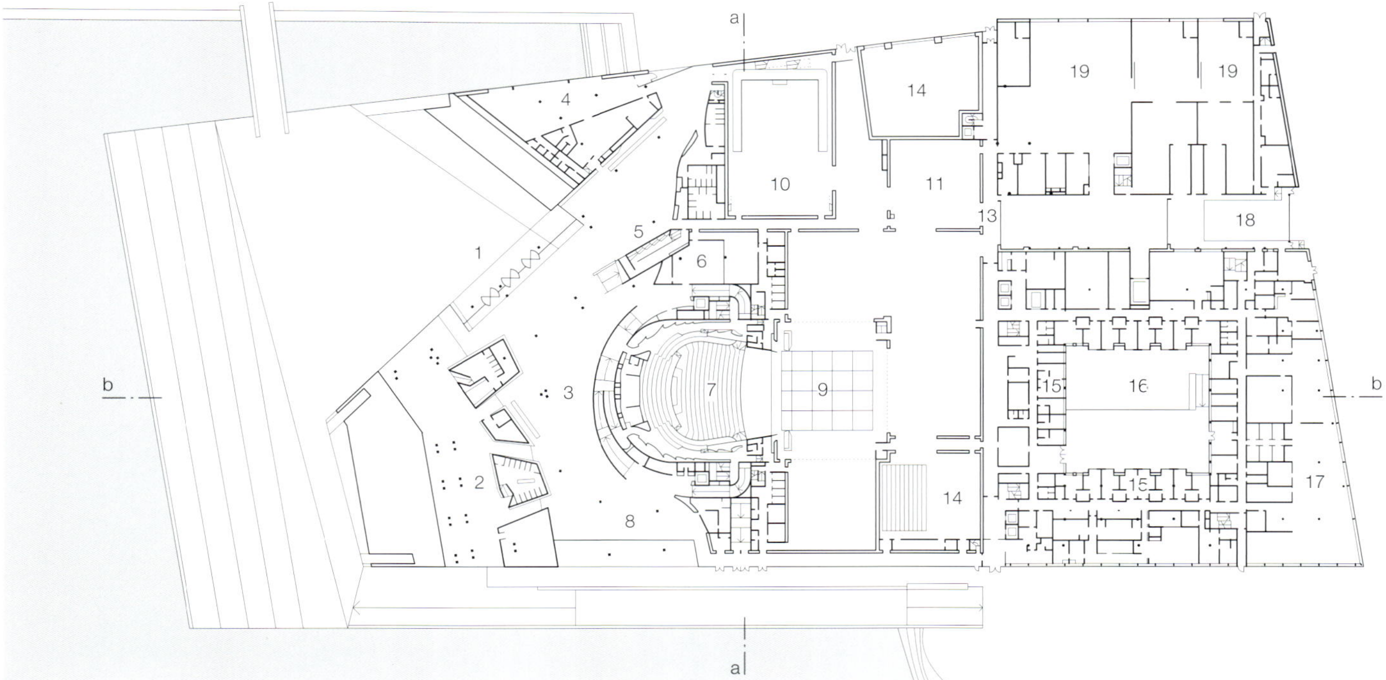

Ground floor

Floor plans
Scale 1:1500

1 Main entrance
2 Cloakrooms
3 Foyer
4 Restaurant
5 Box office
6 Reading room
7 Large auditorium
8 Bistro
9 Stage
10 Small auditorium
11 Assembly shop
12 Void
13 "Opera street"
14 Rehearsal space
15 Dressing rooms
16 Courtyard
17 Costume studio
18 Deliveries
19 Scenery workshop
20 Gallery
21 Upper circle
22 Mechanical services
23 Administration
24 Ballet rehearsals

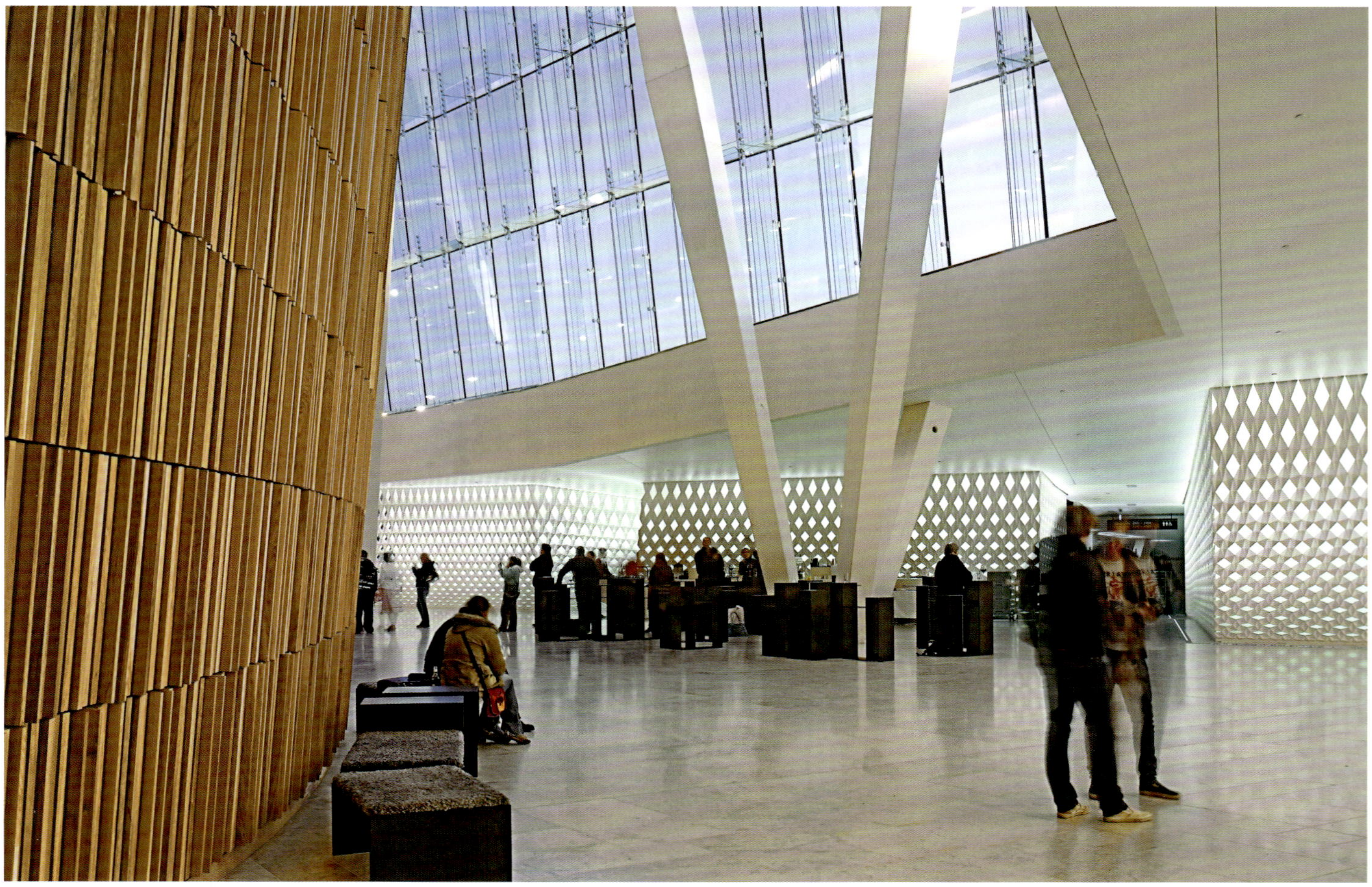

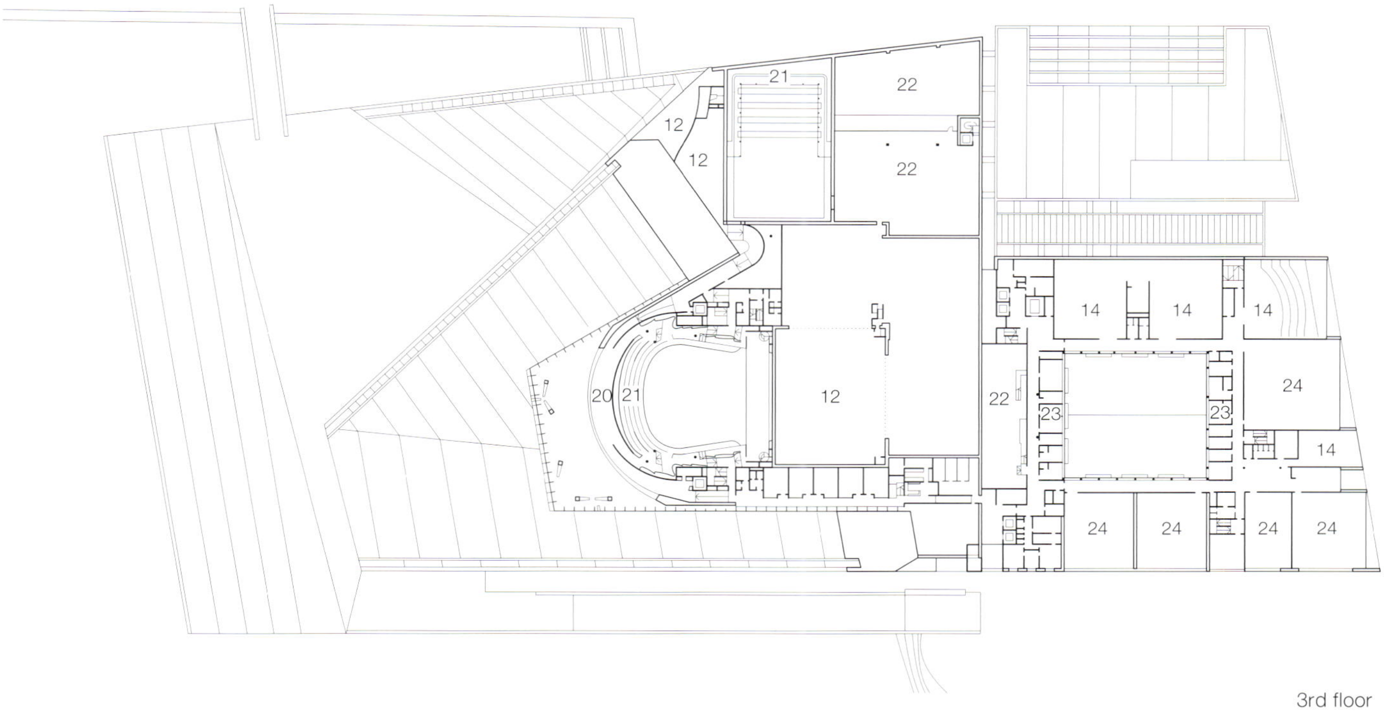

3rd floor

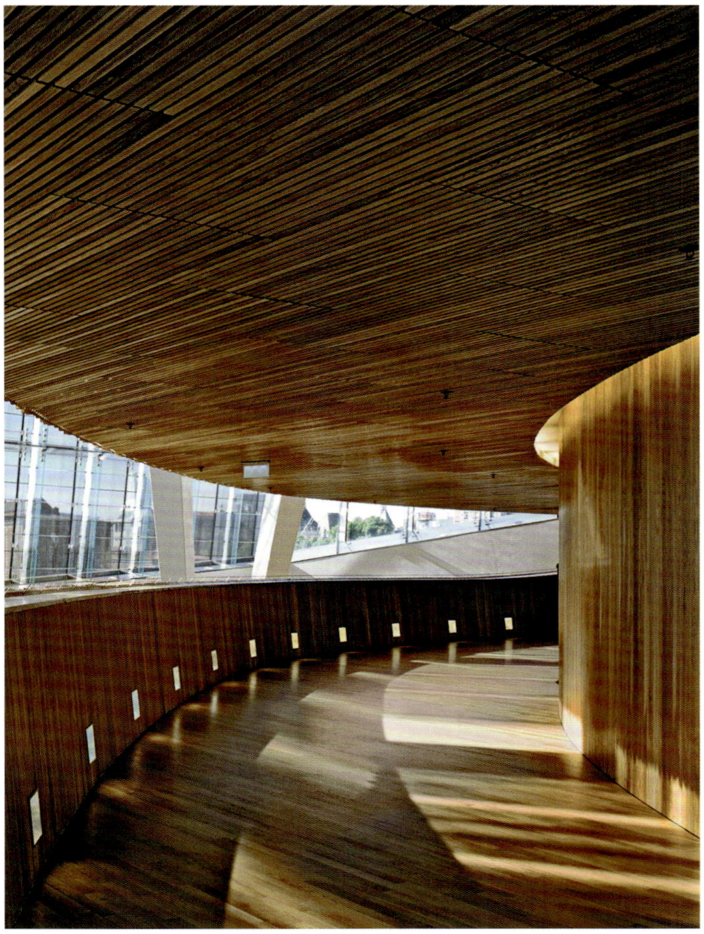

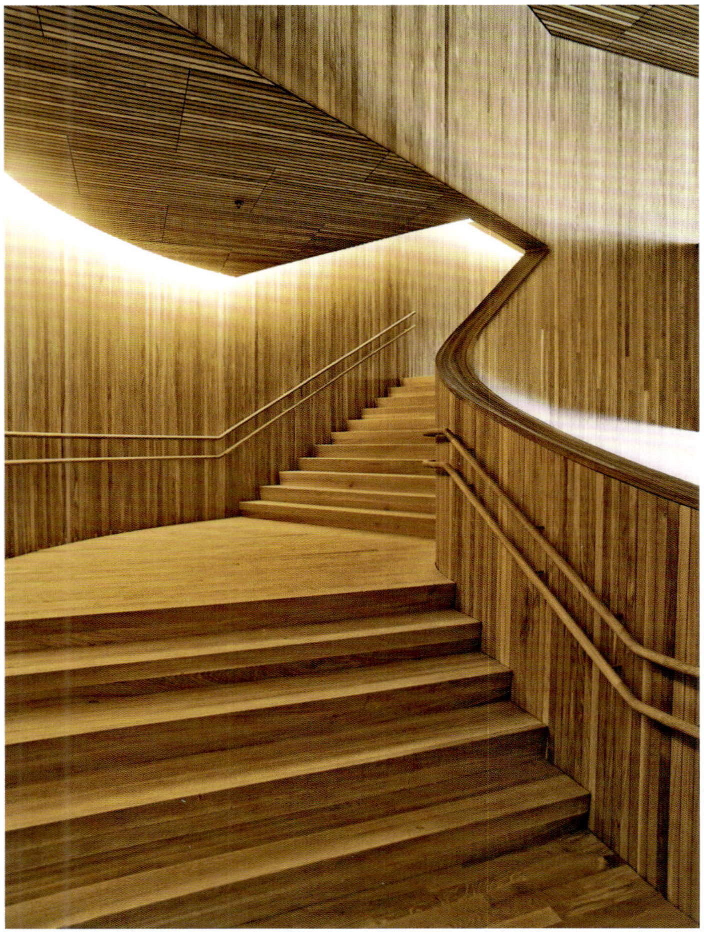

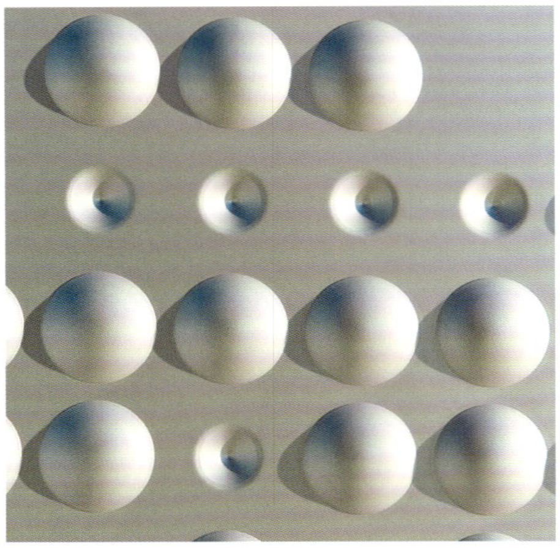

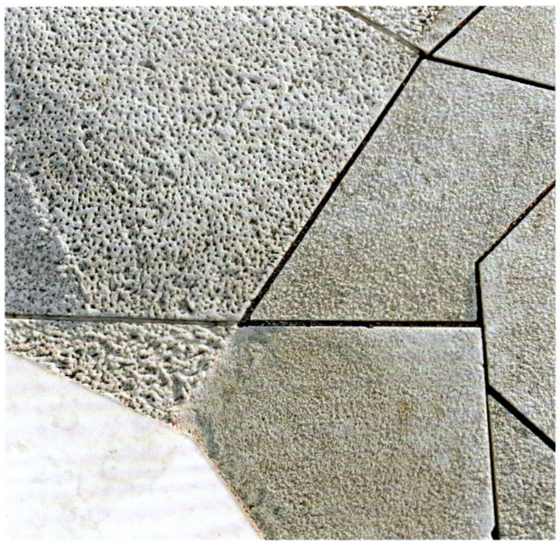

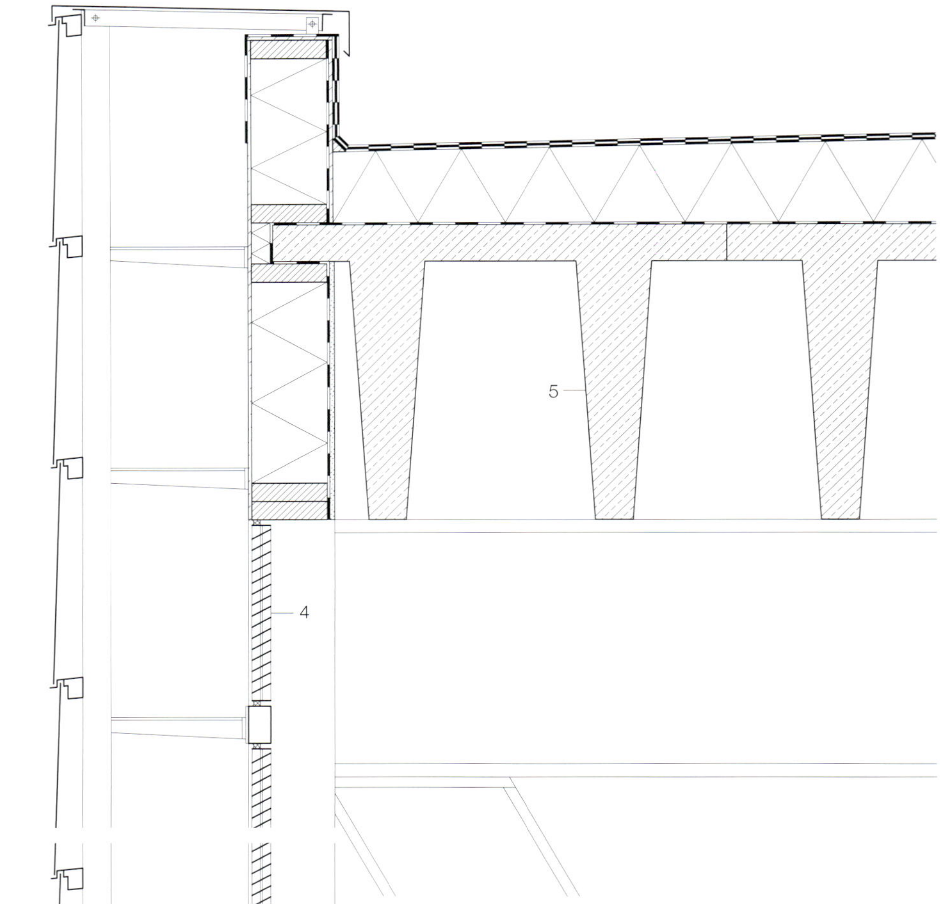
5
4

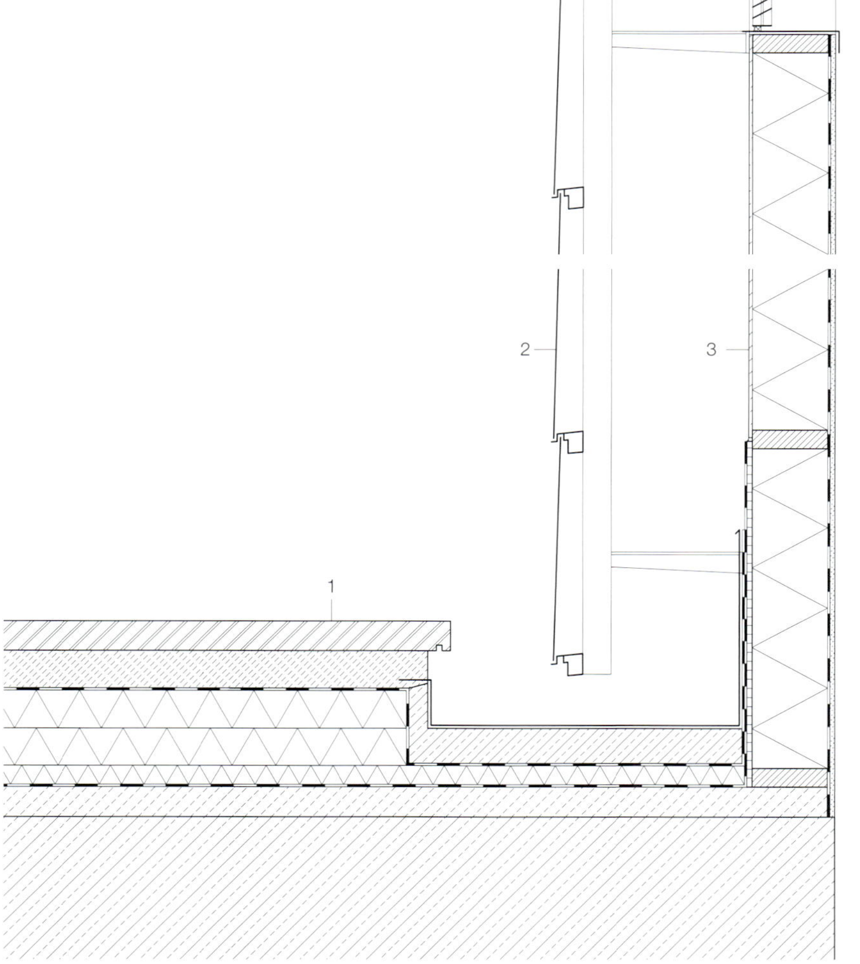
2
3
1

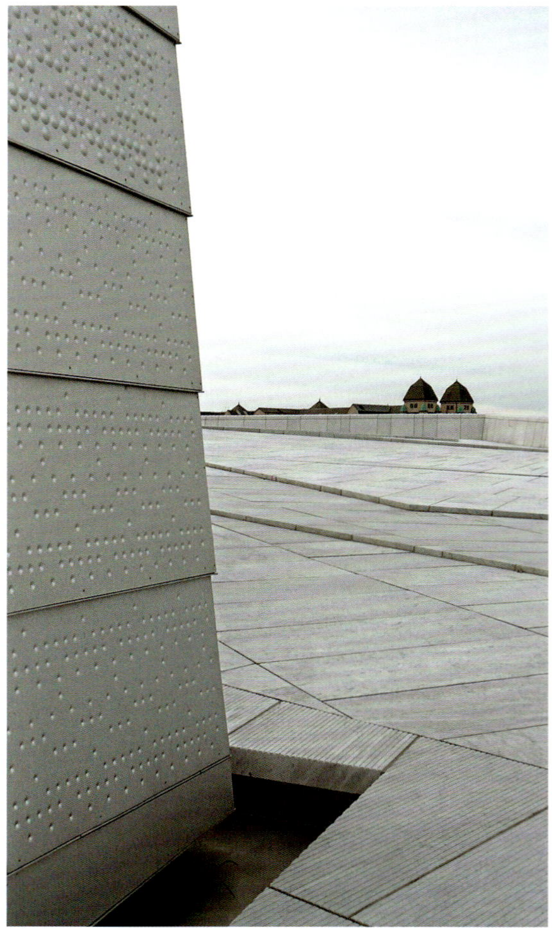

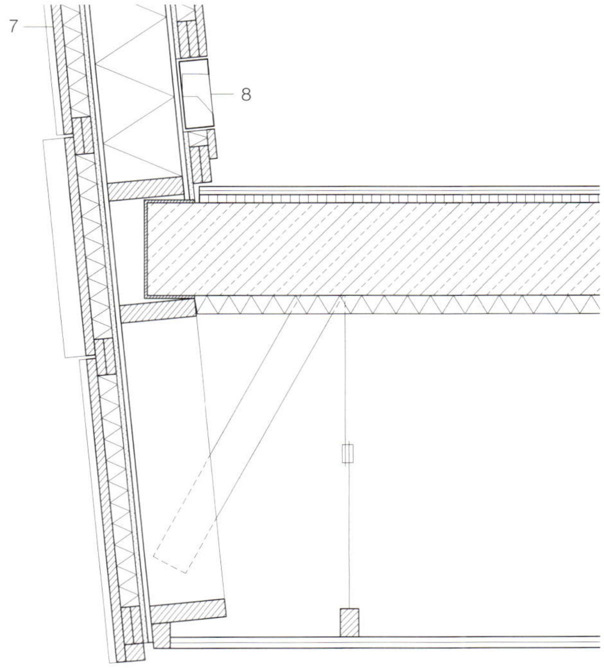

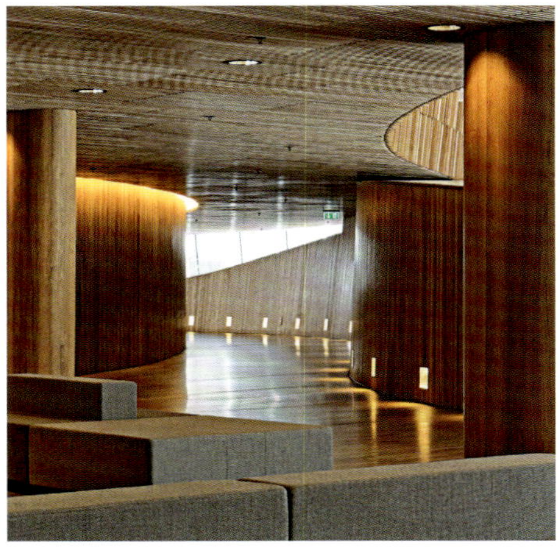

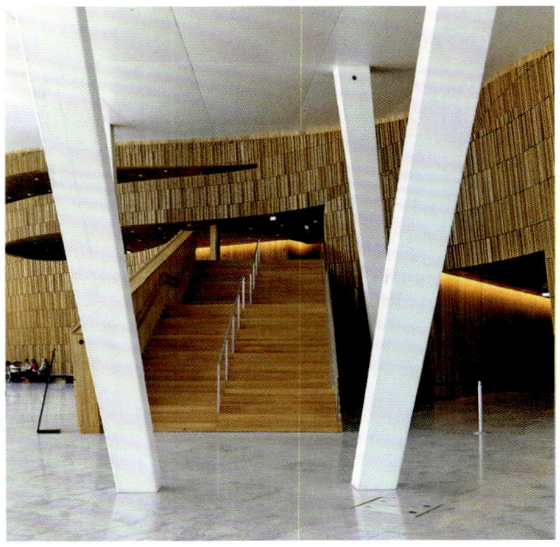

Facade services tower/accessible roof
Vertical section through ramp in "wave wall"
Scale 1:20

1 80 mm white marble
100 mm screed
3 mm felt
2× 100 mm XPS insulation
50 mm mineral wool insulation
3-layer bituminous sheeting
80 mm concrete topping
400 mm precast reinforced concrete unit
50 mm insulation
2 3 mm anodized-aluminium sheeting on
2× 90 mm aluminium hollow sections
3 9 mm fibre-cement sheeting
200 mm mineral wool insulation between
100/200 mm timber studding
vapour barrier
12.5 mm plasterboard
4 Ventilation grating
5 Precast concrete unit:
2× 800 mm T-sections
6 ∅ 12 mm stainless-steel rod
7 Alternating 20/45 mm, 40/45 mm, 60/45 mm
and 80/45 mm oak strips, oiled
50 mm mineral-wool acoustic insulation
13 mm plasterboard
Wood levelling strips
48/198 mm timber framing with insulation,
fixed to steel structure
Wood levelling strips
13 mm plasterboard
50 mm mineral-wool acoustic insulation
20/40 mm oak boarding, oiled
8 Spotlight
9 22 mm oak parquet, oiled
22 mm chipboard
250 mm reinforced concrete
50 mm acoustic insulation
10 Suspended ceiling: 20/28 mm oak louvres,
oiled, with 17 mm spacings
11 14/45 mm oak boarding, oiled
48/98 mm battens
200 mm reinforced concrete
12 4 mm oak veneer, smoke-treated and oiled
50 mm chipboard
48/98 mm timber posts

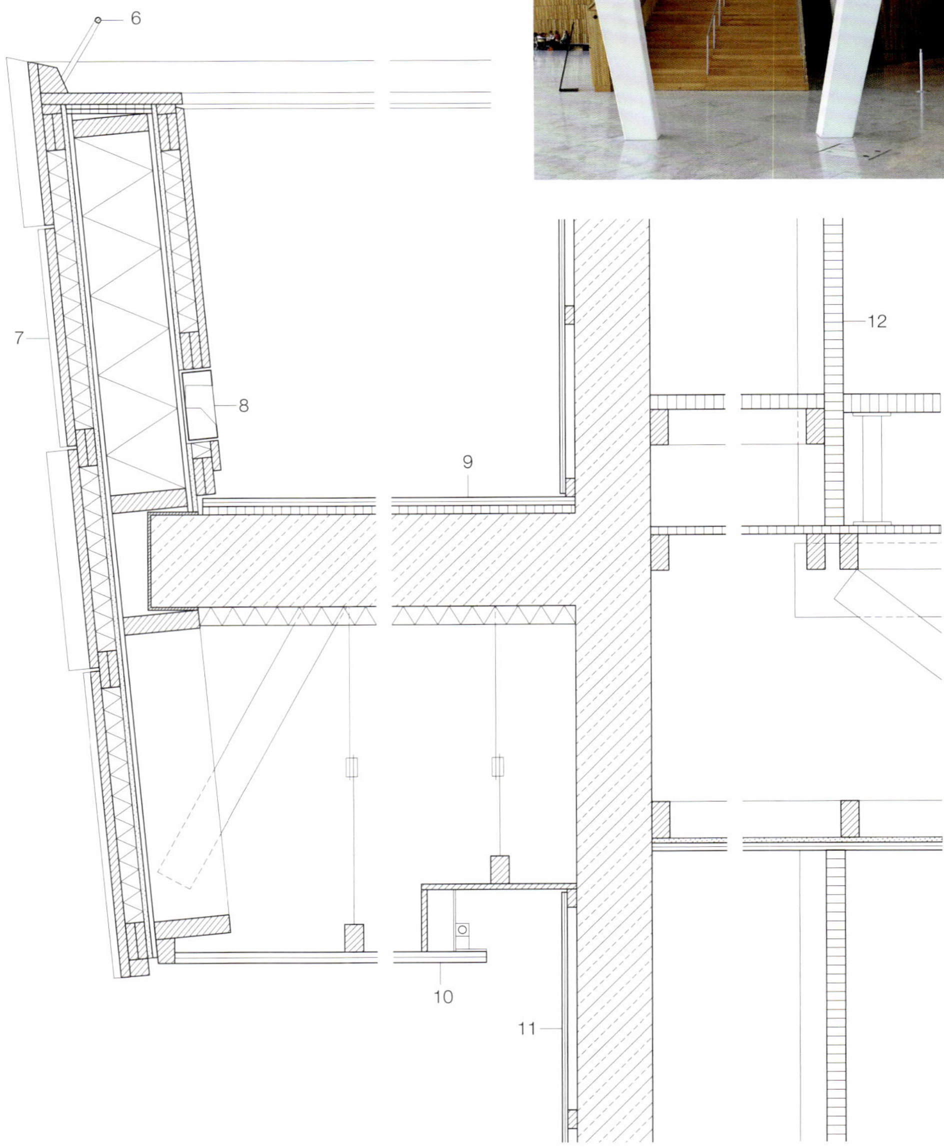

Porsche Museum in Stuttgart

Architects: Delugan Meissl Associated Architects, Vienna

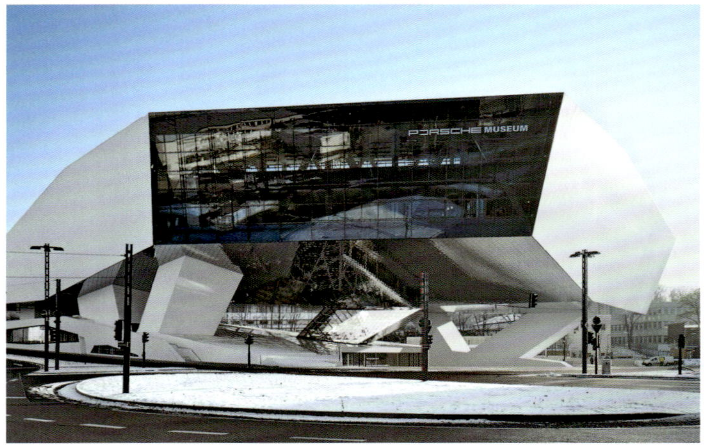

Two escalators flanking a cascading stairway lead visitors through the partially glazed access tract into the exhibition.

Set between production facilities and administration buildings, an arterial road and an S-Bahn station, the Porsche Museum sets a bold architectural accent in the heterogeneous industrial environment of Stuttgart-Zuffenhausen. The polygonal structure that houses the exhibition is clad in white-coated aluminium panels and, resting on only a few points of support, is raised above the flat plinth structure which houses the foyer and restaurant as well as the museum workshop, the interior of which can be viewed from outside. The gently sloping forecourt area provides access to the sunken entrance sheltered by the cantilevered exhibition structure. Much of the sense of weight of the raised steel volume is relieved by the polished stainless-steel sheeting on the underside, which creates changing reflections of the forecourt, plinth and staircase tracts. The view up from the foyer to the web-like glazing overhead heightens this effect, arousing the interest of visitors as they head for the exhibition. A central staircase with two flanking escalators leads through the reflective, partially glazed intermediate zone into the museum landscape. Due to possible differential deformations of approx. 80 mm and a sort of agitating effect of the escalators, the access tract is only supported at two articulated points, in the middle and on the gallery. The connections to this part of the building, which passes through several levels that each deform in different ways, represented a challenge, in particular the movement joint on the glass facade around the escalators.

Exhibition space and circulation

The bright, expansive exhibition space with a total floor area of 5,600 m² resembles a distorted spiral set in the polygonal footprint. The total length of 550 m made it possible to design the circulating route as a very shallow ramp with a pitch of less than 2 %. On this almost level surface, as they pass through the exhibition, visitors overcome a height difference of more than 6 m almost imperceptibly, ending up on the gallery level which leads them back to the descending escalator. The white floors and walls, which are covered with seamlessly joined, mineral-based sheeting, form a neutral background to the presentation of the exhibits – the total of 83 vehicles are almost all in perfect running order. The surrounding 3-metre-deep black strip running around the outer wall acts as a contrasting background for the vehicles and depicts the history of the product from 1948 onwards. This chronological sequence is periodically interrupted by themed areas which explain typical properties of the marque in a variety of different ways with reference to a key exhibit. In addition, an arrangement of themes relates the concept in question to the corresponding cars and prototypes. As a prelude to the content of the exhibition, the "prologue" which winds around the entrance way describes the activities of Ferdinand Porsche prior to establishing the marque in 1948 with early exhibits displayed on landings made of green recycled glass material. Visitors can choose to follow the chronological tour in the form of the abstracted spiral or wander freely through the extensive exhibition, if they wish taking shortcuts via shallow-angled stairs to create their own spatial connections and view the exhibits according to their own preferences. From almost any position within the fluid space, visitors have an overview of the route before them as well as the parts of the exhibition they have already covered. In order to achieve striking spatial effects and varying sightlines, the architects played with alternating narrow and wide footways, rest plateaus with a decelerating effect as well as varying room heights and depths.

The column-free exhibition shell is covered by a self-supporting load-bearing framework with an average span of 60 m and cantilevering out by as much as 50 m, consisting of peripheral trussed girders that extend over the full height of the building, with three-dimensional beam grids in the floor and roof planes. This flexurally rigid upper volume is supported by three cores of reinforced and prestressed concrete with a total of five vertical and raking piers. The use of load-distributing steel inserts and high-strength concrete makes it possible to divert the support forces into the core which, among other things, houses the emergency stairs and a goods/car lift.

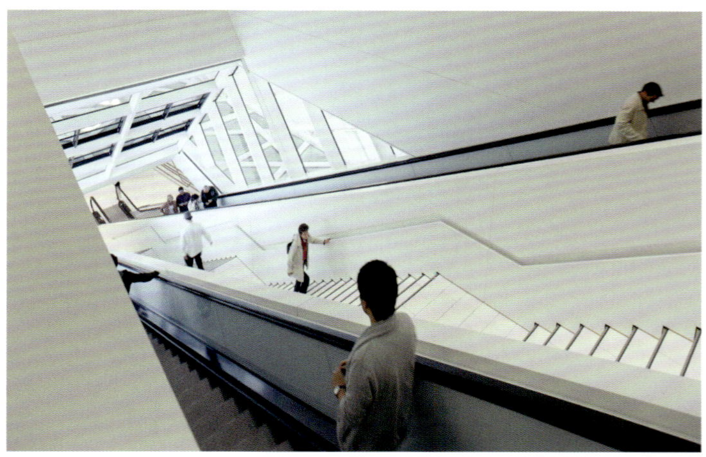

Axonometric view
of access strip
Section
Scale 1:1000

Project data:

Use:	Museum with museum workshop
Access :	Escalators and central cascading stairs, shallow ramps and stairs in the exhibition area
Number of floors:	3 + 2 basement levels
Gross volume:	225,464 m³
Gross floor area:	27,692 m²
Year of completion:	2009
Construction period:	40 months

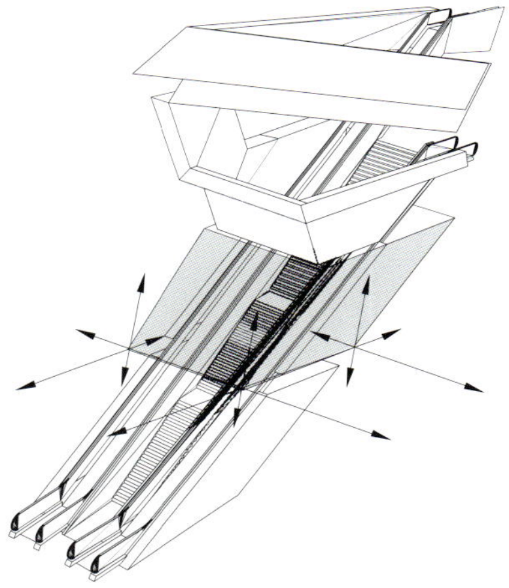

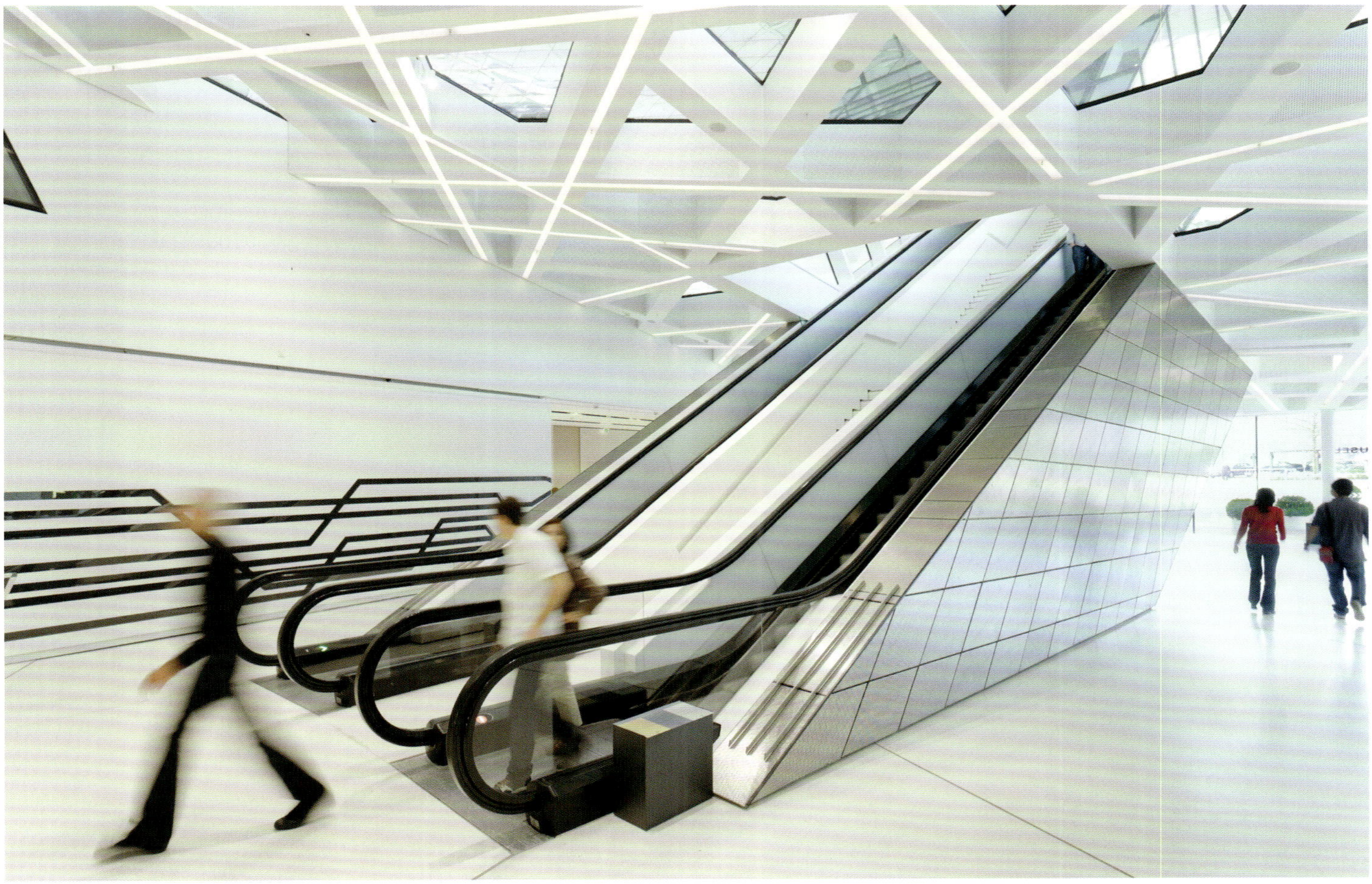

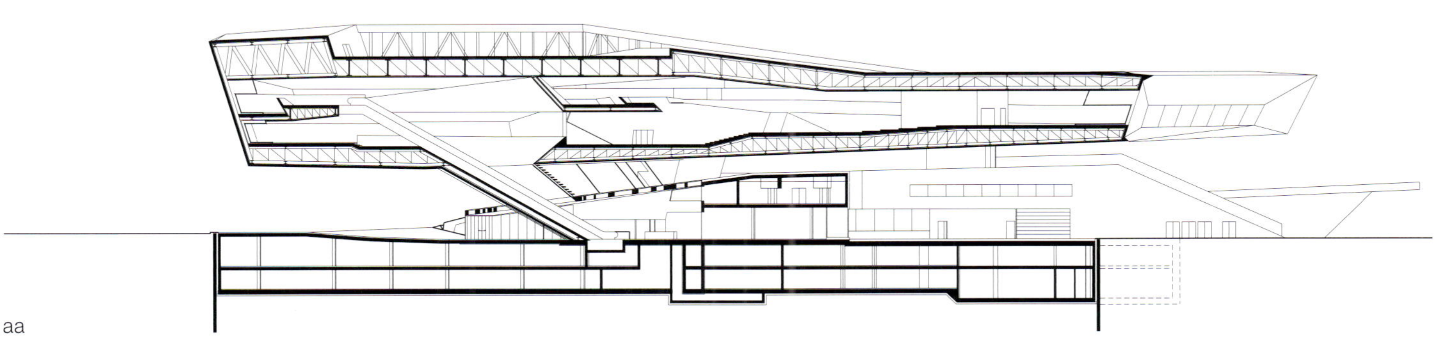

aa

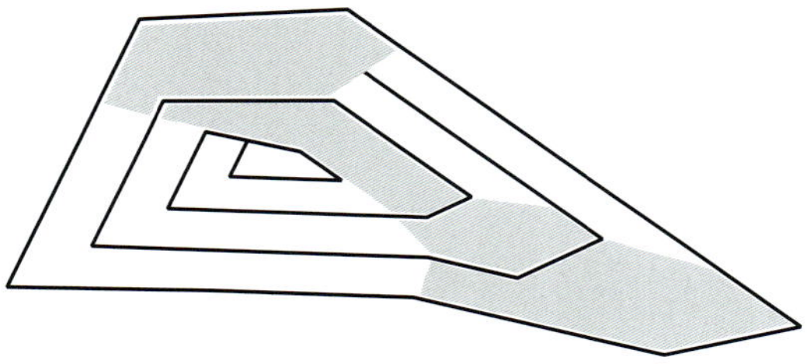

Wide spaces and open plateaus
with facilities for sitting and resting

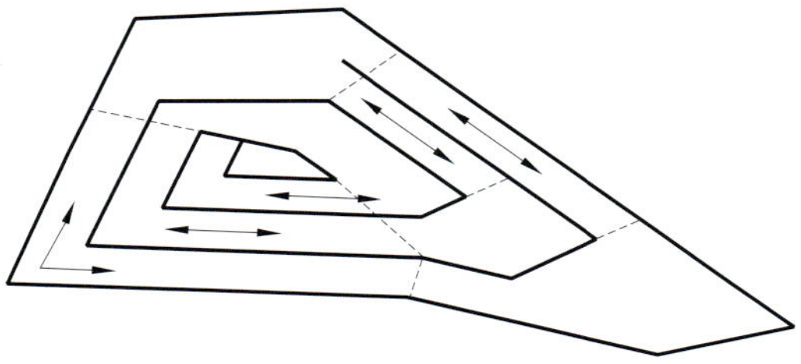

Narrow lanes with an
accelerating effect

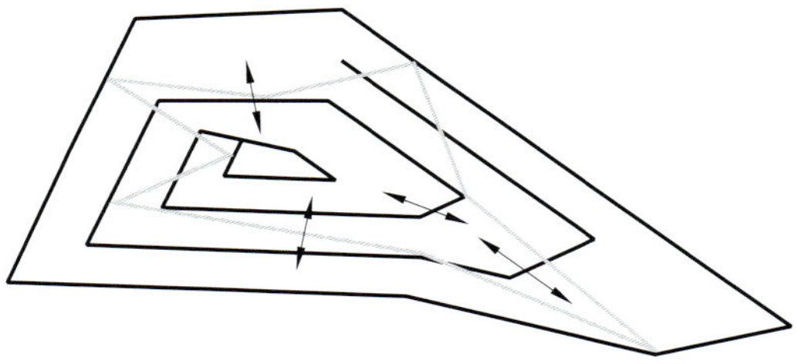

Possible shortcuts via shallow stairs

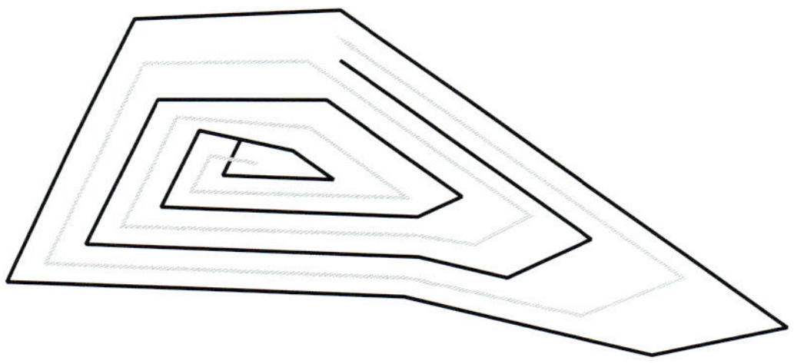

Chronological exhibition
tour following ramp

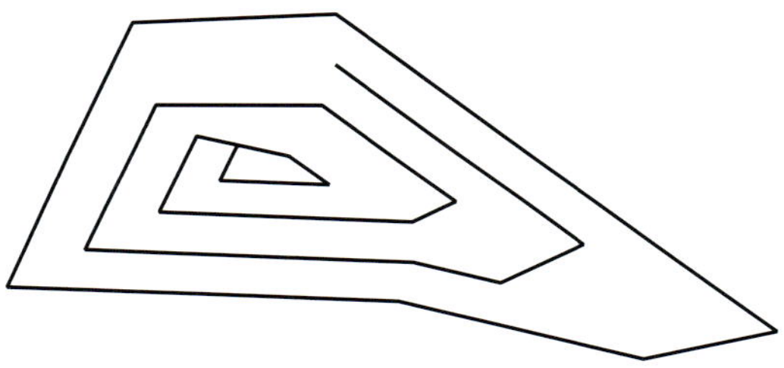

Basic principle: distorted
spiral form as shallow ramp

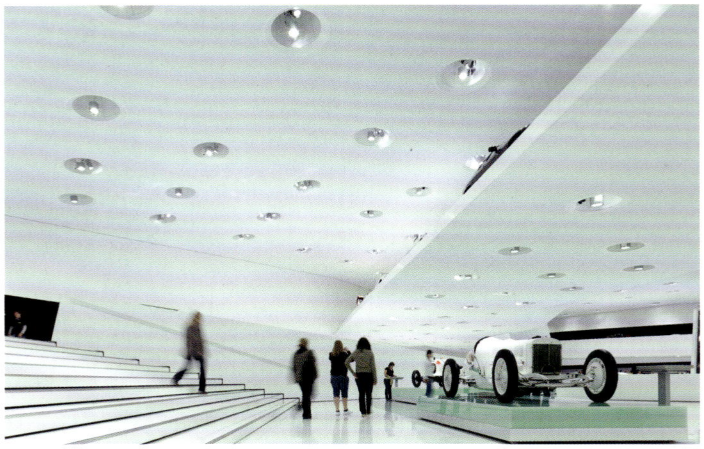

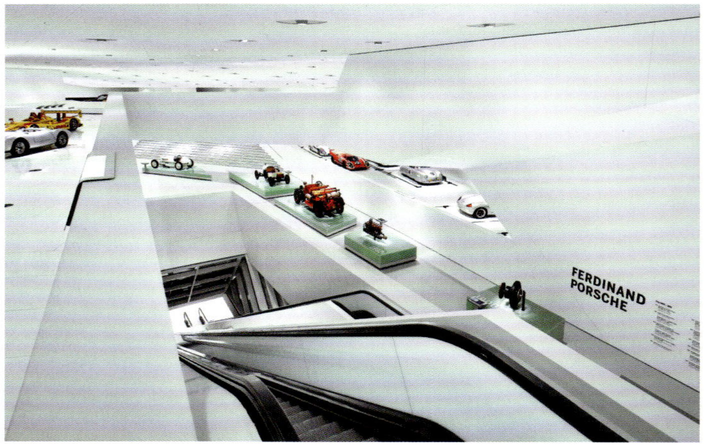

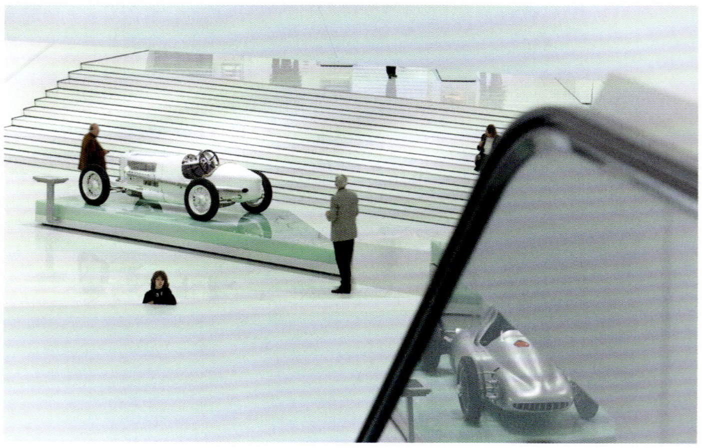

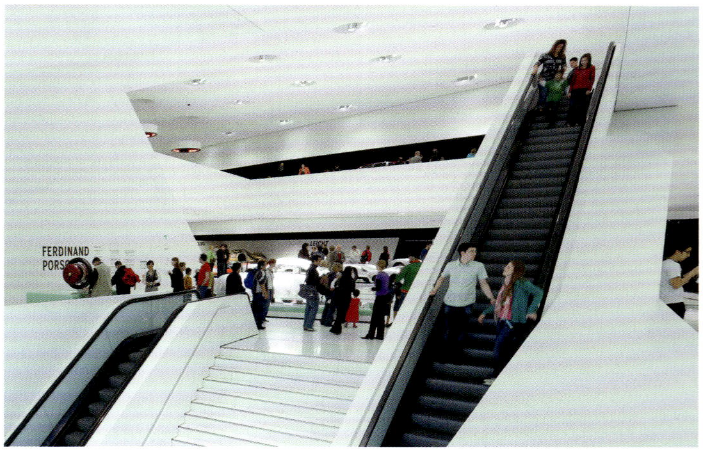

Floor plans
Scale 1:1000

1 Forecourt
2 Foyer
3 Restaurant
4 Café
5 Shop
6 Museum workshop
7 Technical services
8 Deliveries
9 Access to underground car park
10 Exhibition
11 Gallery

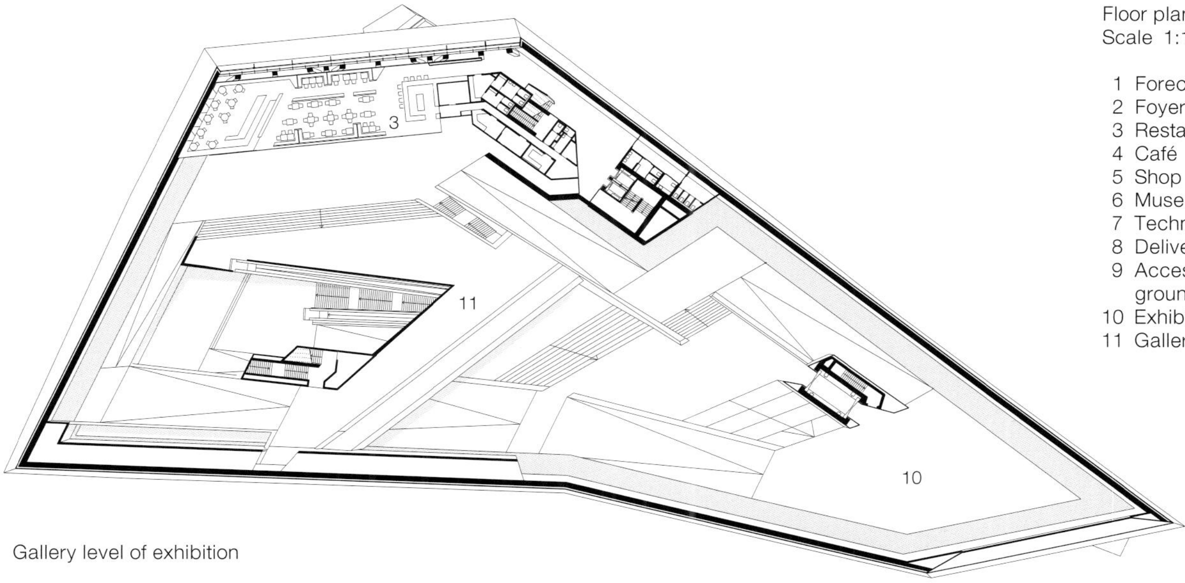

Gallery level of exhibition

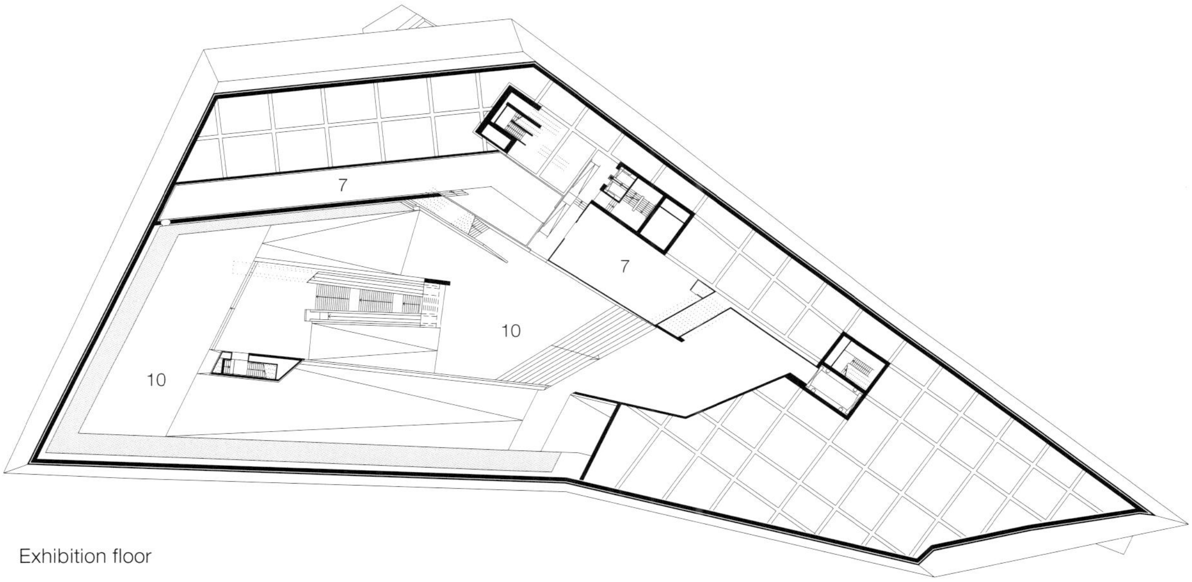

Exhibition floor

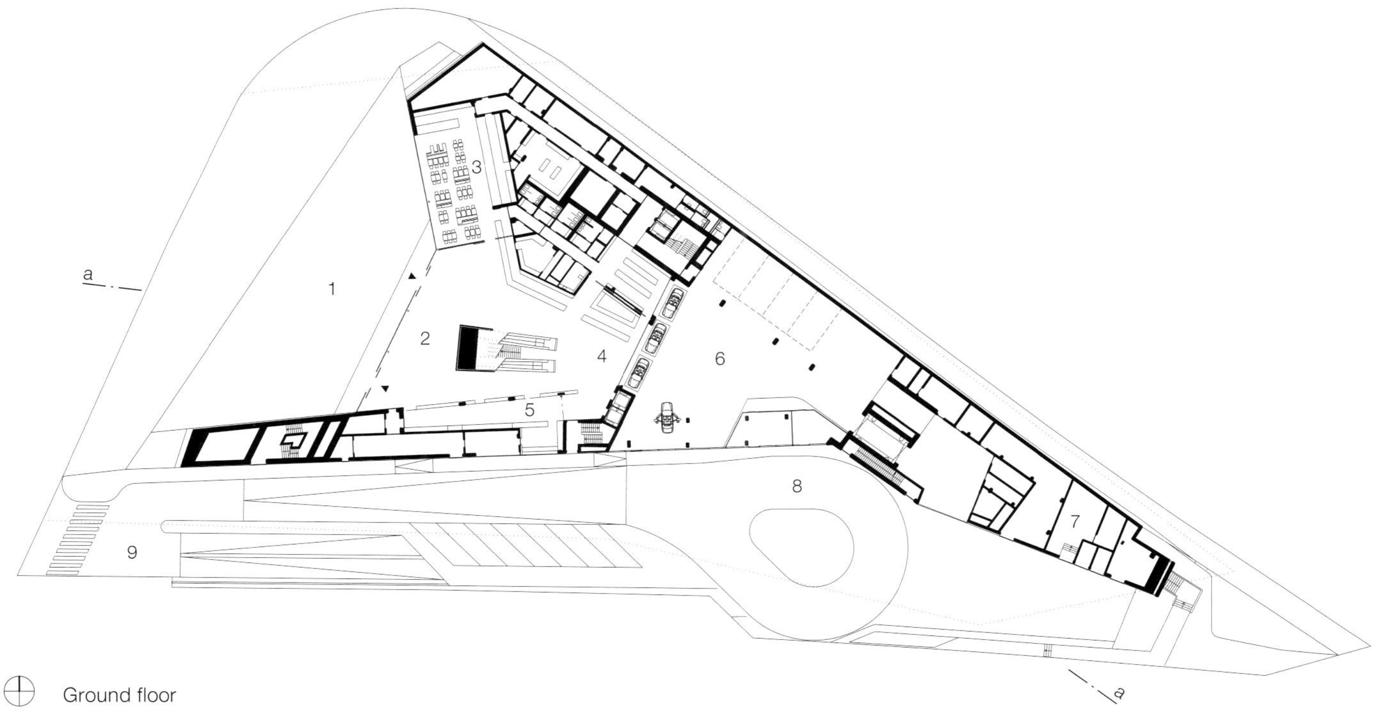

Ground floor

Armani Fifth Avenue in New York

Architects: Doriana and Massimiliano Fuksas, Rome

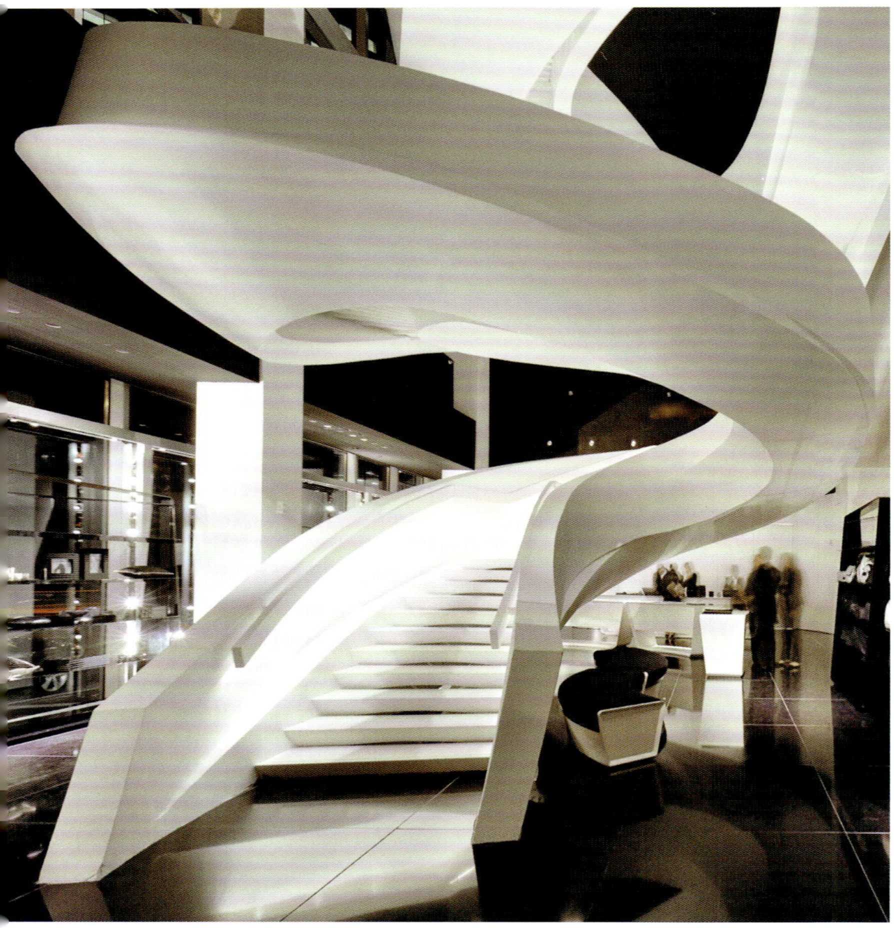

The central element in the four-storey retail space is the free-form sculptural staircase made of rolled-steel sections and glass fibre reinforced concrete.

The Armani Shop on Fifth Avenue in New York, one of the most prestigious addresses in the city, occupies four floors and two blocks at the intersection with 56th Street. Conceived as a continuous space extending over all the sales floors, the interior is dominated by the dynamic, free-form sculptural staircase. Since, due to weight considerations, it was not possible to build the staircase out of concrete, the architects opted for a more lightweight construction of internal rolled-steel sections and steel sheeting. These elements were prefabricated individually and then assembled and clad on site. The structure was then clad with glass fibre reinforced concrete elements at the transitions and the staircase was finished with a high-gloss white coating. The gleaming white finish, intensified by the lighting in the handrail, makes the expressive form of the staircase sculpture stand out against the black marble floors and ceilings. As a spatial core, the staircase connects the different levels on which the products are displayed. The interior design and furnishings are developed from the sweeping curves of the staircase. The side walls and parapets, following different curves in the floor plan, are echoed throughout the space in the display tables and counters of varying lengths. On the second floor there is an exclusive restaurant with a view over Central Park. Another staircase leads down to the basement level, only part of which is used as a retail space.

Sections • Floor plans
Scale 1:400

1 Main entrance
2 Window dressing
3 Changing rooms
4 Staff
5 Other users
6 Void
7 Ladies' fashion
8 Storeroom
9 VIP zone
10 Restaurant
11 Kitchen
12 Cloakroom
13 Men's fashion
14 Underwear

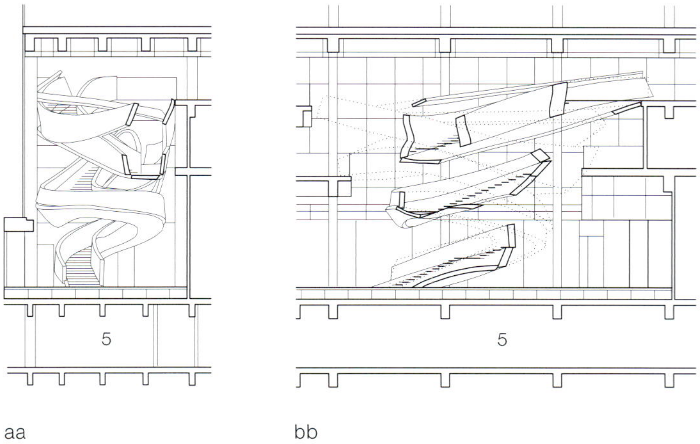

aa bb

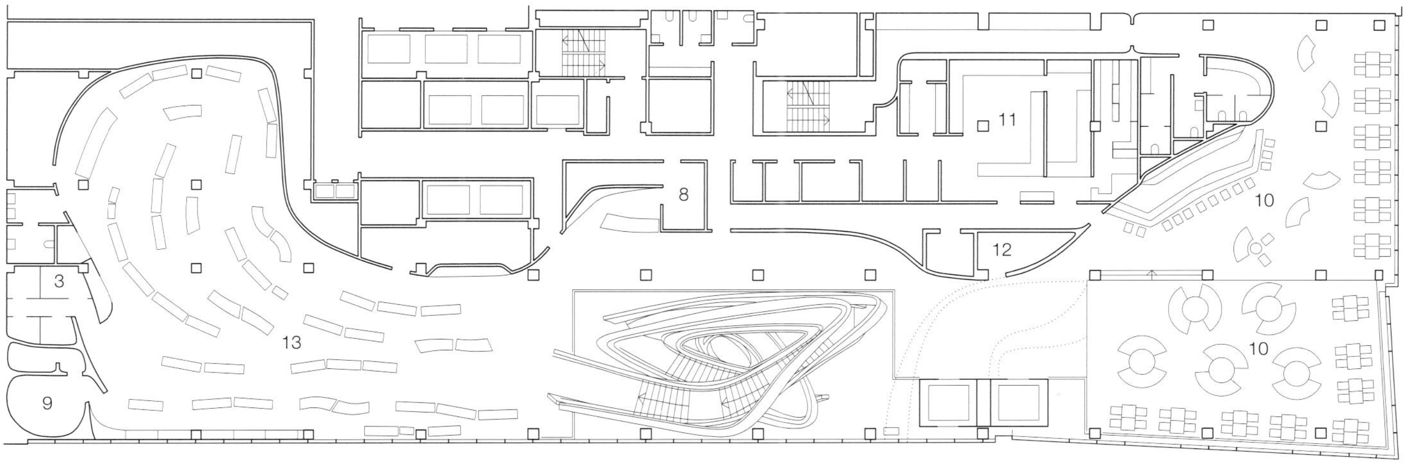

2nd floor

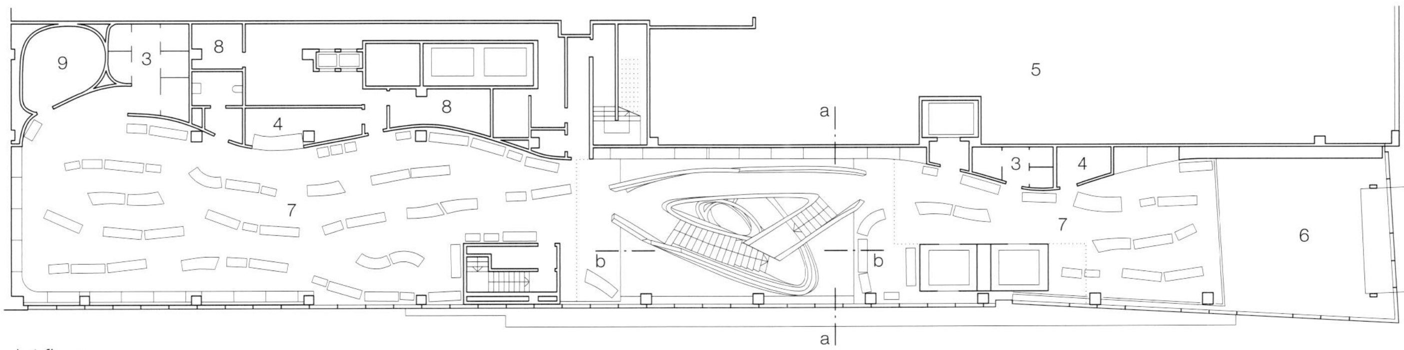

1st floor

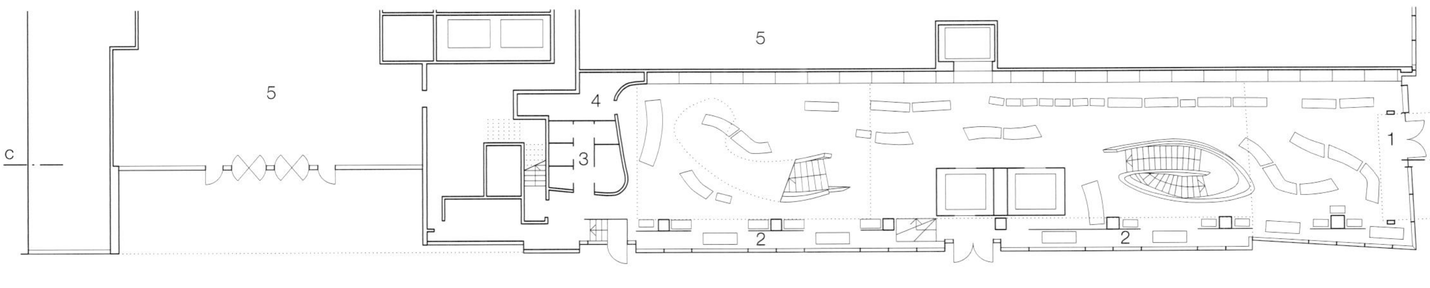

Ground floor

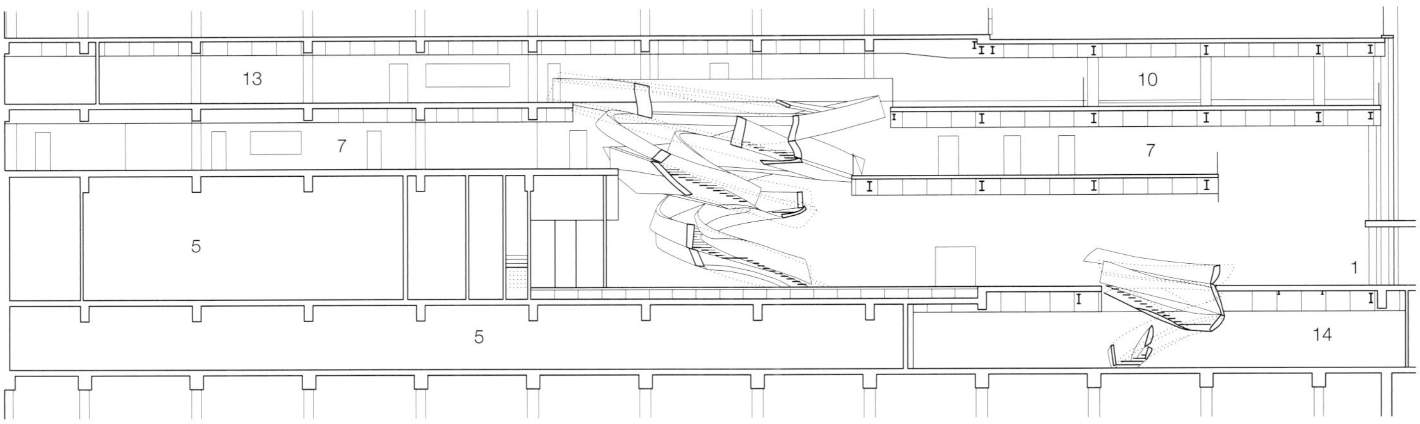

cc

Axonometric view of layered construction
Sectional details Scale 1:20

1 ∅ 168/8 mm tubular steel
2 6 mm sheet steel
3 20 mm glass fibre reinforced concrete, concrete with high-gloss white surface coating, aluminium grille, expanded metal mesh
4 ∅ 127/4 mm tubular steel
5 20 mm white cement modified with plastic coating, polished, 6 mm sheet steel

Project data:

Use:	Retail space with restaurant
Access:	Sculptural staircases, lifts
Number of floors:	3 + 1 basement level
Clear room height:	2.60 – 4.95 m
Gross floor area:	2,500 m^2
Year of completion:	2009
Construction period:	12 months

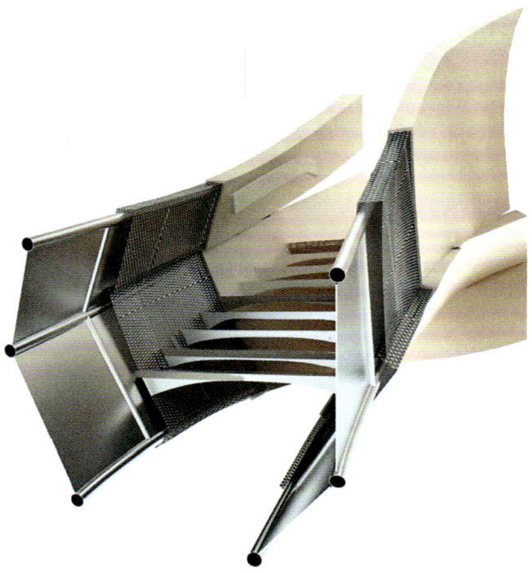

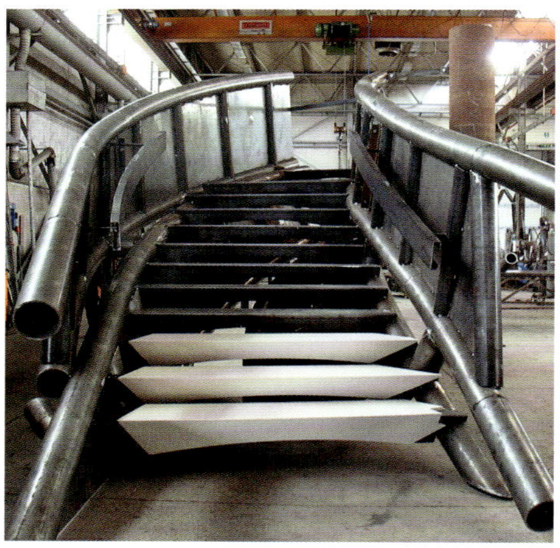

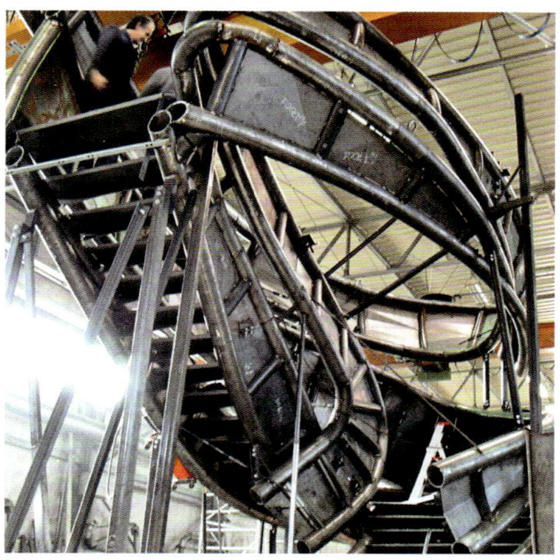

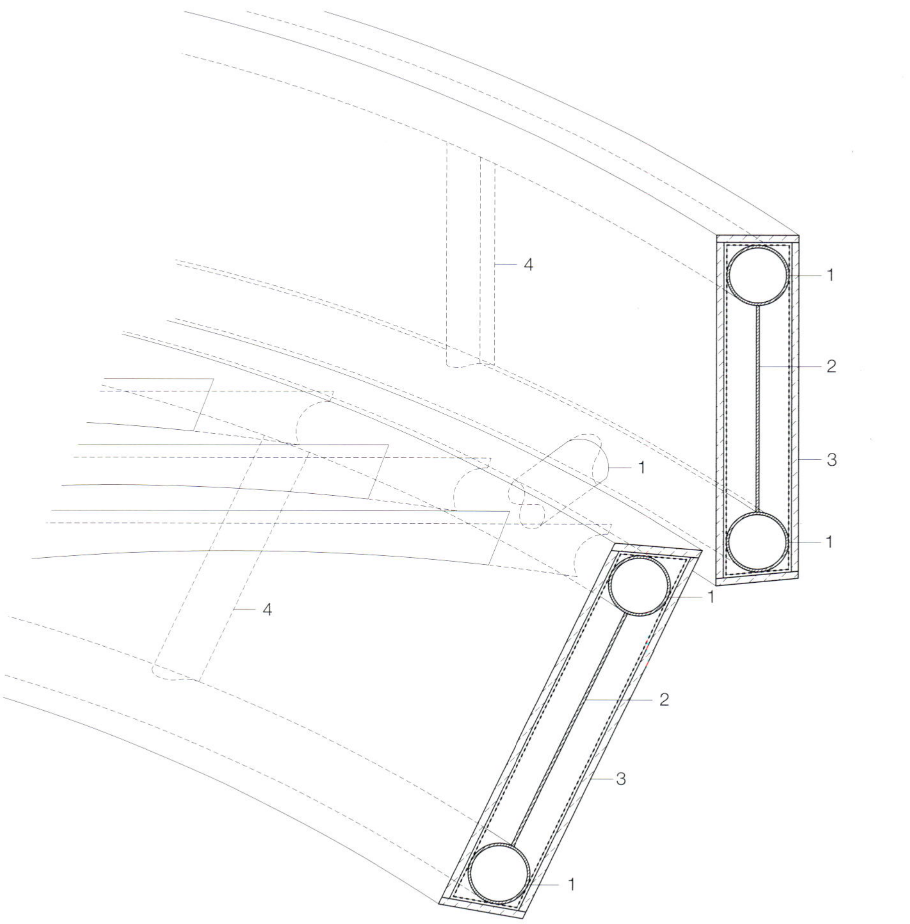

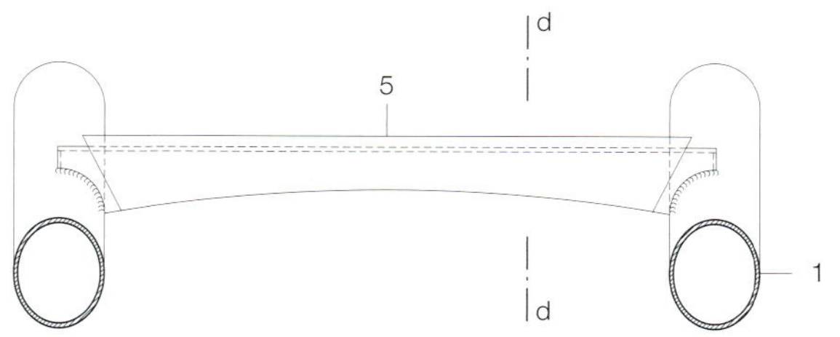

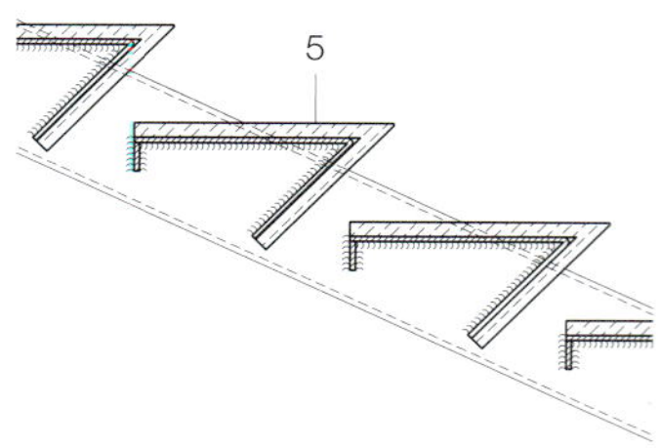

dd

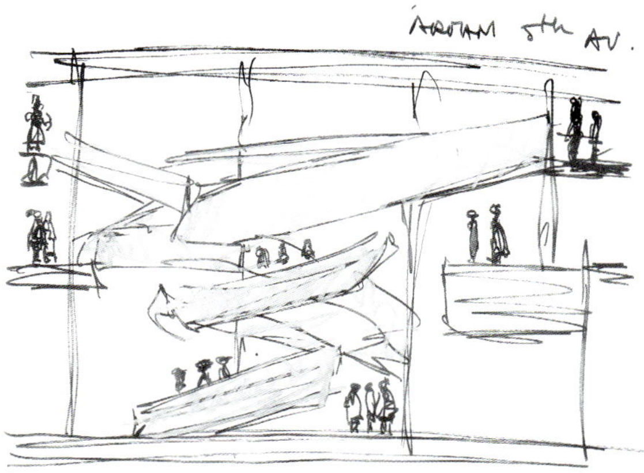
5th AV.

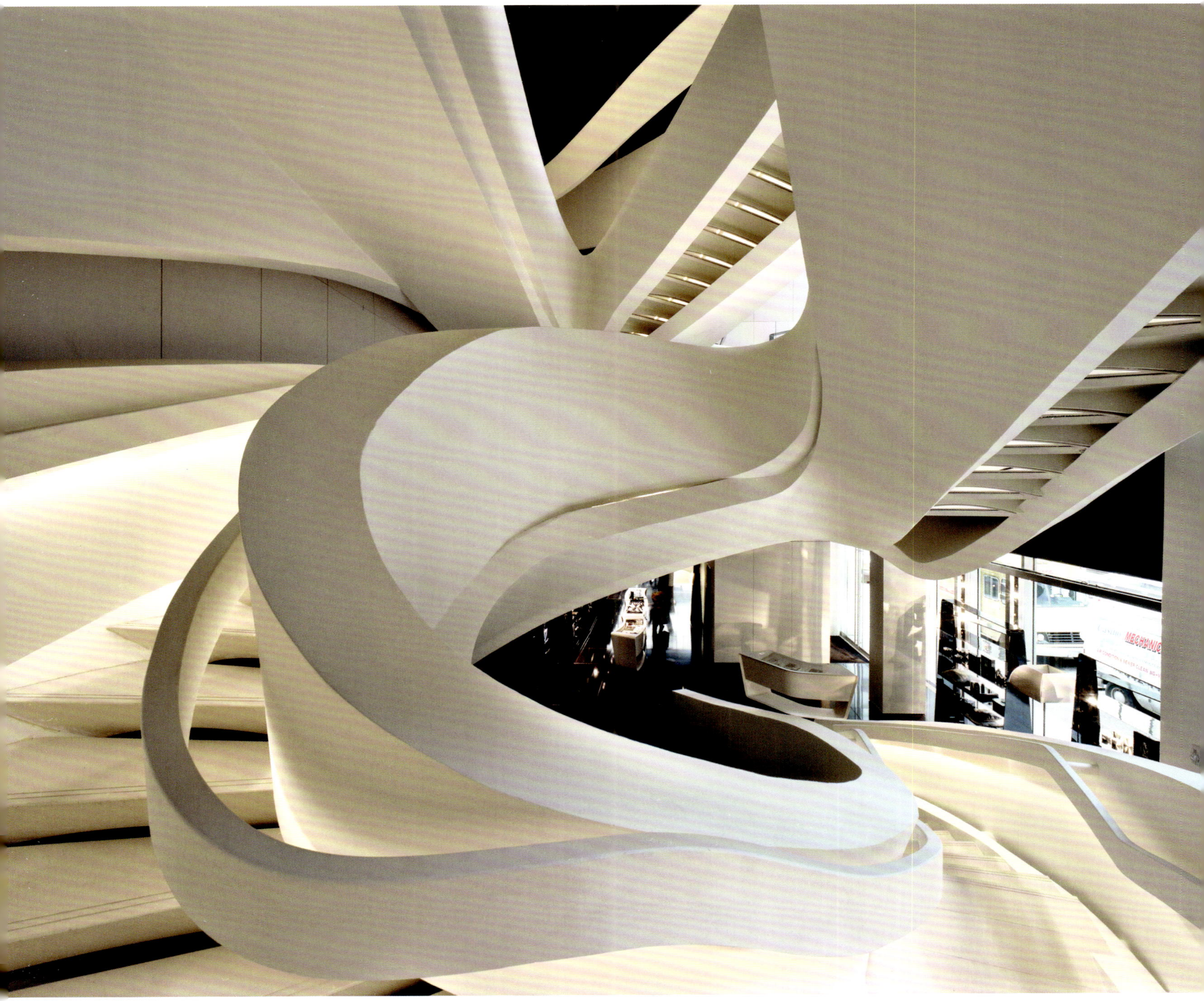

New York Times Building in New York

Architects: Renzo Piano Building Workshop, Paris;
FXFowle Architects, New York

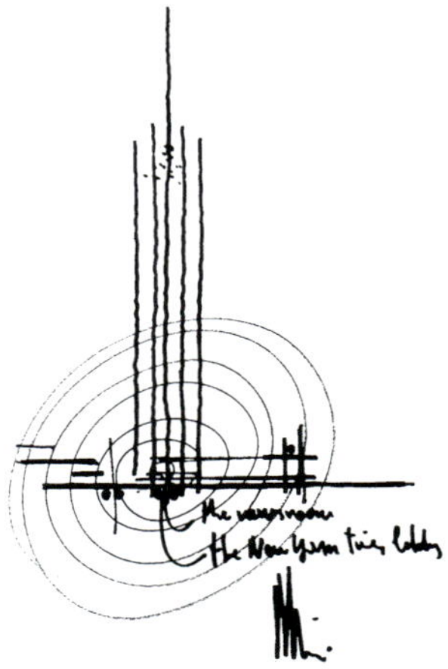

Even from a distance, the glowing orange entrance area of the New York Times Building demonstrates openness through transparency.

After nearly a century the New York Times, which was founded in 1851, has moved out of its high-rise yet conspicuously narrow publishing house on 43rd Street where it had been based since 1913. The editorial department and administration of what is probably America's most important daily newspaper are now housed in a modern skyscraper. In comparison with many other office towers, the new Times Tower is rather modest in appearance. What stands out is the meticulous attention to detail and exemplary quality of openness – something not to be taken for granted in what is Manhattan's first major high-rise building project to be completed in the wake of 11 September 2001. Nonetheless, following the attacks the decision was made to remain true to the spirit of the architectural competition held in 2000 rather than creating a bunker-like fortress.

Axes structure the open ground floor

It was important that the building should appear light and open on the ground floor at street level: this has been achieved by means of an inviting entrance zone. From Eighth Avenue, it is possible to look through between the lift shafts, along the main axis, through the lobby and garden into the auditorium. The ground floor also remains transparent between the two side streets, 40th and 41st Street. The idea of layering is evident here: various glass facades allow complete transparency.
For this reason, all fixtures and fittings in the stores in the shopping zone must remain below eye level. Also, at 6.50 m the ceiling level on the ground floor is relatively high by New York standards, which also contributes to the light and open appearance, yet without seeming arrogant or intimidating.
Fundamentally, the base structure is organised as follows: because two parties share the use of the building – the Times and the tenants in the upper half of the skyscraper – there are two lobbies: one accessed from Eighth Avenue and – because of a requirement imposed by the city authorities – one public connecting lobby between the side streets, which together create a T-shaped foyer. The approximately 9-metre-wide connecting passage running along the main axis between the lift shafts is flanked to the left and right by an art installation consisting of small screens presenting the latest news from the newsroom in a wide variety of forms.
In order to keep the central axis free, the building services installations and escape routes had to be located in the double wall of the lift shafts or routed around these.
The New York Times wanted to concentrate the core of the editorial department, the area containing several hundred journalists in which the newspaper is created, on a maximum of three floors. However, the levels in the tower are not large enough for this purpose. The architects therefore interpreted this newsroom as a three-storey structure surrounding a central, glass covered atrium on the ground floor.

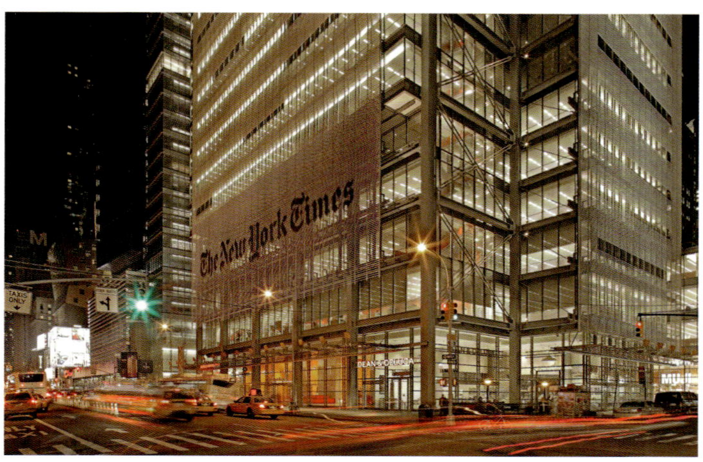

Connecting stairs ensure circulation

Another important element on the office floors of the Times are the connecting stairs made of steel, located on two corners directly behind the facade in order to improve circulation and communication between the different levels and reduce use of the lifts. Not only do they provide the staff with views and thus a certain connection with the surrounding city, they also communicate to viewers on the street something of the inner life of the newspaper. In order to comply with the strict New York fire regulations, these stairs are equipped with horizontal fire protection roller shutters on every second floor, allowing them to be divided into the required fire compartments.
A striking feature of the Times lobby is the wall finished in orange Venetian stucco, combined with shades of red and natural wood tones. This cheerful colour scheme is continued through all of the interiors, including the offices: cherrywood fixtures and red paintwork in the core as well as on the connecting stairs, with vivid red metal panels used in the balustrades. Prominently visible from outside, these add a colourful accent behind the relatively monochrome outer skin.

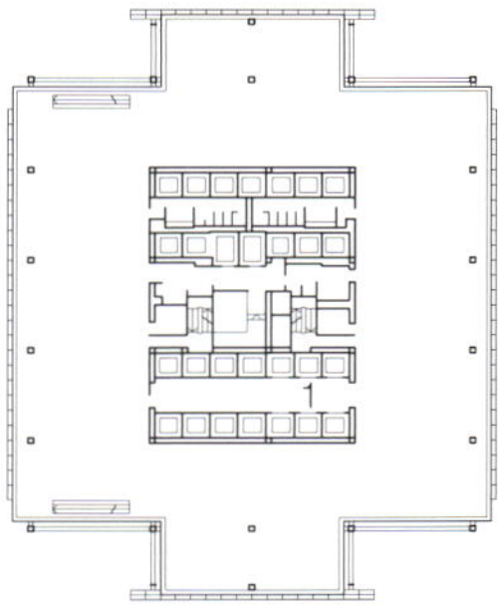

Standard floor New York Times
1.–26. level

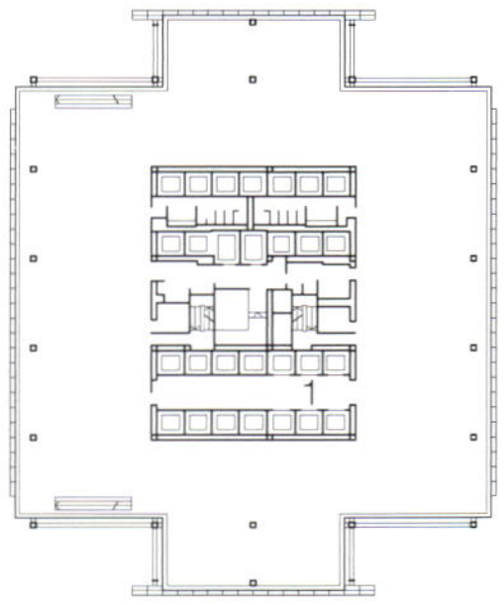

Standard floor tenants
middle levels

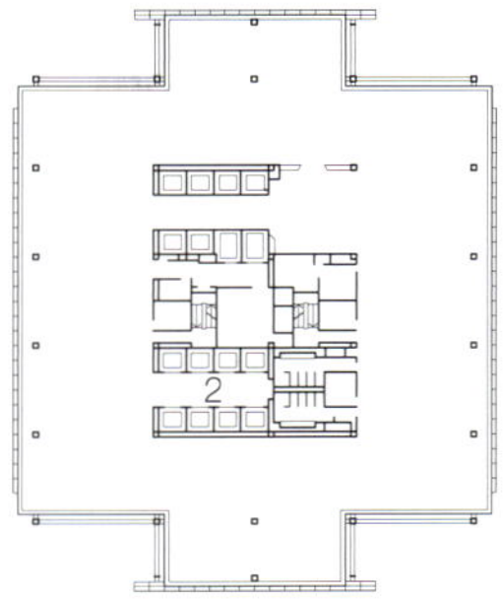

Standard floor tenants
upper levels

Floor plans • Section
Scale 1:1500

1 Lifts to NY Times
2 Tenants' lifts
3 Services storey
4 Cafeteria
5 Lobby
6 Garden
7 Auditorium
8 Newsroom

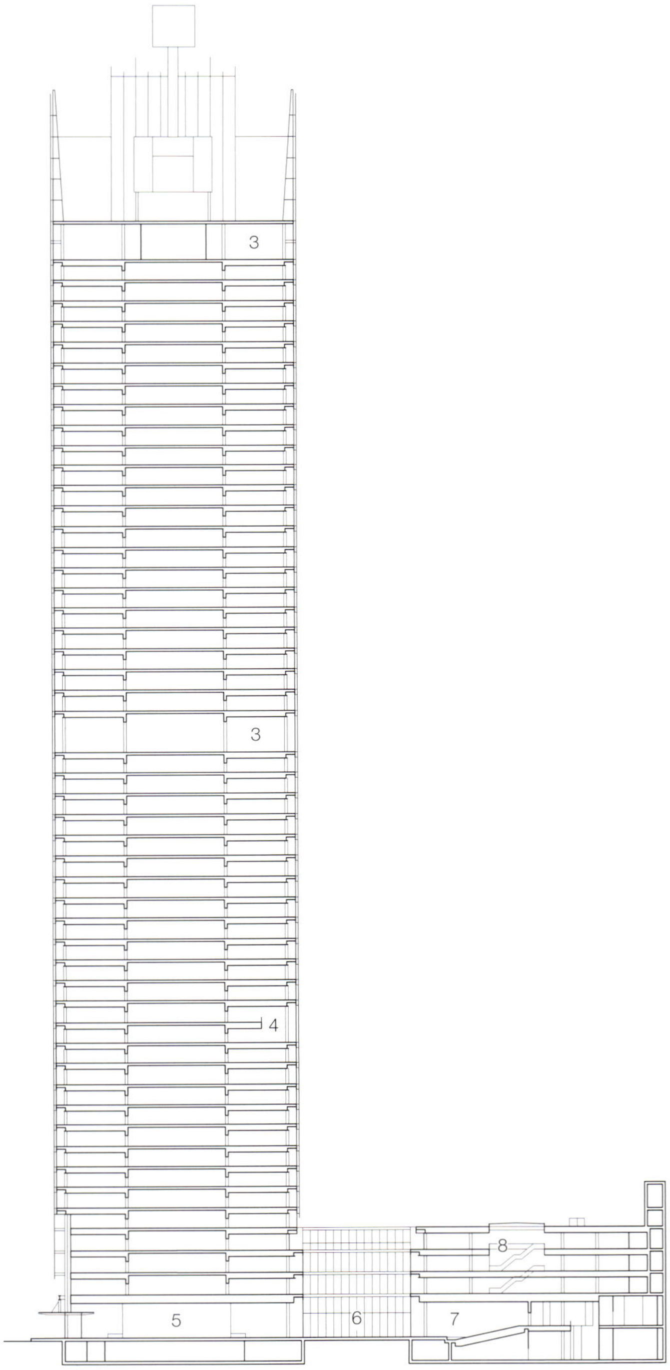

aa

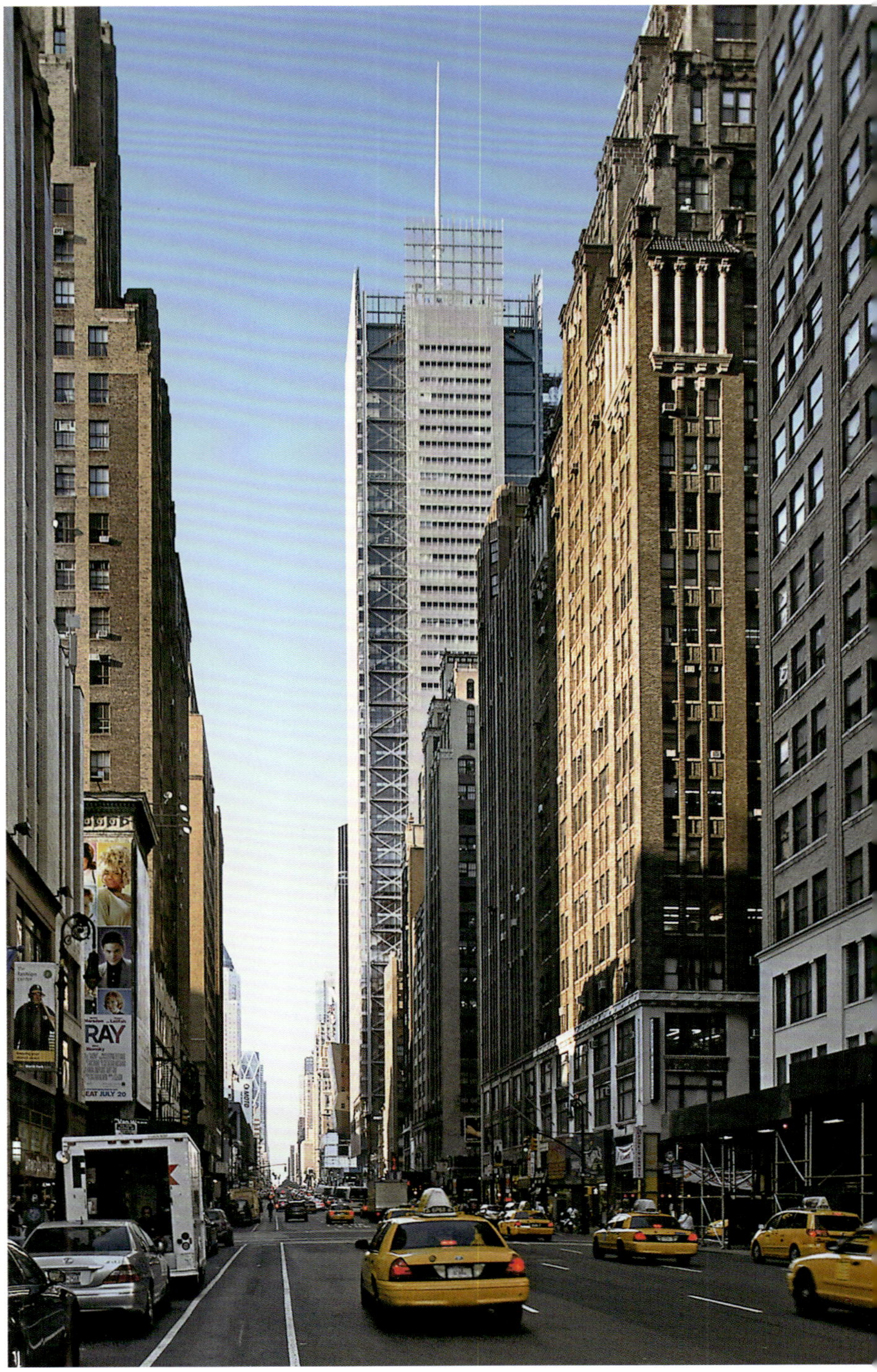

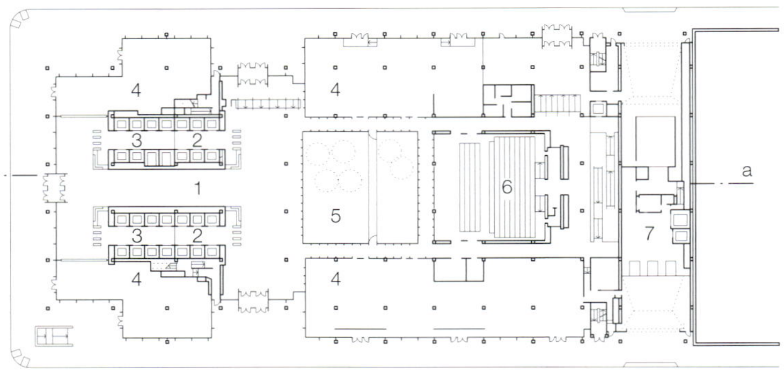

Floor plan
Scale 1:500
Site plan
Scale 1:10000

1 Lobby
2 Lifts to NY Times
3 Tenants' lifts
4 Shops
5 Garden
6 Auditorium
7 Loading zone
8 Hearst Tower
9 Empire State Building

Ground floor

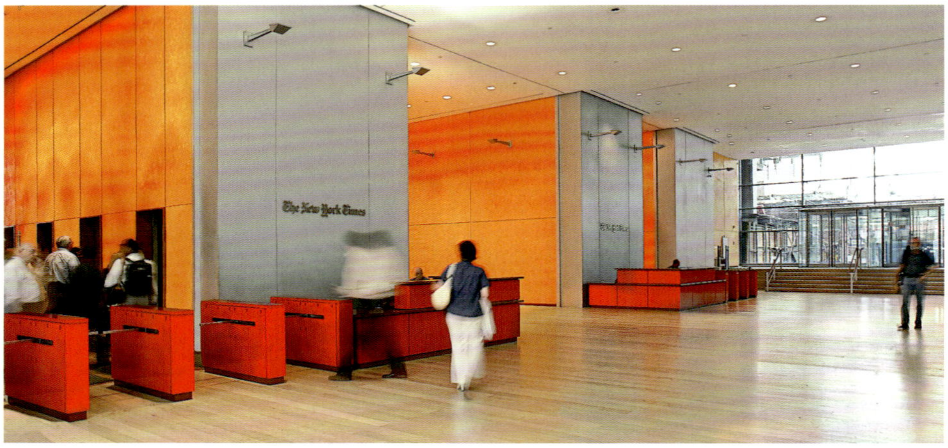

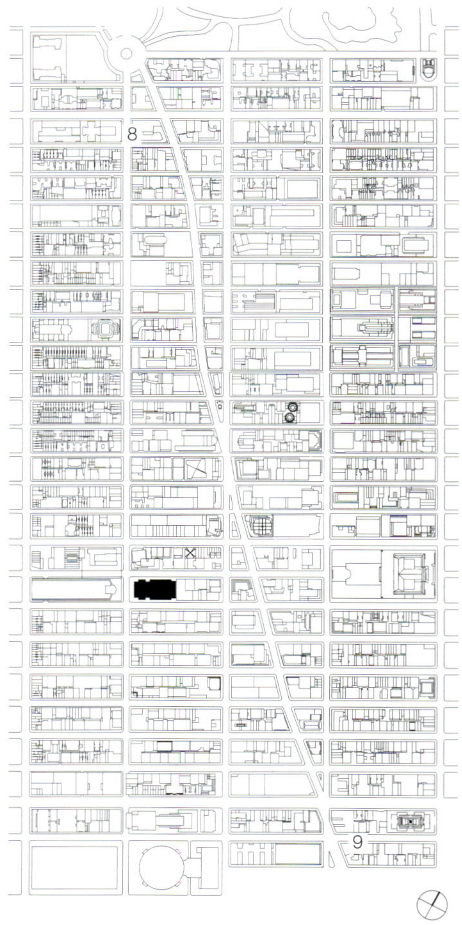

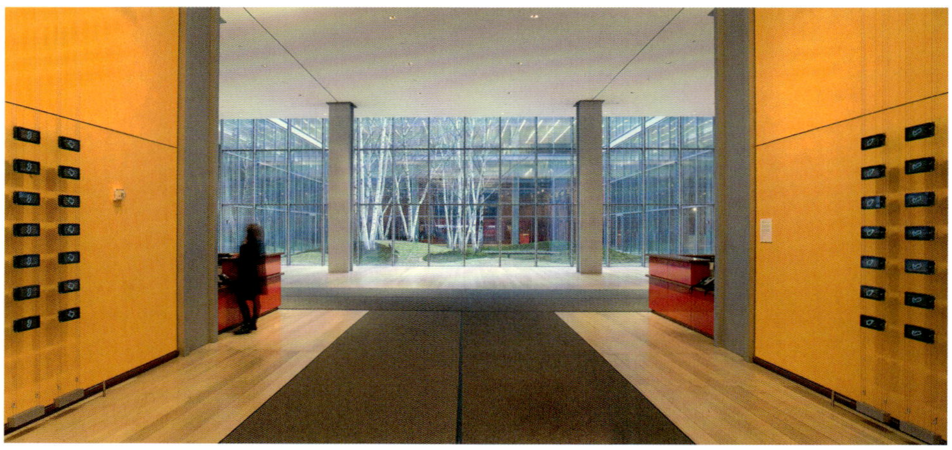

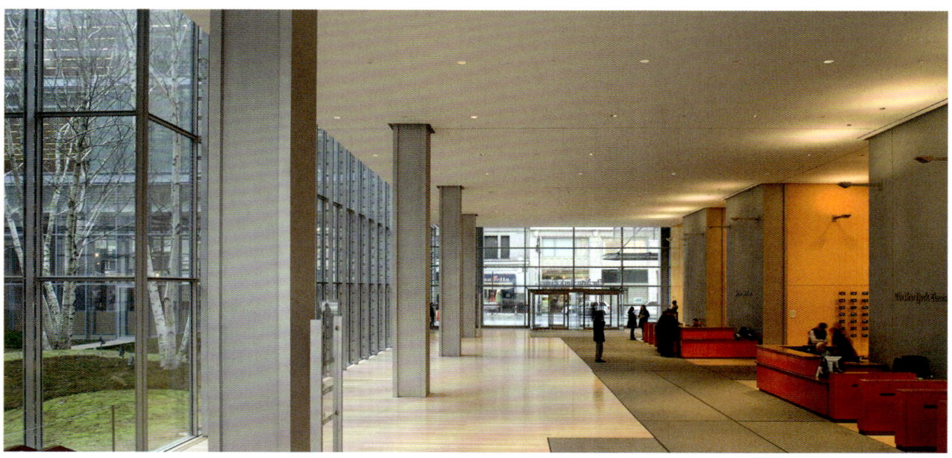

Project data:

Use:	Office/administration
Access:	T-shaped lobby, two lift shafts (12 lifts for New York Times, 16 lifts for tenants)
Number of floors:	52 + 1 basement level
Clear room height:	6.50 m (ground floor, services storey), 2.90 m (standard floor)
Gross floor area:	111,484 m^2
Year of completion:	2007
Construction period:	48 months

Vertical sections: staircase Scale 1:20

10 Double glazing: 6 mm float + 13.2 mm cavity + 6 mm toughened glass (lower storeys: laminated safety glass)
11 Double glazing, printed
12 Outlet vent for heated/cooled intake air
13 I-section ceiling beam with fire-protection cladding
14 Fire bulkhead
15 Automated interior sunblind
16 Balustrade panelling, red painted 6.4 mm sheet steel
17 Stair tread
9.5 mm carpeting, 12.7 mm plywood
6.4 mm elastic underlay
9.5 mm angled sheet steel
18 Cherrywood handrail
19 63.5/19 mm steel section

Floor plan 19th floor with fixtures and fittings
Scale 1:1000

The areas along the facade are organised as open-plan office spaces, individual offices are arranged around the core, with glass walls facing the open-plan area (interior fitting by Gensler according to guidelines by RPBW), connecting stairs are arranged on the corners along Eighth Avenue.

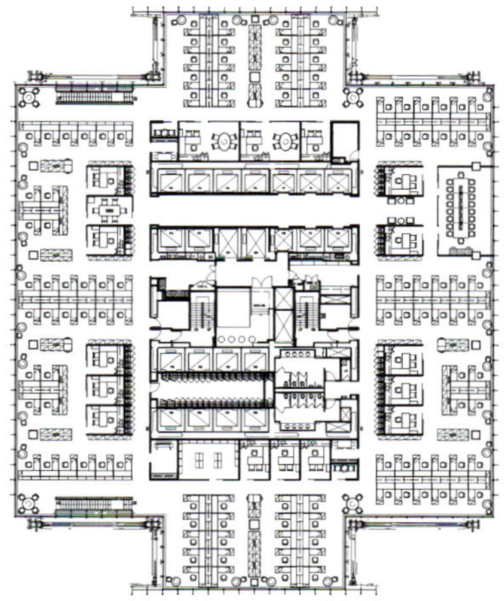

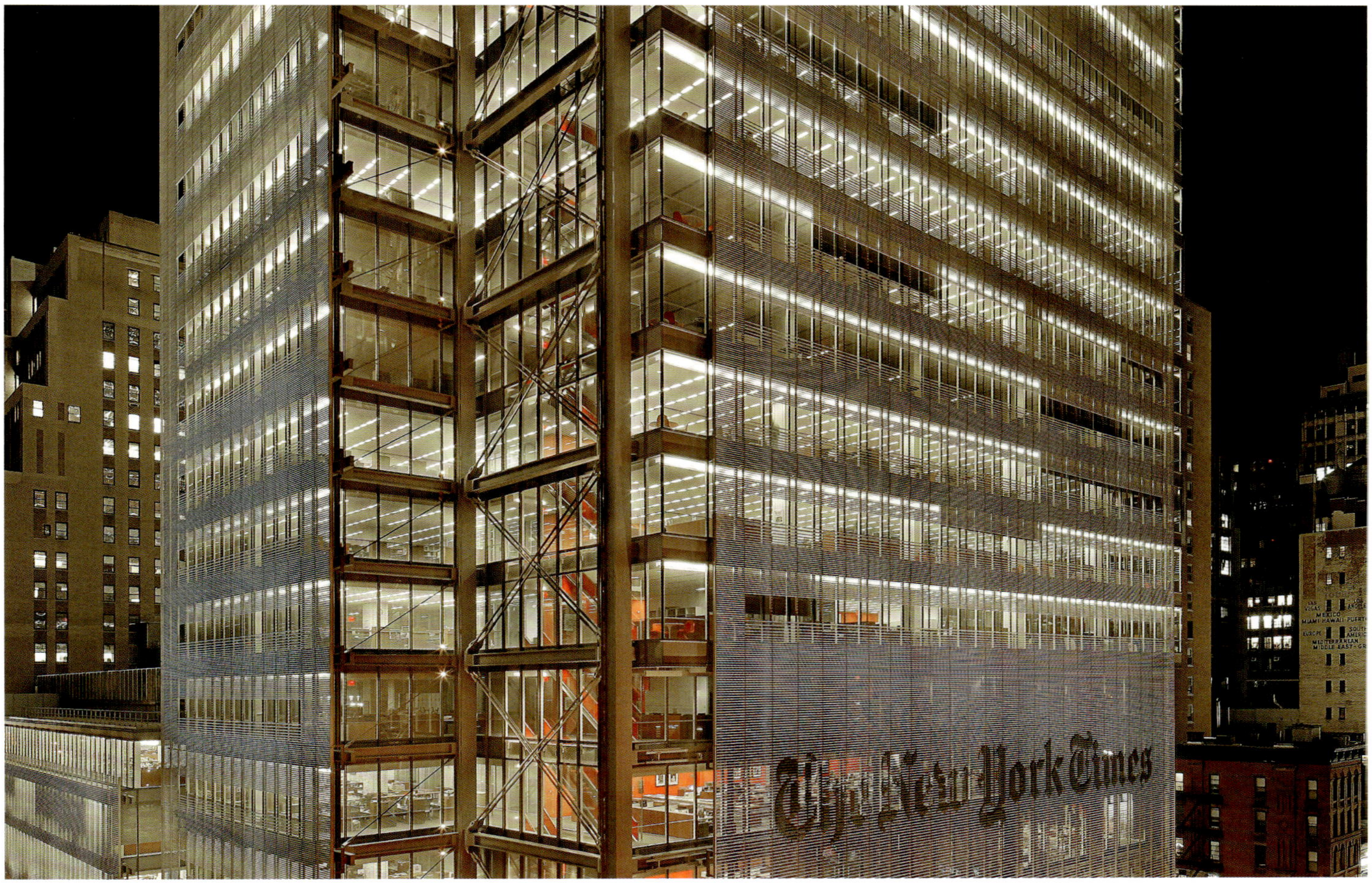

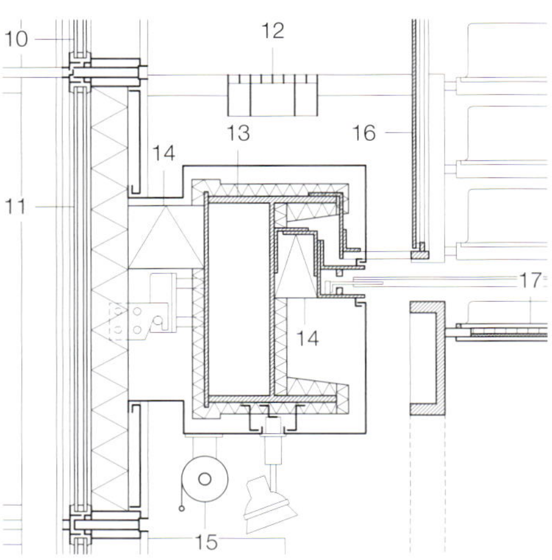

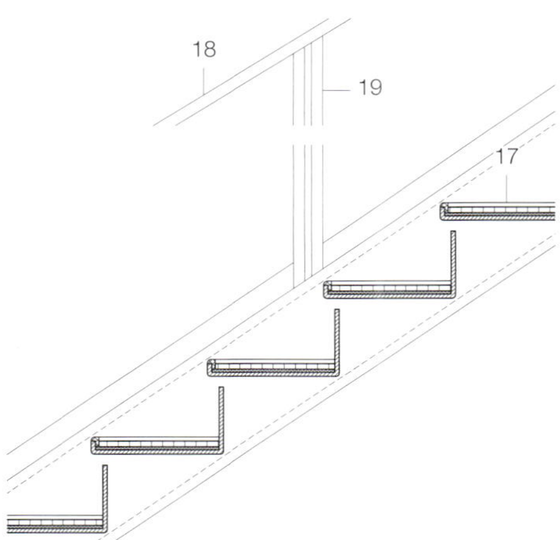

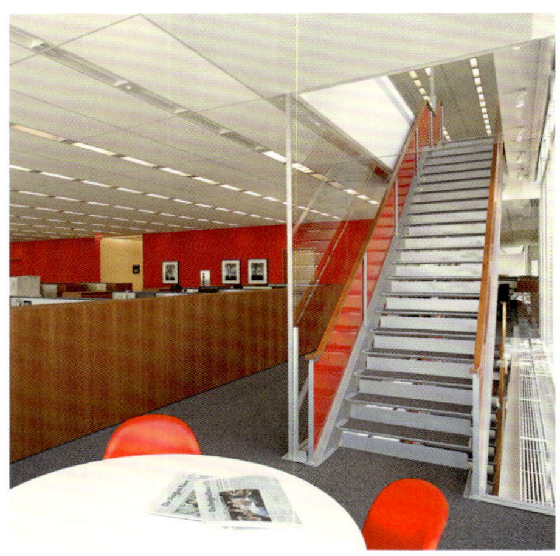

Office building in Sydney

Architects: ingenhoven architects, Düsseldorf;
Architectus, Sydney

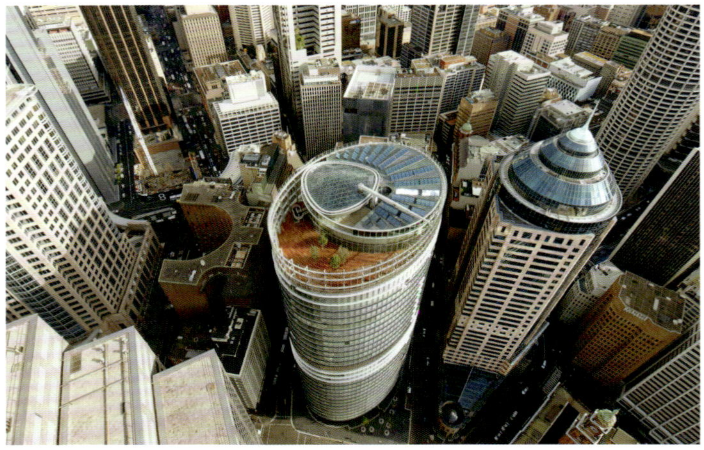

Glass panoramic lifts transform the journey through the building-height atrium into a spatial experience.

The 135-metre-high new office building "1 Bligh" in Sydney's Central Business District, situated around 500 m away from the southern shore of the harbour, forms a kind of hinge in the urban landscape. It is surrounded on all sides by much higher buildings and yet enjoys an unobstructed view in the direction of Sydney Cove and Sydney Harbour Bridge, where generally more low-rise, historical buildings predominate. The elliptical floor plan is angled slightly around the longitudinal axis in relation to the street grid. This creates open spaces around the building – the emphasis here was not on maximum exploitation of the site but on creating added value. A flight of steps in front of the building follows the slope of the terrain and extends Farrer Place, which lies opposite, into the building. It is a popular seating area with passers-by and office workers. Because the building is elevated, the stairs are shaded in summer and pleasantly sunny in winter. On the base level next to the steps is a nursery; below this, four further levels provide parking for cars as well as bike parking, changing rooms and showers for cyclists.

Different lift groups

The publicly accessible ground floor can be reached from Farrer Place via the steps; from Bligh Street it is accessed on ground level via a public entrance area with a planted wall and a kiosk. The glazed facade of the ground floor level is set back in relation to the upper floors. The elements of the facade can be almost completely opened by means of large hydraulically-driven glass gates, drawing the life of the city into the interior of the building. The 13-metre-high foyer contains only a café, a reception desk, a waiting and lounge zone as well as the lift lobby – the only area with access controls.
A central element of the high-rise building is the atrium, which rises the entire height of the building, and makes the journey to the office floors in the glass-walled lift into a spatial experience. The lifts are divided into two groups: the low-rise group, with a speed of 3.5 m/s, serves the floors from the ground up to the 15th, whereas the lifts of the high-rise group travel at 6 m/s, serving floors 15 to 28. Each group of lifts consist of four glass-walled lift cars which circulate inside the atrium and three lifts which run inside enclosed shafts. The drive mechanism and all the mechanical parts of the panoramic lifts remain visible through their glass enclosures.

The 15th floor serves as a transfer point between the two groups of lifts and also contains the reception area of the main tenant, a large law firm. There is also a peripheral exterior terrace created by the recessed facade. The roof terrace on the 28th floor provides a second open space which is not open to the public.

Light and air

In order also to offer the south-facing office spaces an attractive view, rather than creating a central access core the architects located the vertical access routes and usable floor areas on the south-facing facade of the building. This means that both the emergency stairs and the WCs enjoy the unaccustomed luxury of natural lighting and ventilation. The emergency stairwells can also be used within the leased areas as a vertical connection between the levels. Internal connecting stairs located facing the atrium were only installed at one point, between floors 14 and 15, at the request of the tenant. The building-high atrium ensures that even the very deep office spaces on the north side can be lit on both sides; this also assists the natural ventilation of the offices. The 1,630 m² office spaces are only divided by 12 columns, which leaves 1,000 m² completely free of columns, allowing a high degree of flexibility in terms of the way the space is divided. The openness of the building is also intended to encourage better interaction and communication between the users.

Project data:

Use:	Offices
Access:	Lifts in building-high atrium
Number of floors:	28 + 1 base level + 4 basement levels
Clear room height:	2.70 m (standard floor)
Gross floor area:	45,000 m²
Year of completion:	2011
Construction period:	30 months

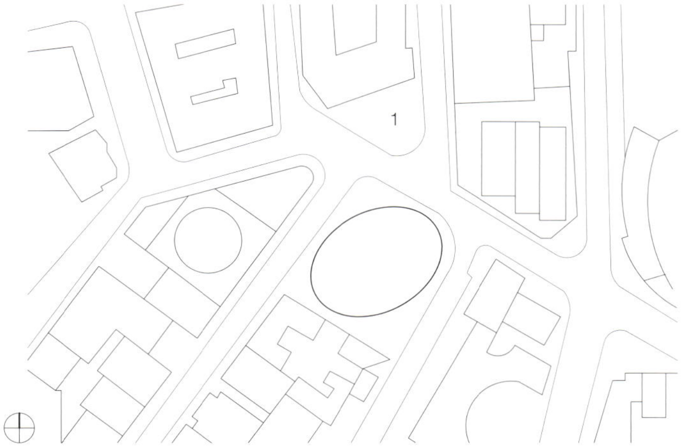

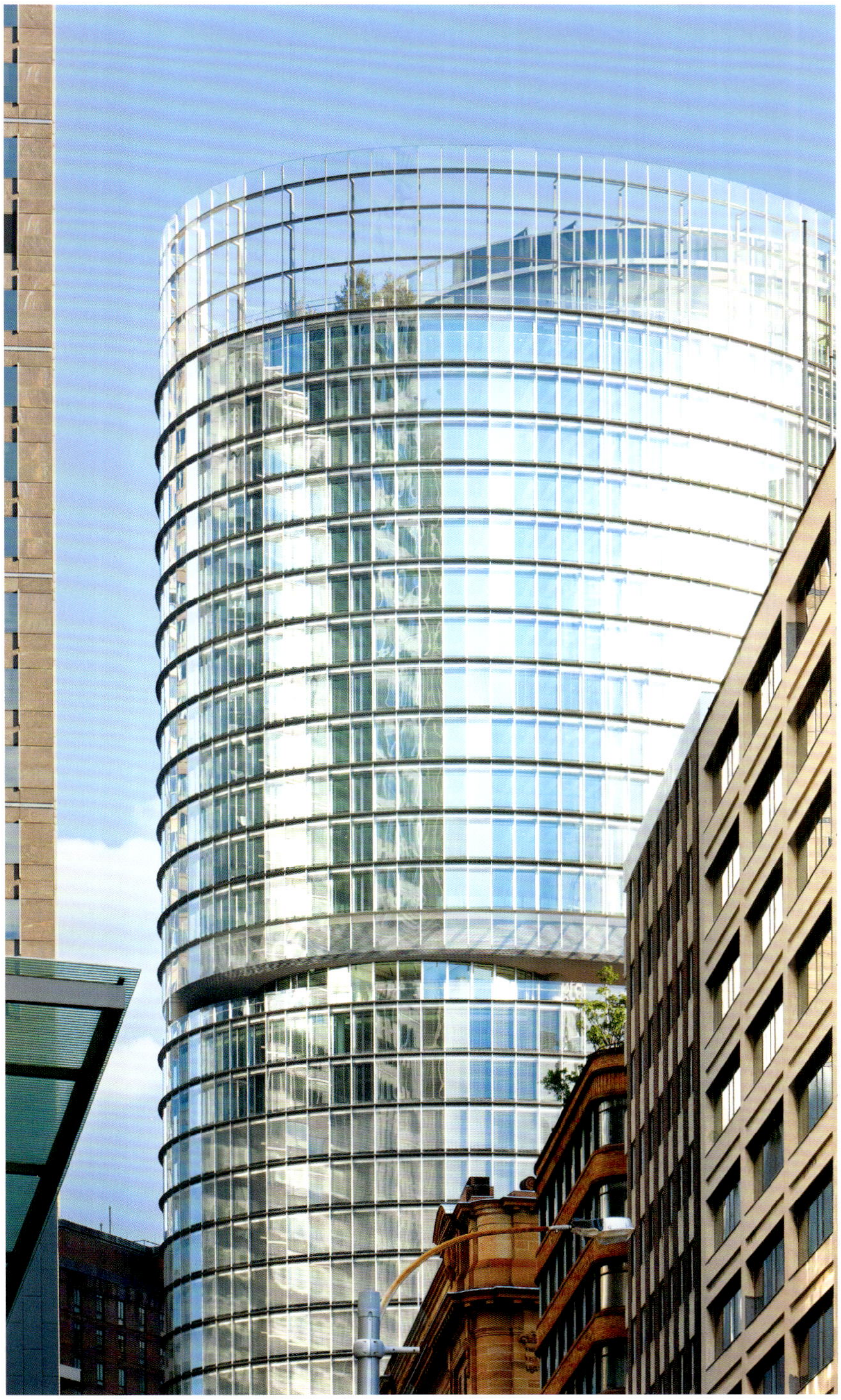

Section Scale 1:1000
Site plan Scale 1:3000

1 Farrer Place

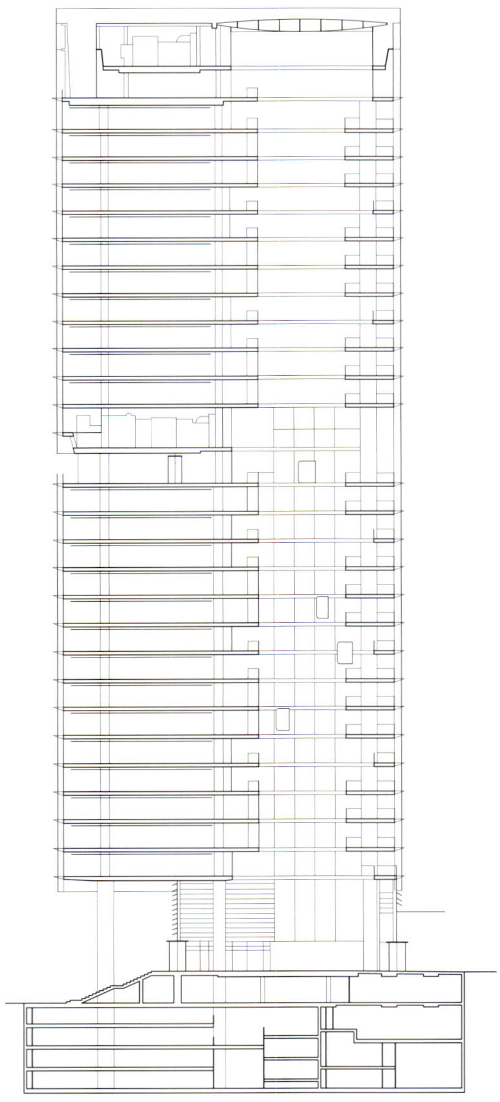

aa

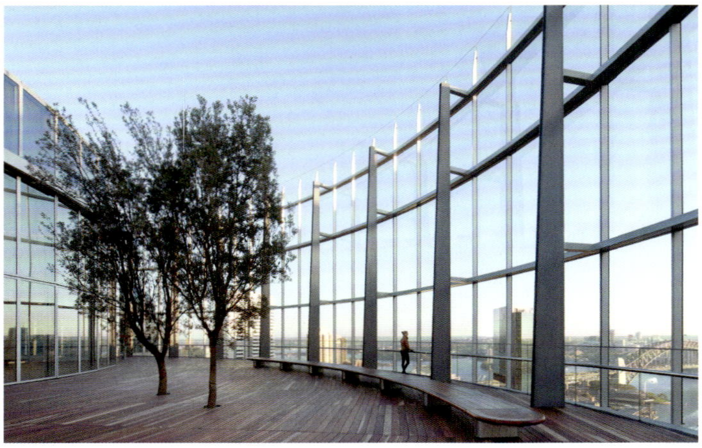

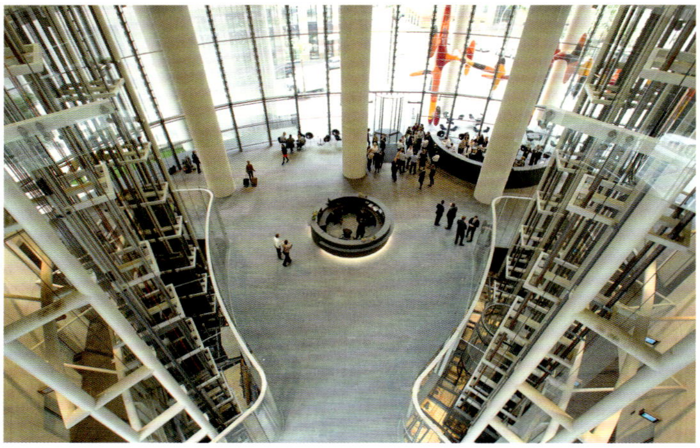

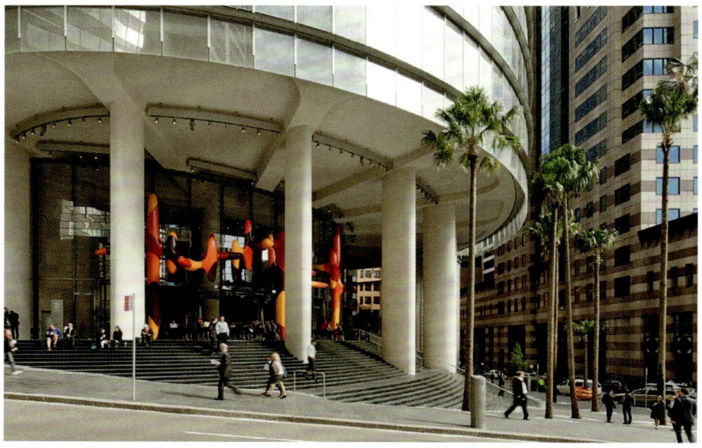

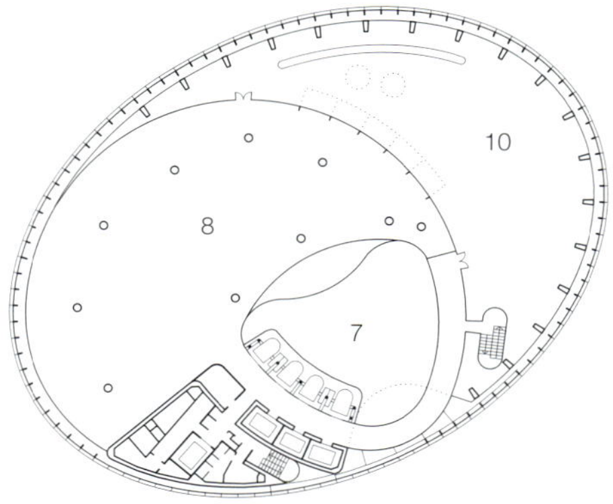

Top floor

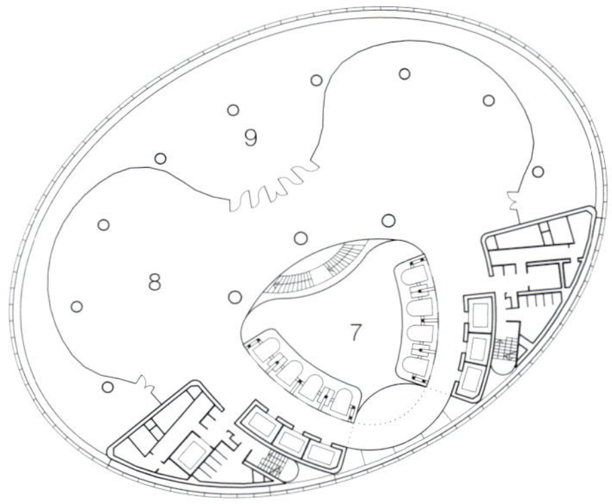

15th floor

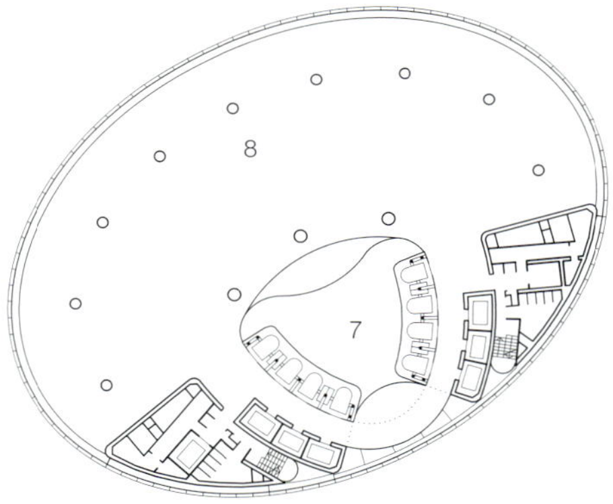

Standard floor

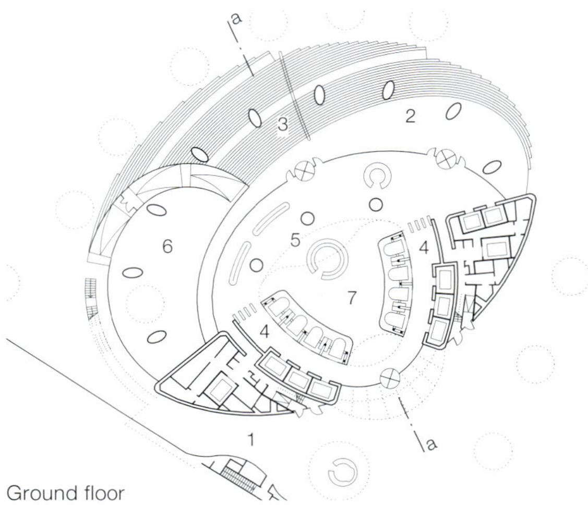

Ground floor

Floor plans
Scale 1:1000

1 Plaza with kiosk and green wall
2 Terrace café
3 Public steps
4 Lift lobby
5 Foyer with reception desk
6 Outdoor area of nursery
7 Atrium
8 Offices
9 Terrace
10 Roof terrace

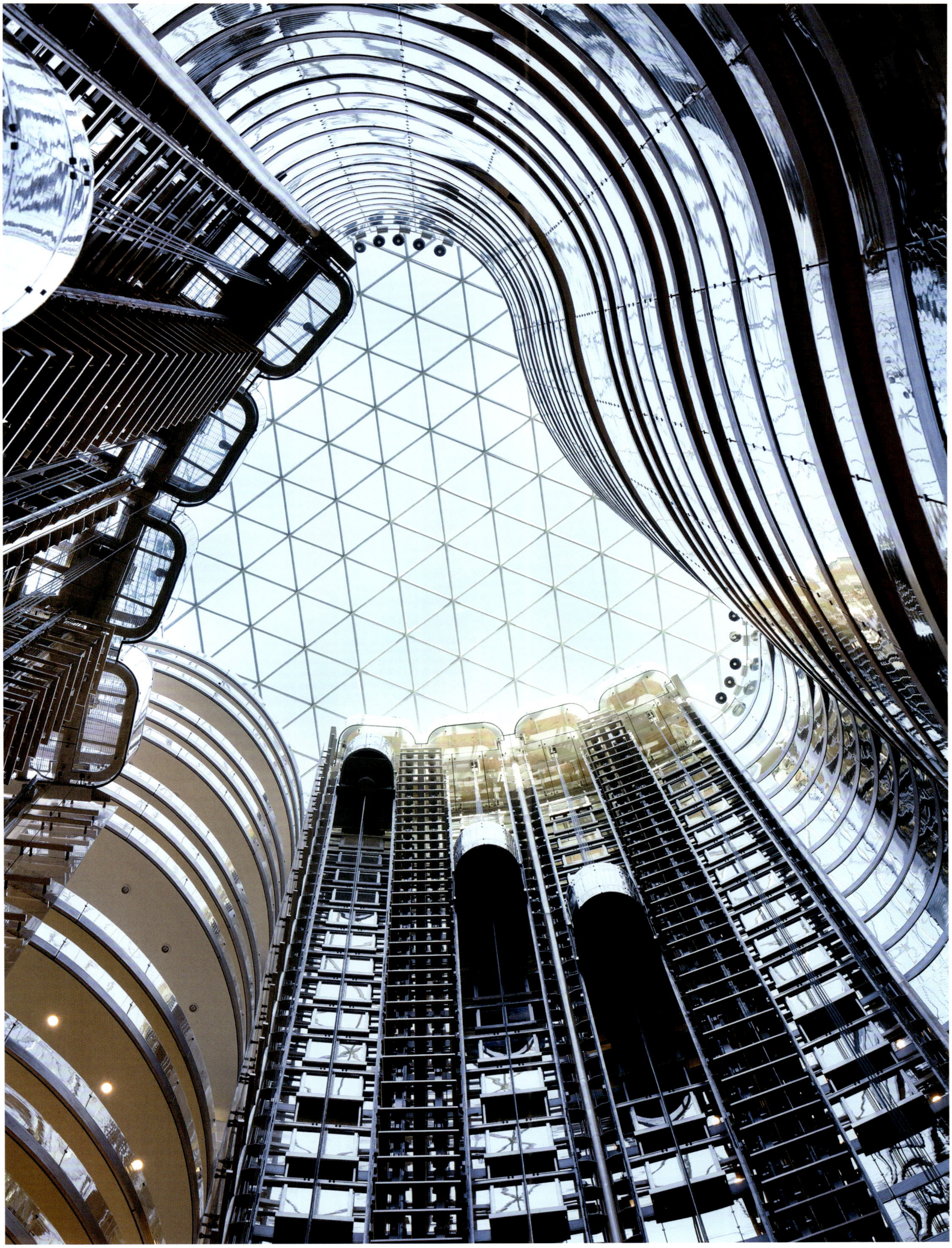

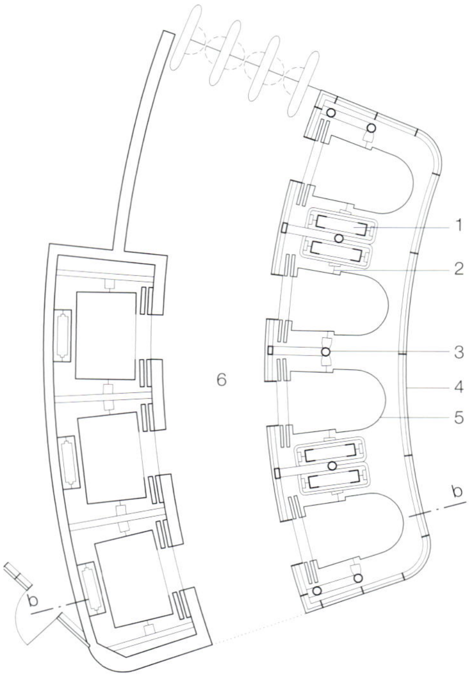

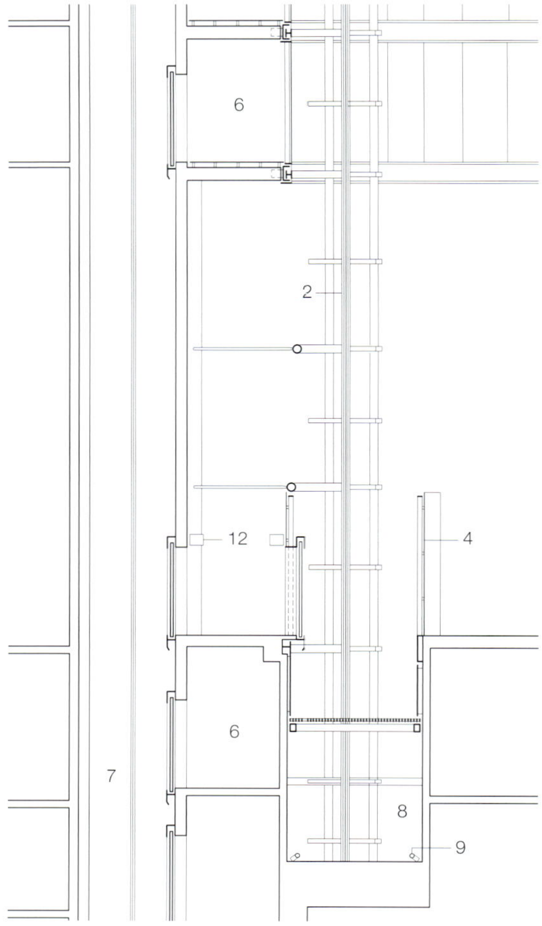

Lift lobby
Floor plan • Section
Scale 1:200

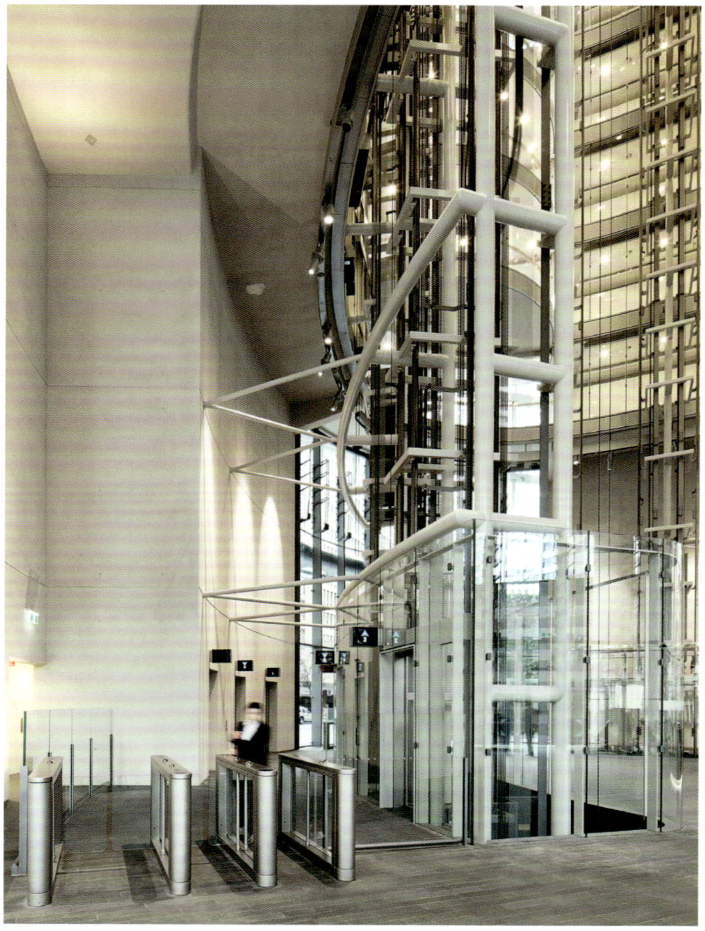

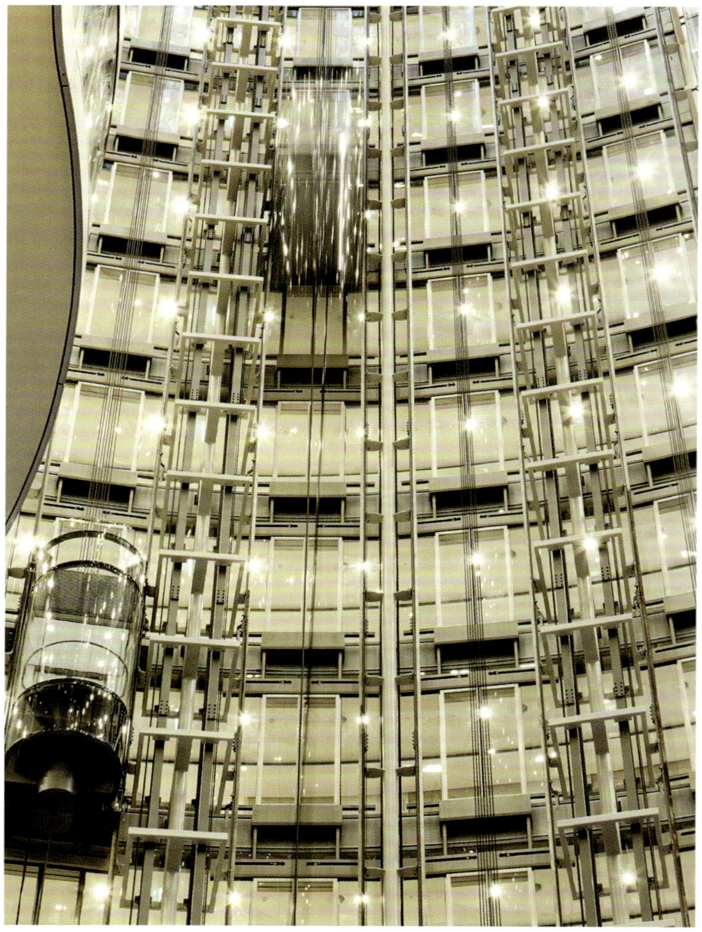

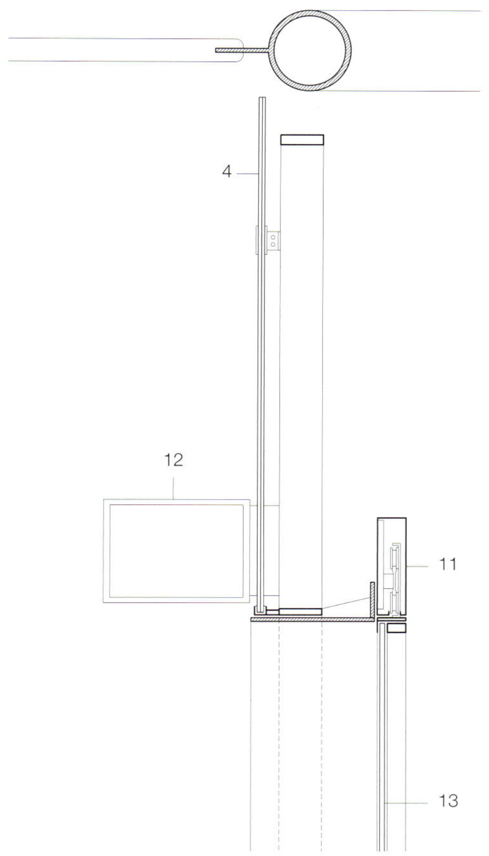

Vertical section of lift lobby/lift shaft
Scale 1:20

1 Counterweight with powder-coated perforated sheet metal covering
2 Guide rails for lift
3 219 mm tubular steel, coated
4 Laminated safety glass wall consisting of 8 mm toughened glass + 1.5 mm PVB + 8 mm toughened glass
Support profile 20/130 mm aluminium, powder-coated
Glass holder 83/50 mm aluminium, powder-coated
5 Panoramic section of lift car glazed with 12.7 mm laminated safety glass
6 Lift lobby
7 Lift shaft
8 Lift underpass
9 Lighting in underpass
10 30 mm granite, Austral Black, R11 slip resistance
70 mm screed with underfloor heating
11 Drive and mechanism for lift door, covering sheet steel, coated
12 Monitor of lift indicator
13 Lift doors of 12.7 mm laminated safety glass, door frames stainless steel, glass bead blasted
14 Acoustic panel with bent aluminium, powder-coated, perforated
15 4 mm composite aluminium panel on steel supporting structure
16 Steel grille, powder-coated, on steel supporting structure

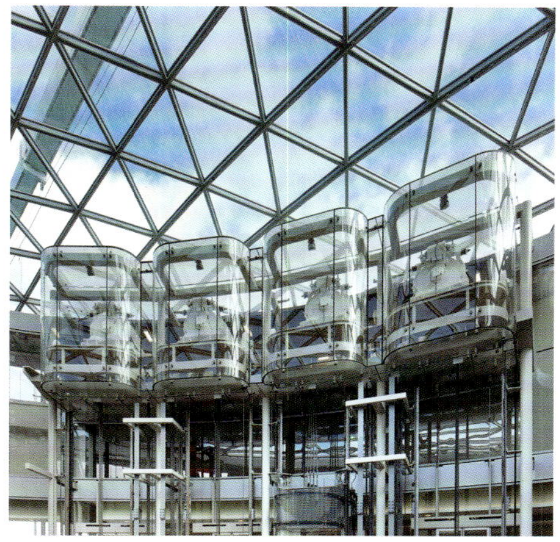

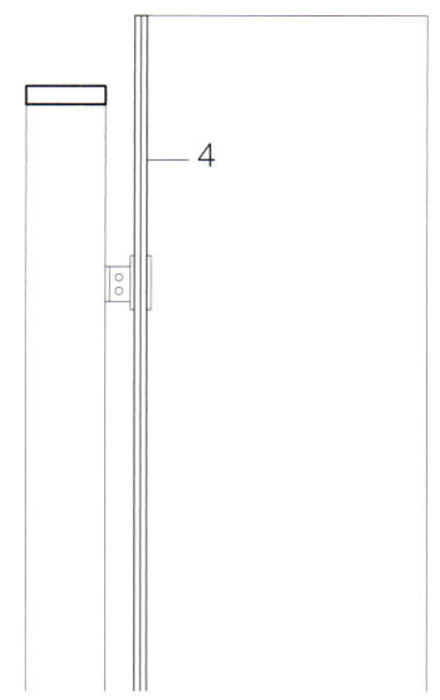

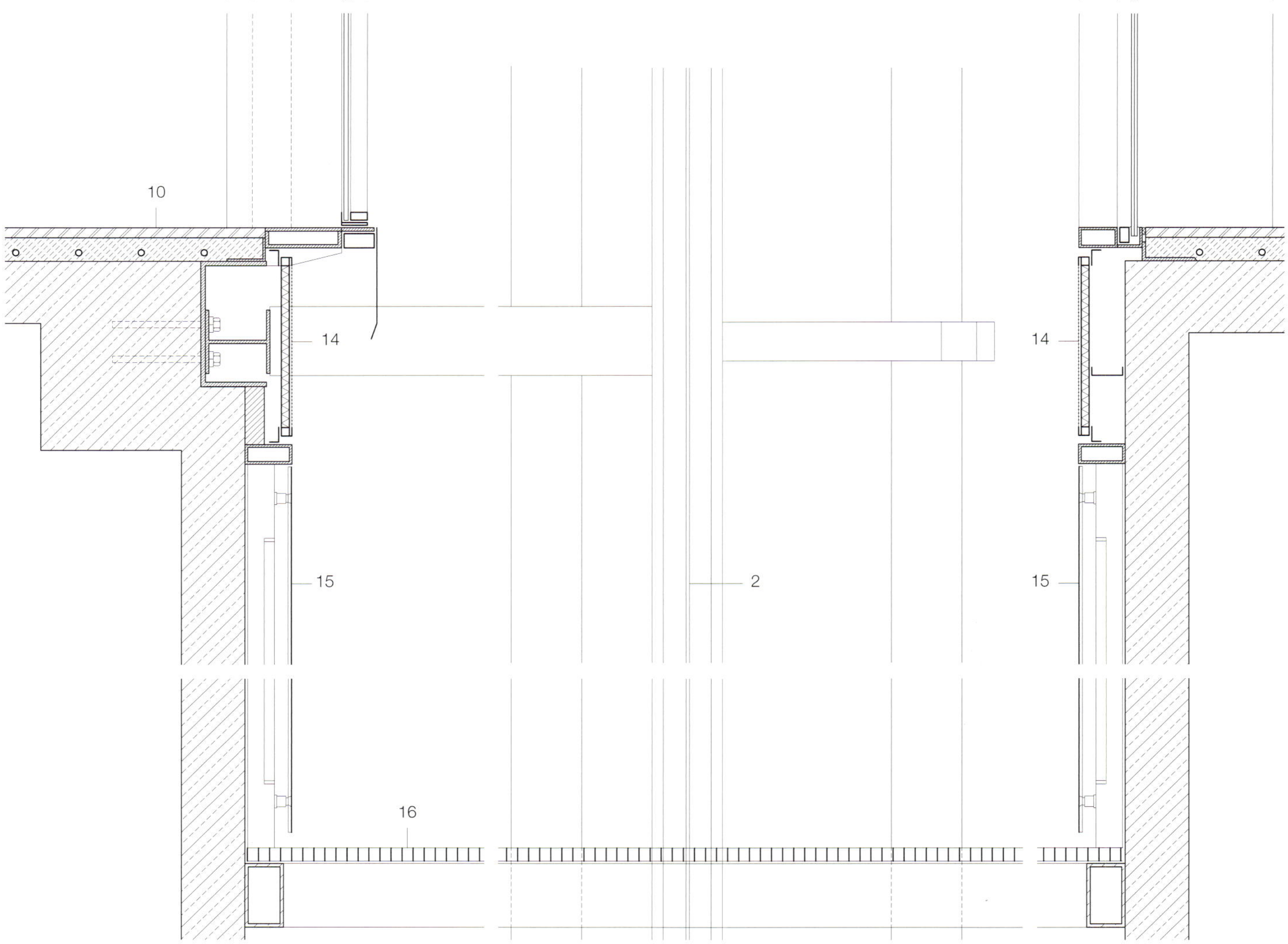

Stadium in Kiev

Architects: gmp • Architekten von Gerkan, Marg und Partner, Hamburg

Despite the constricted urban location and only two main entrances, it was essential to comply with modern safety requirements in terms of access.

In preparation for the European Football Championship in 2012, the Olympic Stadium in Kiev was extensively renovated and expanded in cooperation with local partners. The sports stadium had already been used as a venue for the football tournament during the Olympic Games in Moscow in 1980. Essentially, the finely structured, cantilevered, prestressed concrete upper tier of the stands was retained from the previous structure, which dated from 1968. The supporting structure of the new 43,000 m^2 membrane roof with its translucent ETFE light domes was detached from the existing bowl structure. Because of this transparency, the stadium radiates an impressive glow into the night sky when in operation. A star pattern is created through the superimposed double membrane which follows the lines of force in the stretched fabric. Also, the base structure and the reception building dating from 1948 were retained or reconstructed in modified form.

Access and entry control

Because of the inner-city location of the stadium and the resulting constricted spatial circumstances, coming up with access and evacuation measures which complied with modern safety regulations represented a challenge. Today, the stadium contains over 68,000 seats, 2,300 VIP seats and 40 VIP boxes. At a sporting event like a European Championship game, entry and access to the stadium are governed by strict UEFA regulations. The crowds of spectators have to pass through three security perimeters: the outer ring, around 100 metres before the turnstiles, does not require any structural measures and is formed by a chain of security personnel who carry out a visual control and check tickets. The second consists of the 2.50-metre-high fencing around the stadium site with turnstiles and gates at which the visitors pass through the actual security check and the electronic ticket check. Essentially, the stadium can only be entered from two sides: via Troitzka Square to the west (dimensioned for 42,500 visitors with 70 turnstiles) and from the north via the Palace of Sport (for 22,400 visitors with 32 turnstiles). The number of turnstiles was calculated based on turnstile capacity of 700 persons per hour and the time of two hours stipulated by UEFA during which all spectators must pass through the turnstile system. The third control takes place in the stadium, immediately before one is guided to one's seat by the stewards. In addition to the turnstiles, all the stadium's access points also include disabled-friendly entry controls and an accessible route into the stadium. In the VIP section this is made possible with the aid of appropriately dimensioned lifts running from the underground car park to the boxes. The seating is coloured in different tones of blue and yellow, the national colours of Ukraine, which are distributed over the tiers seat by seat like a pixel image. This makes even areas of empty seats look lively. As the fans stream into the stadium, this picture changes continuously as new colours are added.

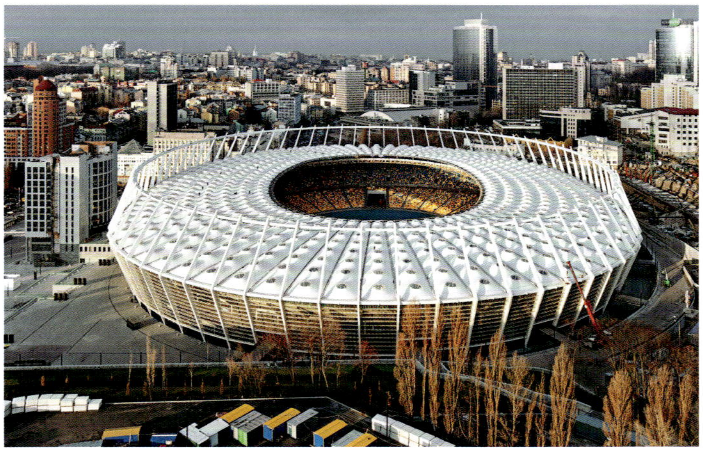

Evacuation and emergency stairs

Evacuation routes and assembly points on and in front of the site must be wide enough in order to allow the flows of spectators to leave the stadium and subsequently the site in a safe and orderly manner in 8 and 16 minutes respectively. A distinction is made here between the evacuation time taken for all spectators to move from the inside of the stadium into a first evacuation zone, a safe place of assembly in which there is no immediate danger from falling parts of the structure – in Kiev this means outside the perimeter line of the outermost edge of the roof – and the time taken for the spectators to reach the second zone, an adequately large area in front of the stadium in which emergency services can provide first aid. A final review was carried out with the aid of a 4-D simulation, following which access stairs were widened at critical points. All outdoor emergency stairs were given a standardised rise of 15 cm. The local building approval procedure required an air lock and positive pressure ventilation in interior emergency stairwells.

Section • Floor plan showing entry controls and spectator flows at a sporting event
Scale 1:3000

1 Accessible main entrance
2 Entry for VIP cars and team busses
3 Additional catering facilities
4 Palace of Sport
5 Pitch
6 Parking
7 Entry for OB vans

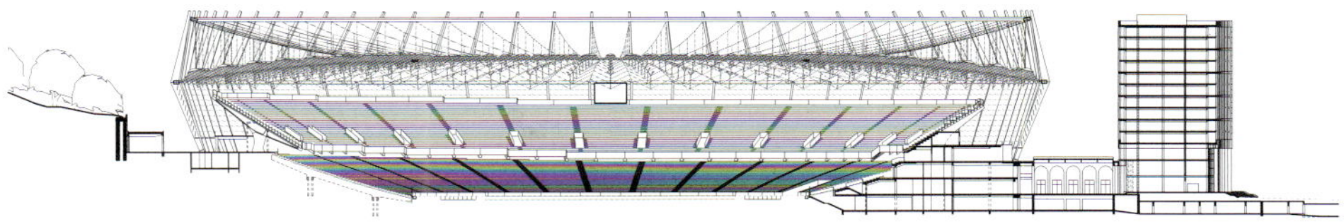

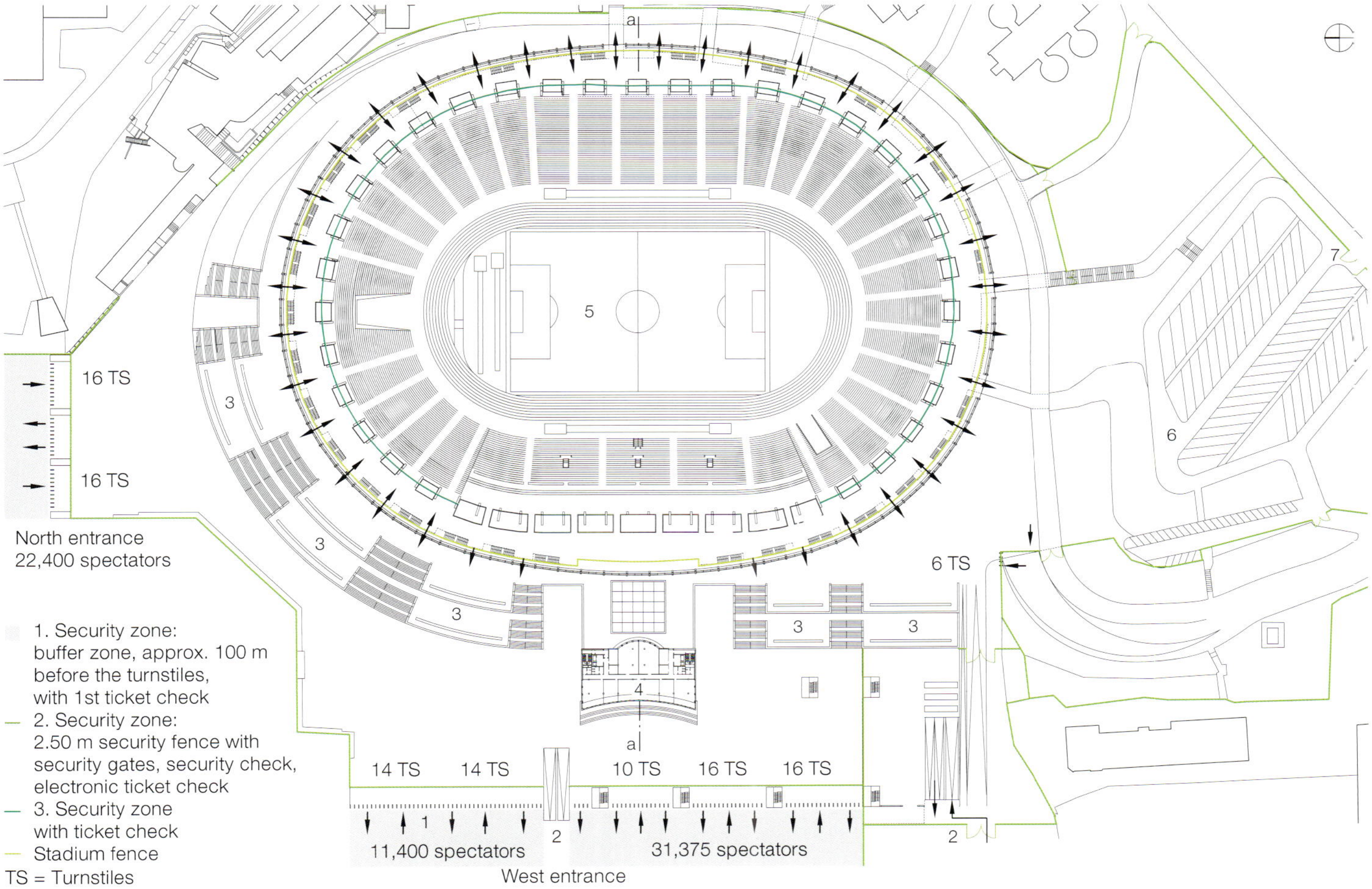

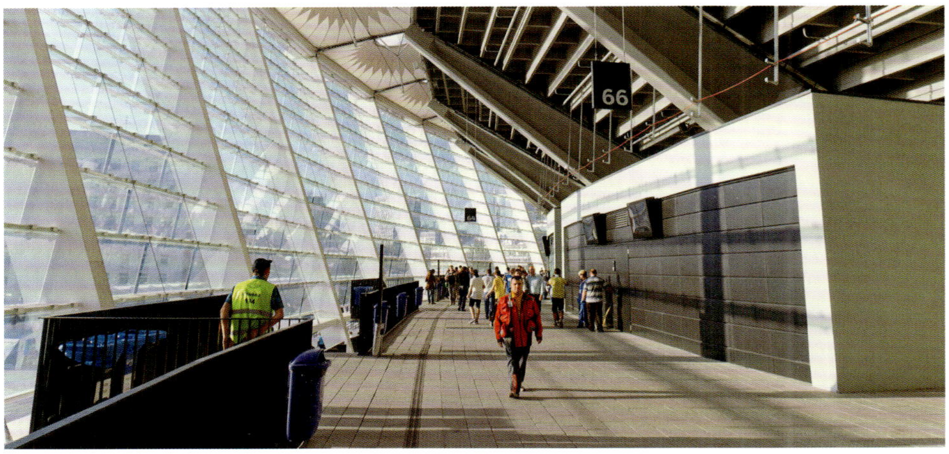

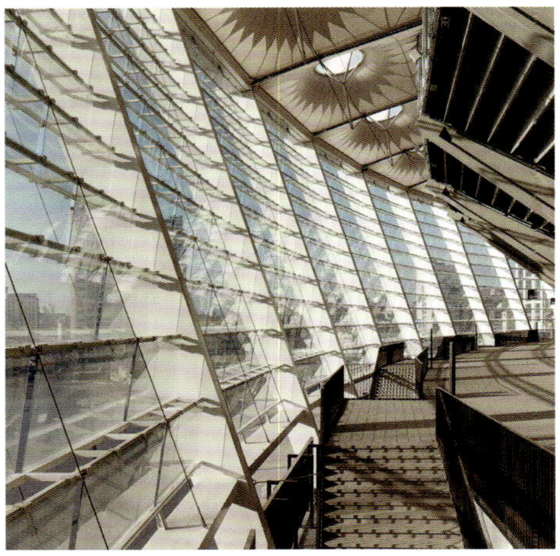

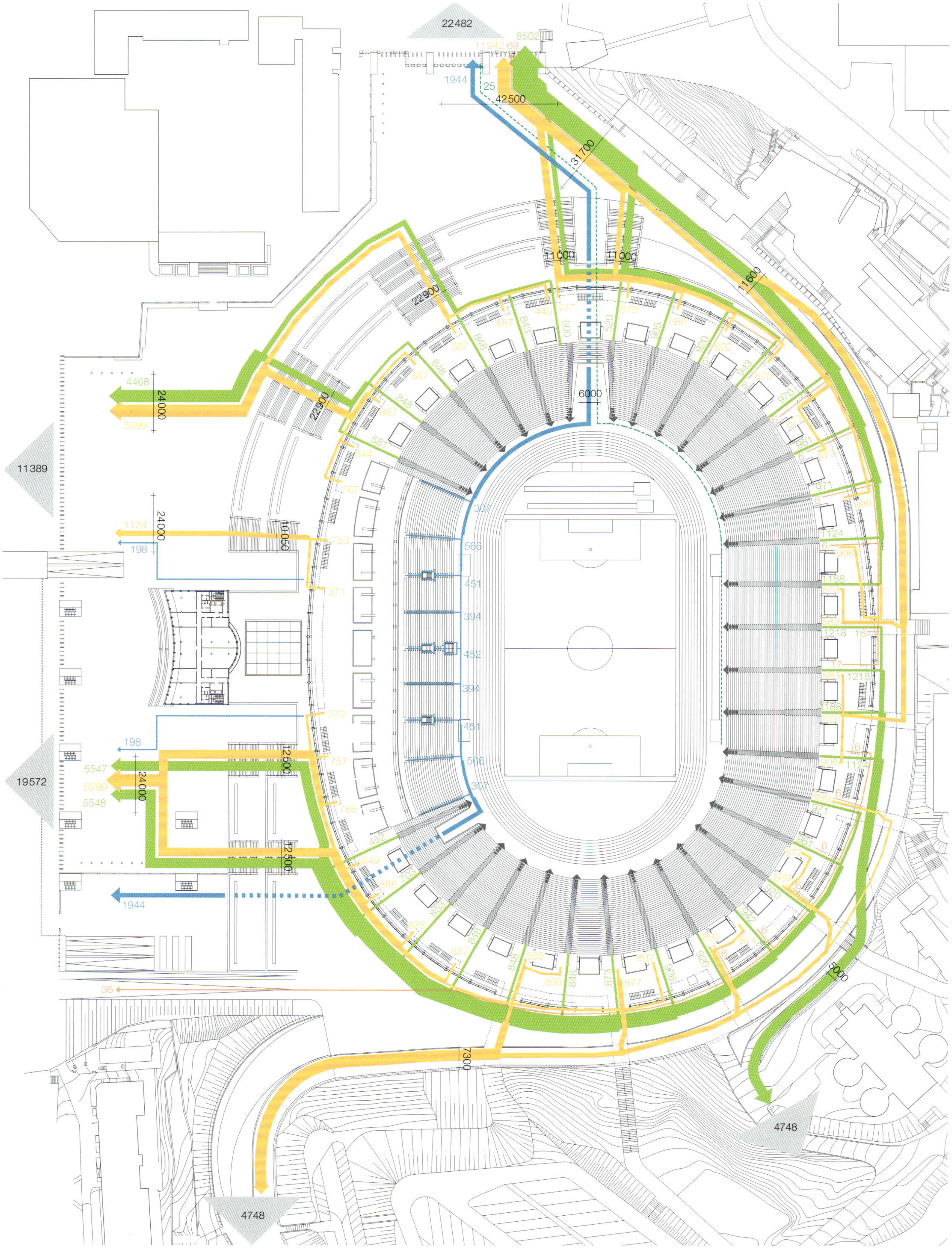
22482
8502
11942
69
1944
25
42500
31700
11000
11000
11600
22900
22900
4468
24000
5599
11389
6000
1124
24000
198
10050
198
19572
5547
6299
5548
24000
12500
12500
1944
36
7300
5000
4748
4748

Evacuation diagram
Scale 1:2000

- Escape route, lower tier
- Escape route, upper tier
- Disabled-friendly escape route
- Escape route from football pitch
- VIP escape route
- → Additional “panic exits”/emergency exits
- ↔ Necessary width of passage way

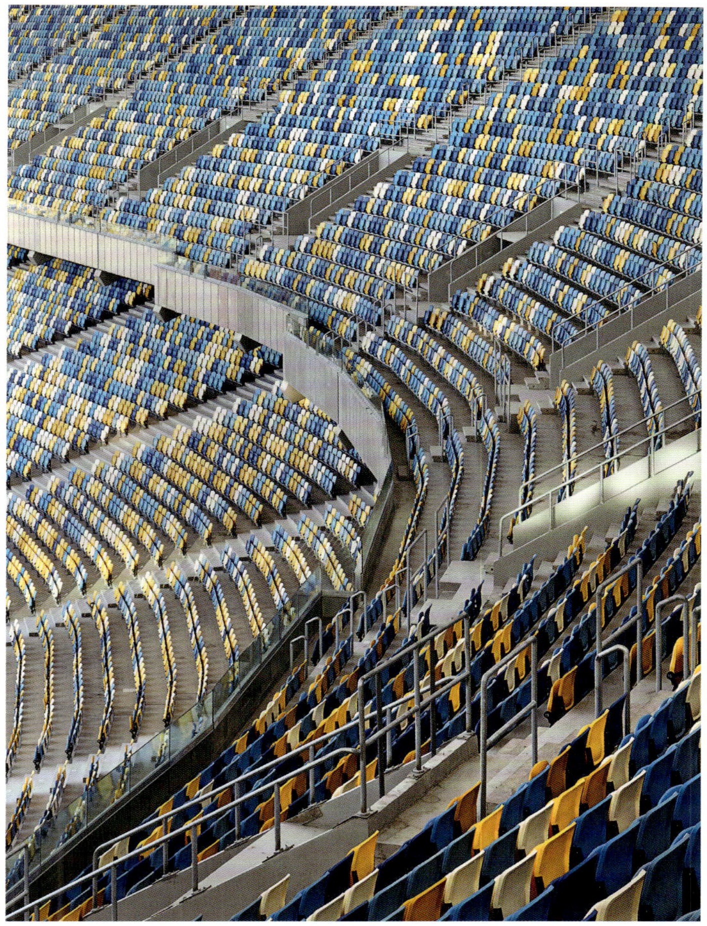

Project data:

Use:	Sports stadium
Access:	3 security zones with turnstile systems
Seats:	68,000 seats, 2,300 VIP seats, 40 VIP boxes
Roof surface area:	43,000 m²
Year of completion:	1968/2011
Construction period:	35 months

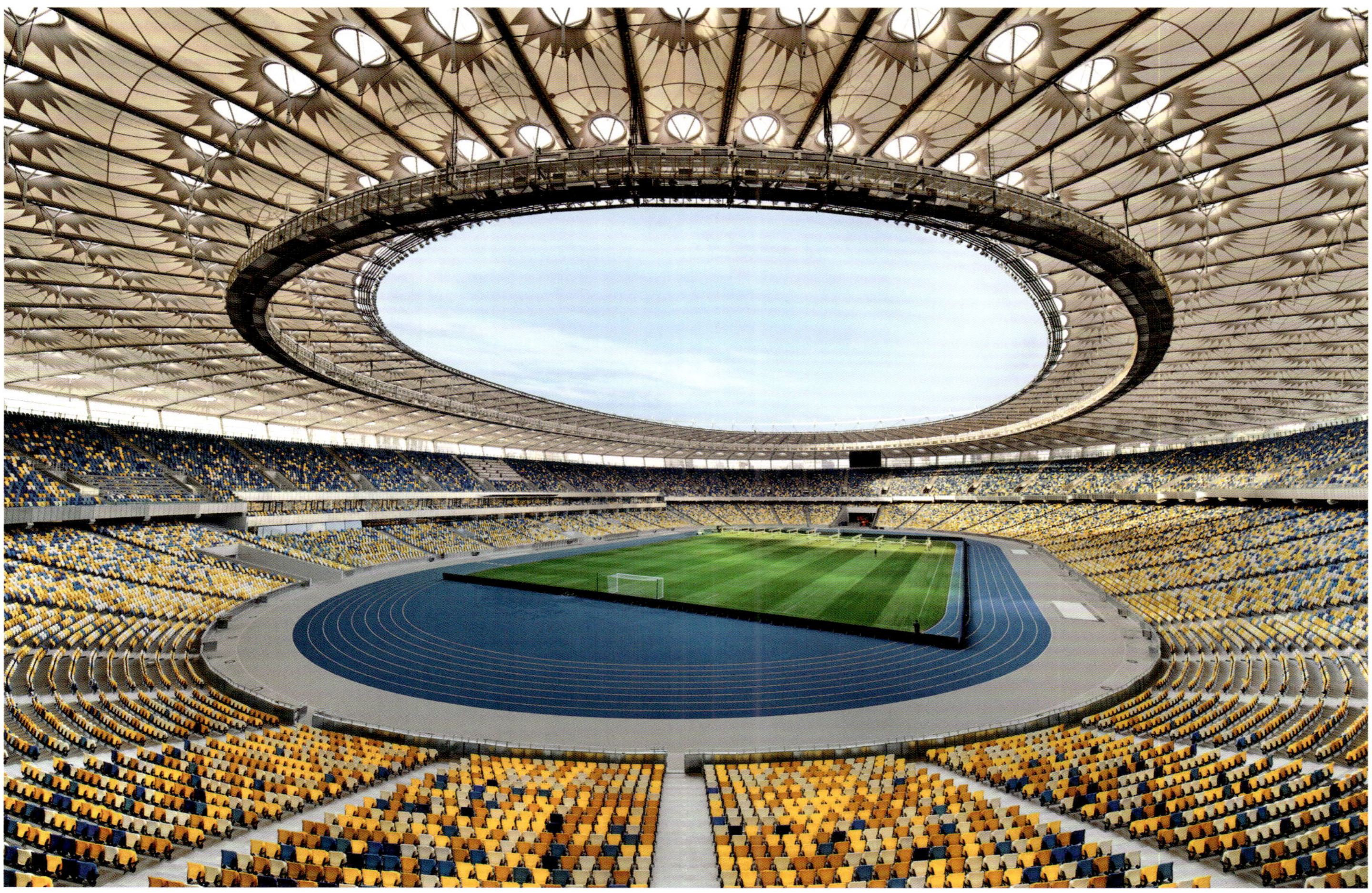

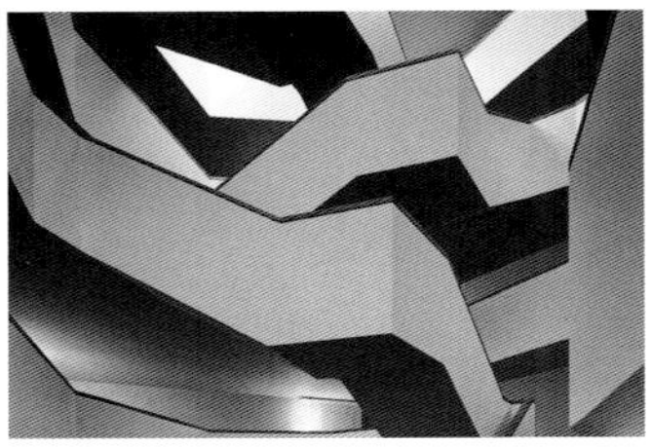

Apartment building in Zurich

Client: Rondo GmbH, Zurich
Architects: Graber Pulver Architekten, Zurich/Bern
Contributors: Manuel Gysel, Jonas Ringli, Susana Elias Robles, Yvonne Urscheler Lofteröd, Marcel Weiler
Project management: Alexander Huhle
Construction: Implenia Generalunternehmung AG, Dietlikon
Building installations: B & G Ingenieure AG, Zurich
Facade consultants: Prometall Engineering AG, Zurich
Year of completion: 2007

www.graberpulver.ch
arch@graberpulver.ch

Marco Graber
Born 1962 in Bern; studied architecture at the ETH Zurich; 1989: obtained degree; 2006–2008: visiting lecturer at the ETH Zurich.

Thomas Pulver
Born 1962 in Bern; studied architecture at the ETH Zurich; 1989: obtained degree; 2006–2008: visiting lecturer at the ETH Zurich.

1992: formation of Graber Pulver Architekten.

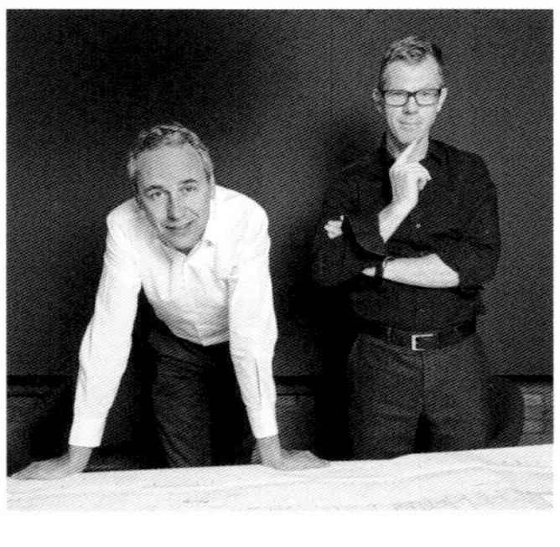

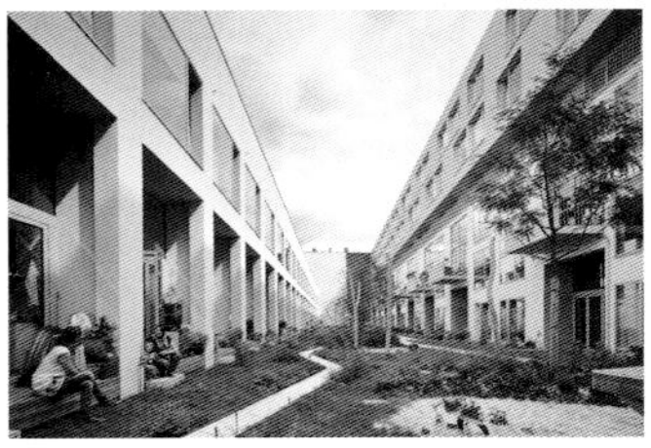

Residential development in Berlin

Client: Bauherrengemeinschaft Zelterstrasse 5 GbR, Berlin
Architects: zanderroth architekten, Berlin
Sascha Zander, Christian Roth
Contributors: Kirka Fietzek, Diana Gunkel, Guido Neubeck, Konrad Scholz, Lutz Tinius
Structural planning: Ingenieurbüro für Statik, Konstruktion und Bauphysik, Berlin, Andreas Leipold
Technical installations, electrical planning: Ingenieurbüro Norbert Lüttgens, Berlin
Fire protection consultants: hhpberlin Ingenieure für Brandschutz GmbH, Berlin
Landscaping: herrburg Landschaftsarchitekten, Berlin
Year of completion: 2010

www.zanderroth.de
kontakt@zanderroth.de

Sascha Zander
Born 1968 in Düsseldorf; studied architecture at the RWTH Aachen, the Academy of Art in Düsseldorf, the Bartlett School of Architecture in London; 1995: obtained degree in architecture at the Bartlett School; 1997: obtained degree in urban planning at the RWTH Aachen.

Christian Roth
Born 1970 in Wiesbaden; studied architecture at the RWTH Aachen, the Academy of Art in Düsseldorf, the E.T.S.A. in Madrid; 1998: obtained degree in architecture at the RWTH Aachen.

1999: formation of zanderroth architekten.

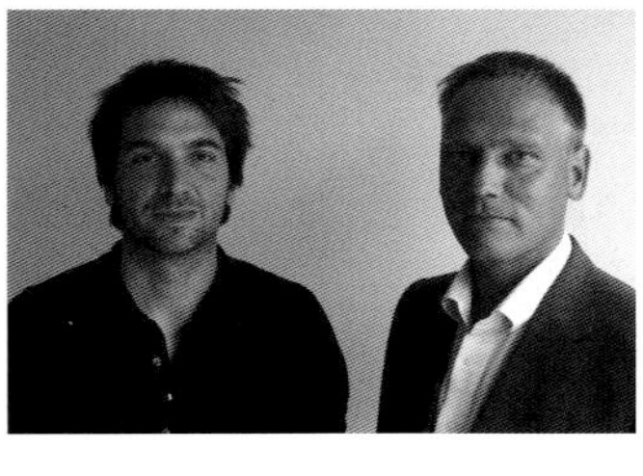

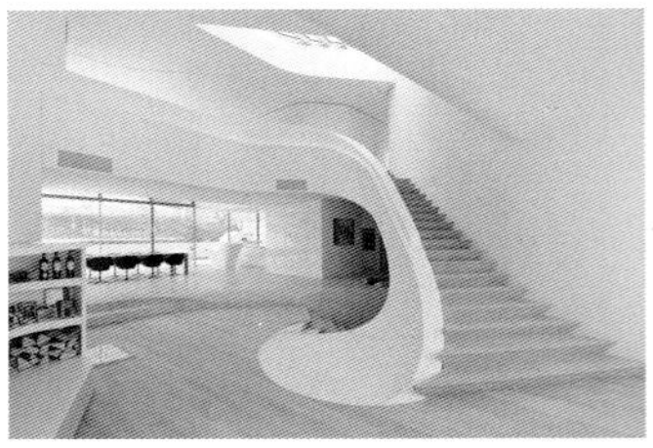

Haus am Weinberg in Stuttgart

Client: private
Architects: UNStudio
Ben van Berkel, Caroline Bos
Contributors: Astrid Piber, René Wysk, Kirsten Hollmann-Schröter, Cynthia Markhoff, Christian Bergmann, Jan Schellhoff, Iris Pastor, Rodrigo Cañizares, Albert Gnodde, Beatriz Zorzo Talavera, Shany Barath, Esteve Umbert Morits, Hannes Pfau
Supervision of works: G + O Architekten GmbH, Leinfelden-Echterdingen
Structural planning: Bollinger + Grohmann GmbH, Frankfurt am Main; Kraft Baustatik, Besigheim
Lighting consultants: ag licht GbR, Bonn
Landscaping: Atelier Dreiseitl GmbH, Überlingen
Year of completion: 2011

www.unstudio.com
info@unstudio.com

Ben van Berkel
Born 1957 in Utrecht; 1979–1987: studied architecture at the Rietveld Academy in Amsterdam and the Architectural Association in London; 1988: formation of Van Berkel & Bos Architectenbureau in Amsterdam, since 1998 UNStudio; professorship at the Städelschule in Frankfurt am Main.

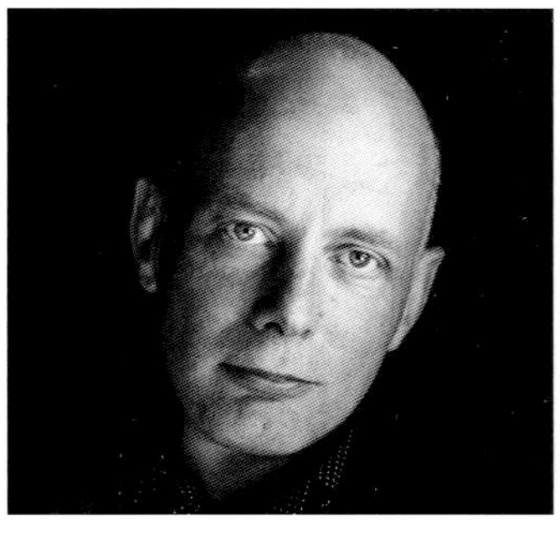

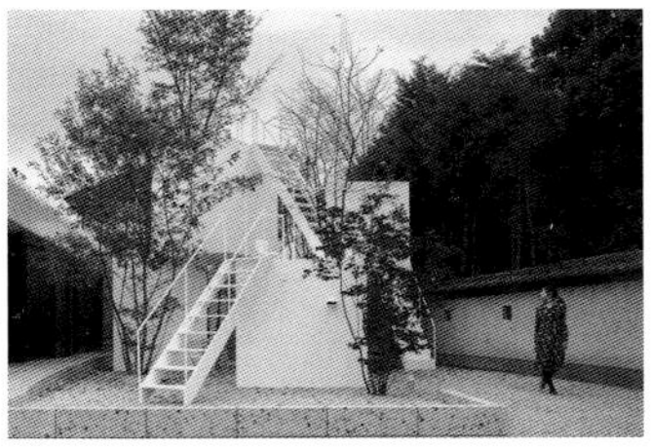

"House before House" in Utsunomiya

Client: Tokyo Gas, Tokyo
Architects: Sou Fujimoto Architects, Tokyo
Contributors: Yasushi Yamanoi
Structural planning: Jun Sato Structural Engineers, Tokyo; Naotake Koyama
Supervision of works: Sou Fujimoto Architects, Tokyo
Construction: Toyota Woodyou Home Corporation, Utsunomiya City
Environmental engineering: Kankyo Engineering Inc., Takafumi Wada
Lighting design: Sirius Lighting Office, Tokyo, Hiroshi Totsune
Year of completion: 2008

www.sou-fujimoto.net
project@sou-fujimoto.net

Sou Fujimoto
Born 1971 on Hokkaido; studied architecture at the University of Tokyo; 1994: obtained bachelor's degree; 2000: formation of Sou Fujimoto Architects.

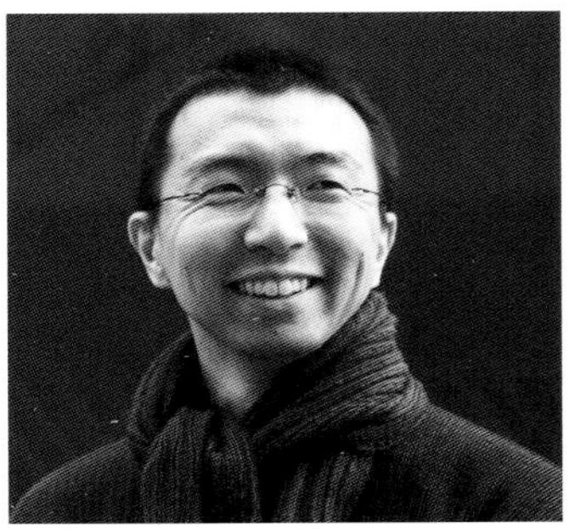

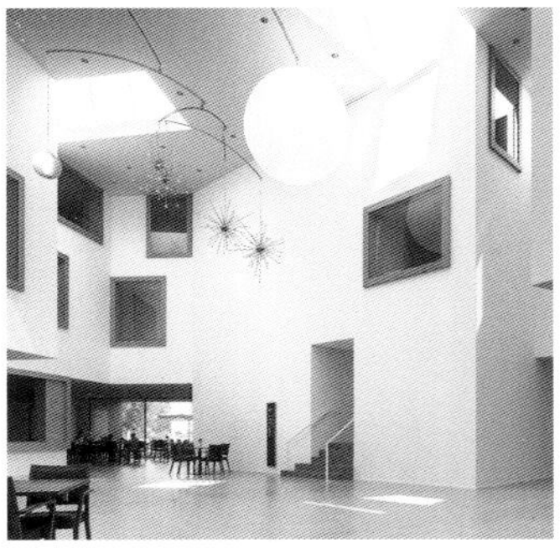

Centre for senior citizens in Maienfeld

Client: Stiftung Alterszentrum Bündner Herrschaft, Maienfeld
Architects: Arbeitsgemeinschaft Isler Gysel, Zurich, bhend.klammer architekten, Zurich
Contributors: Kim Sneyders, Claudia Wunderlich
Structural planning: Edy Toscano AG, Chur
Landscape architects: Schweingruber Zulauf Landschaftsarchitekten, Zurich
Signage: Bodara, Büro für Gebrauchsgrafik, Zurich
Year of completion: 2011

www.islergysel.ch
info@islergysel.ch
www.bhend.klammer.ch
info@bhend.klammer.ch

Dominik Isler
Born 1974; 1995–2001: studied architecture at ETH Zurich and EPFL in Lausanne.

Manuel Gysel
Born 1975; 1995–2001: studied architecture at ETH Zurich and EPFL in Lausanne.

2009: formation of Isler Gysel Architekten.

Christof Bhend
Born 1967; 1994–1999: studied architecture at ETH Zurich.

Sergej Klammer
Born 1974; 1994–2001: studied architecture at ETH Zurich.

2003: formation of bhend.klammer architekten.

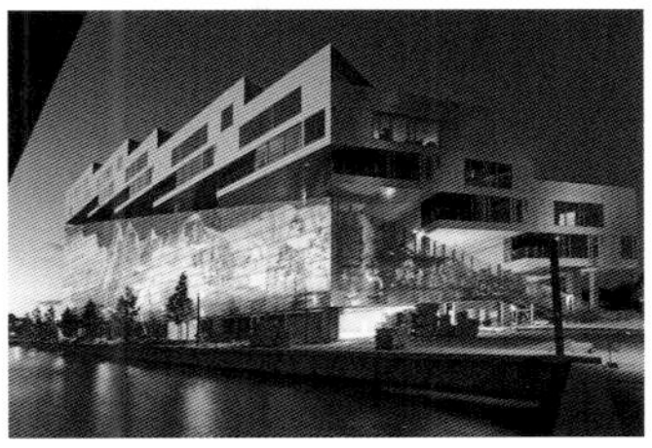

Apartment mound in Copenhagen

Client: Høpfner A/S, Copenhagen; Danish Oil Company A/S, Copenhagen
Architects: BIG – Bjarke Ingels Group, Copenhagen; JDS Architects, Copenhagen/Brussels/Belo Horizonte
Project management: Bjarke Ingels, Jakob Lange, Finn Nørkjær, Jan Borgstrøm
Contributors: Henrik Poulsen, Annette Jensen, Dariusz Bojarski, Dennis Rasmussen, Eva Hviid-Nielsen, João Vieira Costa, Jørn Jensen, Karsten V. Vestergaard, Karsten Hammer Hansen, Leon Rost, Louise Steffensen, Malte Rosenquist, Mia Frederiksen, Ole Elkjær-Larsen, Ole Nannberg, Roberto Rosales Salazar, Rong Bin, Sophus Søbye, Søren Lambertsen, Wataru Tanaka
Structural planning: Moe & Brødsgaard A/S, Copenhagen
Year of completion: 2008

www.big.dk
big@big.dk
www.jdsa.eu
office@jdsa.eu

Bjarke Ingels
Born 1974 in Copenhagen; studied architecture at the Royal Academy of Arts in Copenhagen and the School of Architecture of Barcelona (ETSAB); worked at OMA in Rotterdam; 2001: formation of PLOT Architects; 2005: formation of BIG – Bjarke Ingels Group.

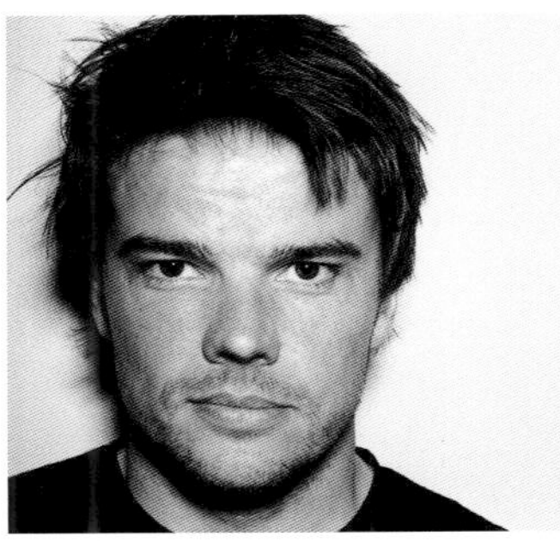

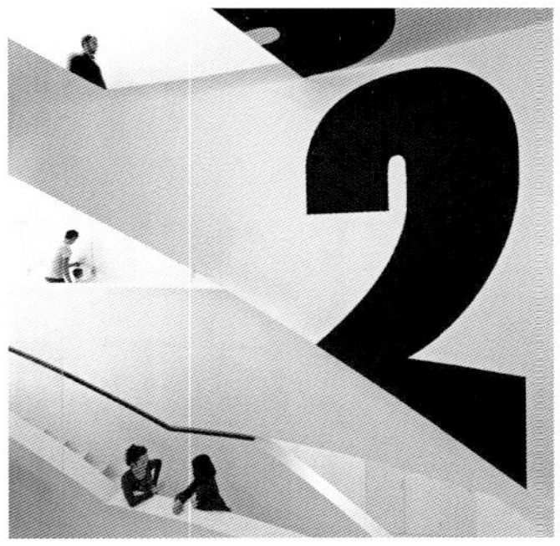

Hostel in Split

Client: SAFIR d.o.o., Split
Architects: STUDIO UP, Zagreb
Lea Pelivan, Toma Plejić
Contributors: Antun Sevšek, Iva Denona Vusić, Jelena Martić, Domagoj Jurić, Ivan Grubišić Tasić, Robert Tičić, Paula Prkačin, Jasna Hrga
Structural planning: Darko Fadić, Split
Building installations: Nikša Nižetić, Davor Lučin, Split
Electrical planning: Petar Trumbić, Split
Signage: Damir Gamulin, Zagreb
Year of completion: 2010

www.studioup.hr
info@www.studioup.hr

Lea Pelivan
Born 1976 in Split; studied architecture at the Faculty of Architecture in Zagreb; 2001: obtained degree.

Toma Plejić
Born 1977 in Rijeka; studied architecture at the Faculty of Architecture in Zagreb; 2001: obtained degree.

2003: formation of STUDIO UP.

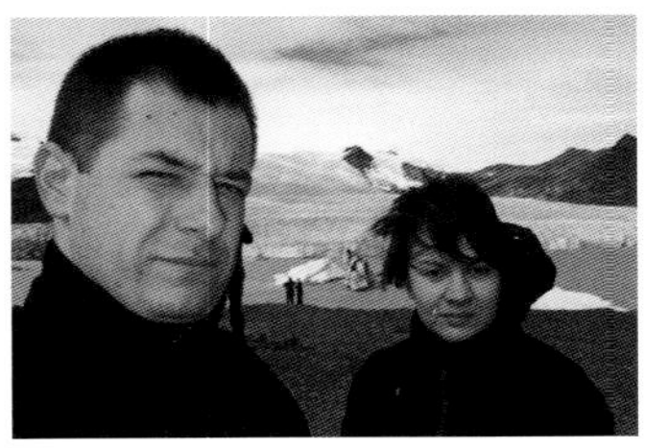

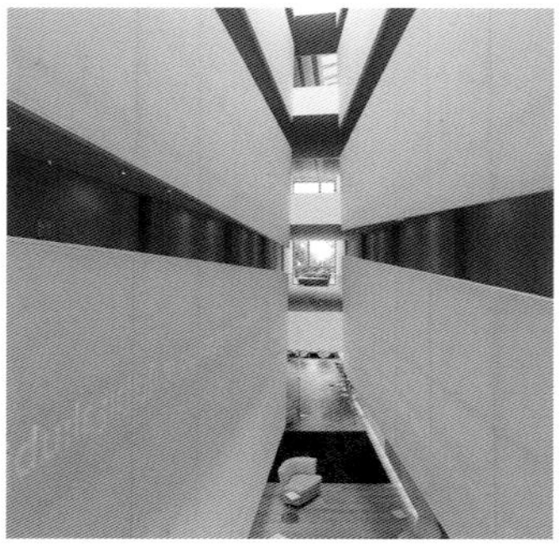

Hotel am Domplatz in Linz

Client: Diözesane Immobilien-Stiftung (formerly Stiftung St. Severin)
Architecture: hohensinn architektur, Graz
Project management: Erich Ganster (hotel and car park); Karlheinz Boiger (design)
Contributors: Ognjen Persoglio, Klemens Mitheis, Mario Mayrl, Thomas Klietmann, Franz Jelisitz
Project supervision: Jastrinsky GmbH & Co. KG, Salzburg
Structural planning: Peter Pawel, Linz; Praher & Schuster ZT GmbH, Linz
Soil mechanics: Sigma Consult GmbH, Linz
Building physics: Dr. Pfeiler GmbH, Graz
Technical installations: TB Freunschlag GmbH, Linz
Lighting design: Licht-Innovativ GmbH, Innsbruck
"Integration" art project including "Lautschrift" illuminated lettering in the atrium panels: Isa Stein, Christoph Fürst, Linz
Local supervision of works: Hinterwirth Architekten ZV OG, Gmunden
Year of completion: 2009

www.hohensinn-architektur.at
office@hohensinn-architektur.at

Josef Hohensinn
Born 1956 in Upper Austria; studied architecture at the Höhere Technische Bundeslehranstalt in Linz and at the TU Graz; 1990–1995: assistant lecturer at the TU Graz; 1995: establishment of an independent architectural practice; 1996–1997: practice partnership with Hubert Riess; 1998: formation of hohensinn architektur in Graz.

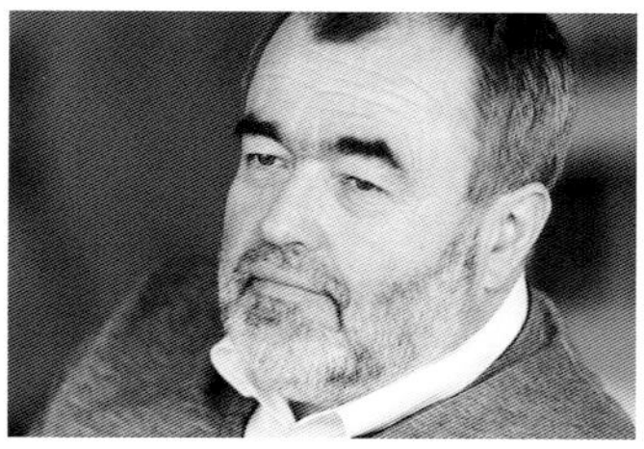

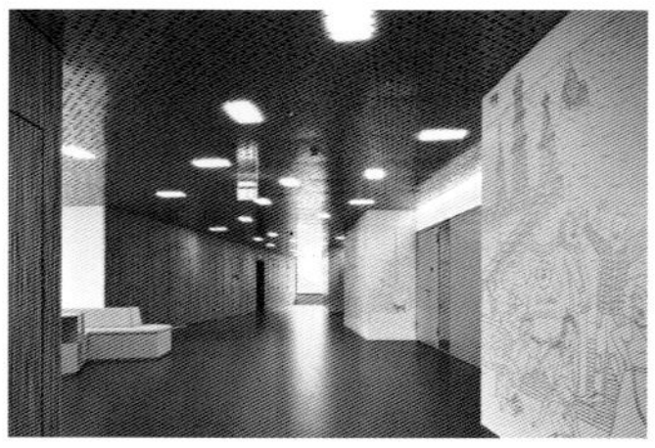

Children's hospital in Basel

Client: Cantons of Basel-Stadt and Basel-Landschaft
Architects: Stump & Schibli Architekten BSA AG, Basel
Structural planning: Schnetzer Puskas Ingenieure AG, Basel
Construction management, supervision of works: Proplaning AG, Basel
Facade consultants: Emmer Pfenninger Partner AG, Münchenstein
Landscape architects: Bechthold. Lenzin, Liestal
Signage: Integral Ruedi Baur, Zurich
Year of completion: 2009

www.stumpschibliarch.ch
info@stumpschibliarch.ch

Yves Stump
Born 1962 in Aarau; 1978–1982: apprenticeship as draughtsman; 1982–1985: studied architecture at Fachhochschule beider Basel (FHBB); 1986–1991: worked at various architectural practices; 1992–1994: established own practice; assistant lecturer at the EPFL in Lausanne and visiting lecturer at the FHBB.

Hans Schibli
Born 1964 in Wettingen; 1981–1985: apprenticeship as draughtsman; 1985–1989: studied architecture at Fachhochschule beider Basel (FHBB); 1989–1995: worked at various architectural practices.

1995: formation of Stump & Schibli Architekten (since 1999 BSA, since 2006 AG).

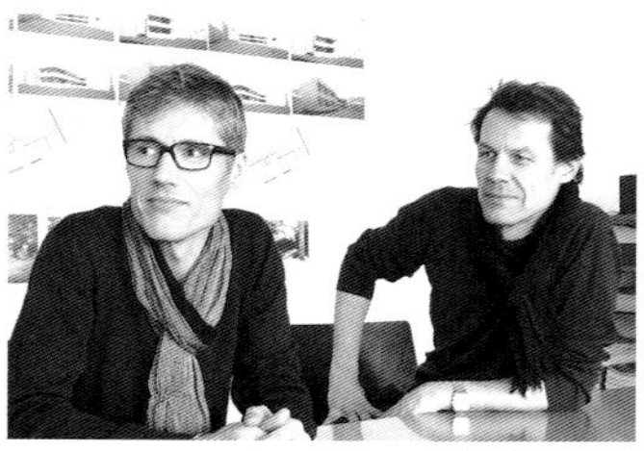

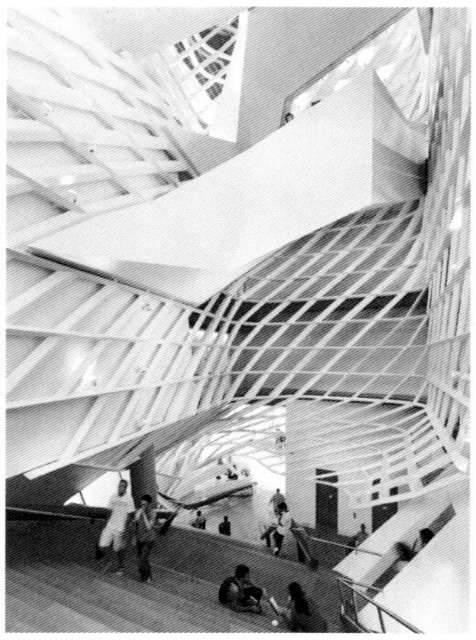

Cooper Union in New York

Client: The Cooper Union for the Advancement of Science and Art, New York
Architects: Morphosis Architects, Culver City
Design director: Thom Mayne
Project manager: Silvia Kuhle
Project team: Pavel Getov, Chandler Ahrens, Jean Oei, Natalia Traverso Caruna, Go-Woon Seo, Irena Bedenikovic, Salvador Hidalgo, Marcin Kurdziel, Debbie Lin, Kristina Loock
Contributors: Sean Anderson, Charles Austin, Dominique Cheng, Guiomar Contreras, Ben Damron, Patrick Dunn-Baker, Graham Ferrier, Mauricio Gomez, Brock Hinze, Eui Yeob Jeong, Mark Johnson, Jennifer Kasick, Amy Kwok, Michelle Siu Lee, Shannon Loew, Mark McPhie, Greg Neudorf, Mike Patterson, Michael Sargent, Reinhard Schmoelzer, Christin To
Partner architects: Gruzen Samton, New York, Peter Samton
Structural planning: John A. Martin Associates, Los Angeles; Goldstein Associates, New York
Technical installations, electrical planning: IBE Consulting Engineers, Sherman Oaks; Syska Hennessy Group Inc., New York
Lighting design: Horton Lees Brodgen Lighting Design, New York
Signage: Pentagram, New York
Year of completion: 2009

www.morphosis.com
studio@morphosis.net

Thom Mayne
Born 1944; studied architecture at the University of Southern California and the Harvard Graduate School of Design; 1972: formation of Morphosis Architects; various teaching appointments.

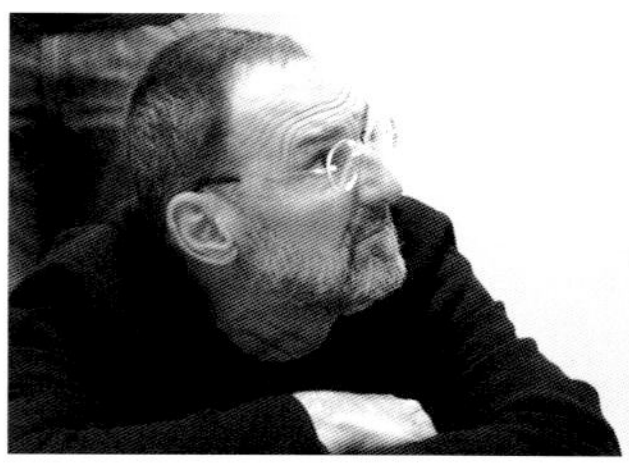

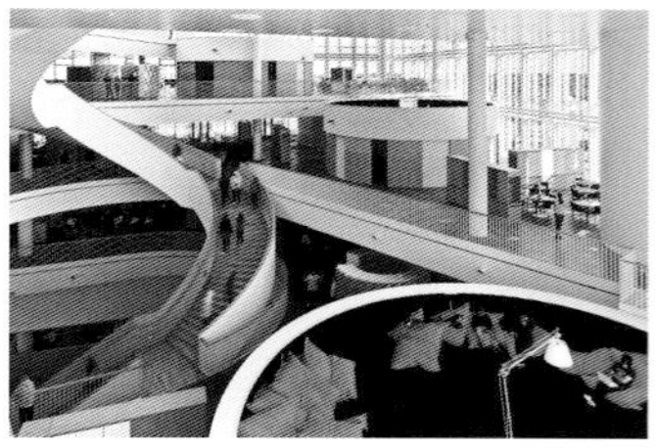

High school in Copenhagen

Client: Copenhagen municipal authorities
Architects: 3XN, Copenhagen
Project management: Kim Herforth Nielsen
Structural planning, technical installations, electrical and acoustic planning: Søren Jensen Consulting Engineers, Copenhagen
Landscaping: 3XN, Copenhagen
Supervision of works: 3XN, Copenhagen
Pedagogical consultants: Helle Mathiasen, Aarhus
Year of completion: 2007

www.3xn.dk
3xn@3xn.dk

Kim Herforth Nielsen
Born 1954; graduated from the Aarhus School of Architecture; 1986 co-founder of 3XN.

Bo Boje Larsen
Born 1951; graduated from the Royal Danish Academy of Fine Arts in Copenhagen; partner at 3XN since 2003.

Jan Ammundsen
Born 1972; graduated from the Aarhus School of Architecture; worked at 3XN since 2005, partner since 2007.

Kasper Guldager Jørgensen
Born 1976; graduated from the Aarhus School of Architecture; obtained master's degree at the Southern California Institute of Architecture; since 2007 head of GXN, the innovation division of 3XN.

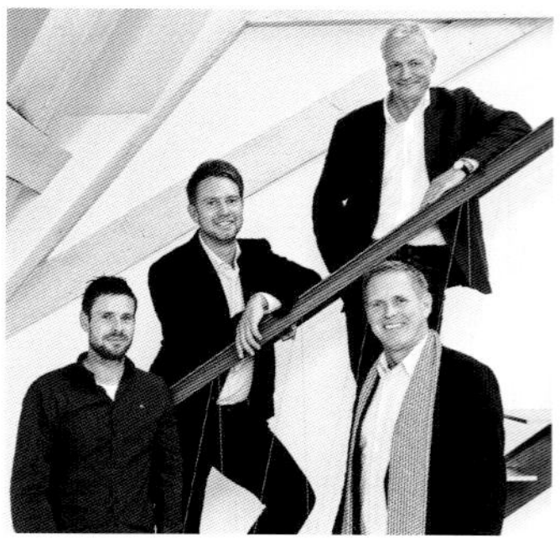

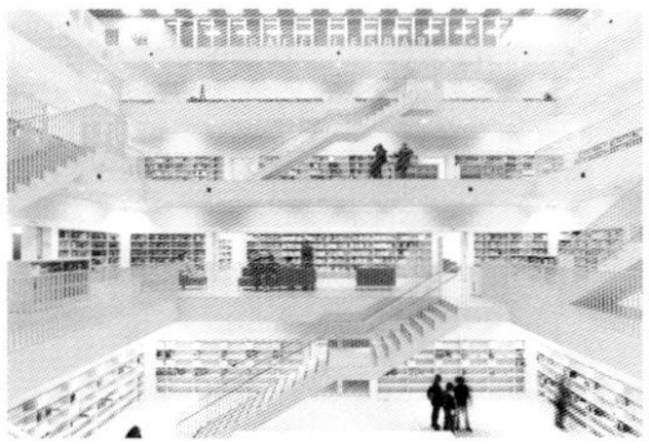

Municipal library in Stuttgart

Client: Landeshauptstadt Stuttgart
Architect: Eun Young Yi, Cologne
Project management:
Drees & Sommer, Stuttgart
Structural planning: Boll und Partner, Stuttgart
Technical installations: Rentschler + Riedesser, Filderstadt
Electrical planning: Conplaning, Ulm
Planning of conveyor systems: Werner Schwarz, Stuttgart
Building physics, energy technology: EGS-Plan, Stuttgart
Fire protection consultants: HHP Nord-Ost, Braunschweig; HHP Süd, Ludwigshafen
Landscaping: Gänßle + Hehr, Esslingen
Facade consultants: R + R Fuchs, Munich
Building security: Höhler + Partner, Aachen
Year of completion: 2011

www.yiarchitects.com
info@yiarchitects.com

Eun Young Yi
Born 1956 in Daecheon; 1976–1983: studied architecture at Hanyang University, 1984–1990 at the RWTH Aachen; 1990: worked at Oswald Mathias Ungers architects; 1991–1992: worked as freelance architect in Cologne; 1993: worked at Joachim Schürmann architects; 1994: formation of Yi Architects in Cologne and Seoul; 1994–1999: research assistant at the RWTH Aachen; 2000–2010: professor at Hanyang University.

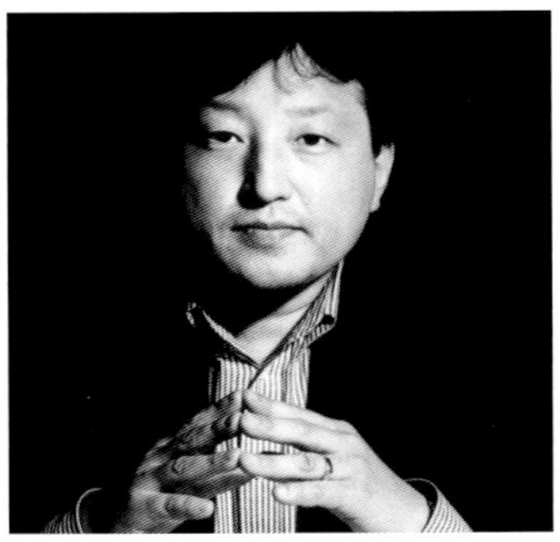

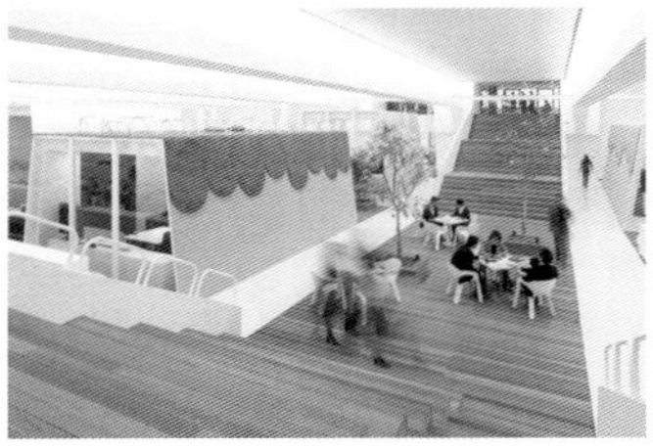

Advertising agency in Tokyo

Client: TBWA\HAKUHODO
Architects: Klein Dytham architecture, Tokyo
Contributors: Yukinari Hisayama, Yoshinori Nishimura, Joe Keating, Mayumi Ito, Nazuki Konishi, Makiko Okano, Hiroshi Ohsu
Structural planning: Arup Japan, Tokyo
General contractor: D. Brain, Masaya Tanimoto, Tokyo
Year of completion: 2007

www.klein-dytham.com
kda@klein-dytham.com

Astrid Klein
Born 1962 in Varese; 1986: obtained bachelor's degree in interior design at the École des Arts Décoratifs in Strasbourg; 1988: obtained master's degree in architecture at the Royal College of Art in London; 1988: worked at Toyo Ito Architects & Associates in Tokyo; lectureships at various universities in Tokyo, at the University of California in Berkeley and the California Polytechnic State University in San Luis Obispo.

Mark Dytham
Born 1964 in Northamptonshire; 1988: obtained master's degree in architecture at the Royal College of Art in London; 1988: worked at Skidmore Owings & Merrill in London; 1988: worked at Toyo Ito Architects & Associates n Tokyo; since 1999 lectureship at the Tokyo Science University; lectureships at the Hosei University in Tokyo and the University of California in Berkeley.

1991: formation of Klein Dytham architecture.

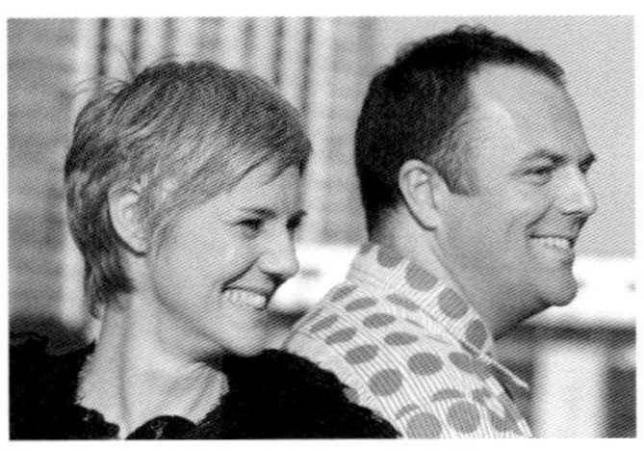

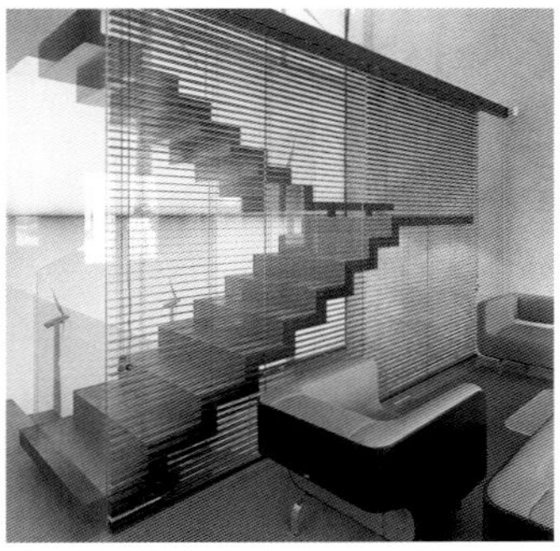

Office building in Bolzano

Client: FRI-EL Green Power, Rome
Architect: Markus Scherer Architekt, Meran
Contributors: Markus Scherer, Adriana Mangiacotti, Elena Mezzanotte
Structural planning: Ingenieur P. Schmidt, Bolzano
Construction: Stuefer Bau, Ritten
Metalwork: Condin Robert, Tramin
Glass construction: Seyr Glas, Bruneck
Rendering: Giuvanni Moling, San Michele
Year of completion: 2007

www.architektscherer.it
info@architektscherer.it

Markus Scherer
Born 1962 in Vienna; studied architecture in Vienna and Venice, obtained degree in 1990; has worked freelance since 1991, initially in a partnership, since 2001 with his own practice.

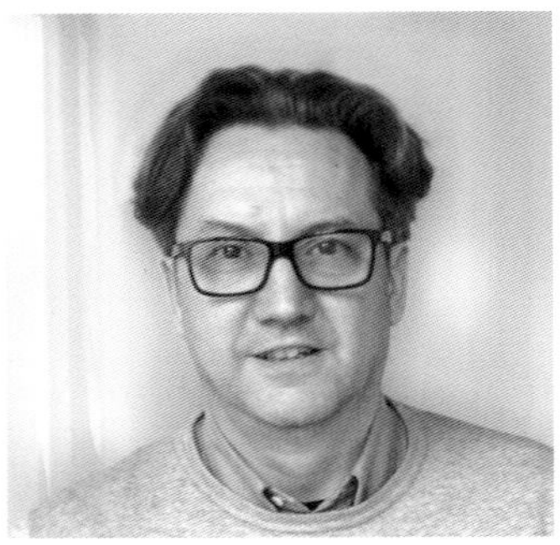

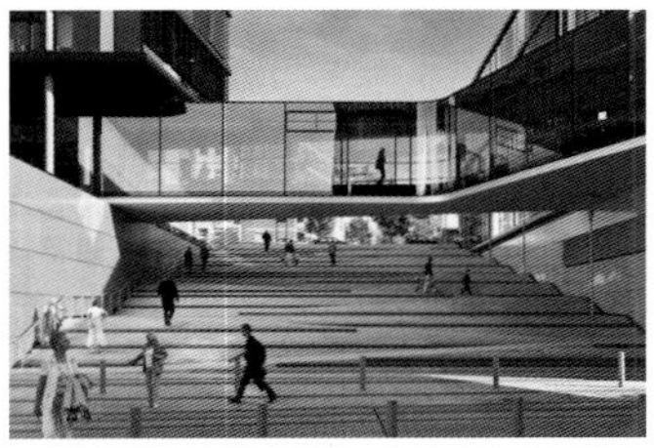

AachenMünchener headquarters in Aachen

Owner: Generali Deutschland Immobilien GmbH
Architects: kadawittfeldarchitektur GmbH, Aachen
Project management: Stefan Haass, Jascha Klusen
Contributors: Sebastian Potz, Michael Tremmel, Frank Berners, Gaby Inden, Roswitha van der Kooi, Susanne Lüschen, Christoph Schlaich, Julia Therstappen, Andrea Thörner, Sascha Thomas, Daniel Trappen, Eva Strotmeier
Structural design: Dr. Binnewies, Ingenieurgesellschaft mbH, Hamburg
Contractor: Alpine Deutschland GmbH with subcontractor in the construction phase: Nattler Architekten, Essen
Project management: Ernst & Young Real Estate GmbH, Troisdorf
Landscaping: Club L94 landscape architects, Cologne
Year of construction: 2010

www.kadawittfeldarchitektur.de
office@kadawittfeldarchitektur.de

Klaus Kada
Born 1940 in Leibnitz; 1971: graduated from TU Graz; 1971–1985: partnership with Gernot Lauffer; since 1988 practice in Graz; 1995–2006: professor at RWTH Aachen; since 1996 practice in Aachen.

Gerhard Wittfeld
Born 1968 in Moers; 1995: graduated from RWTH Aachen; since 1994 work with Klaus Kada, Graz; 1997–2004: teaching post at RWTH Aachen; since 1999 partner at kadawittfeldarchitektur.

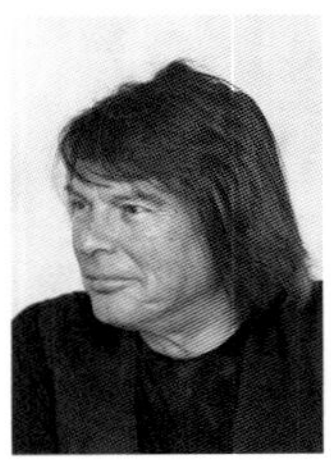

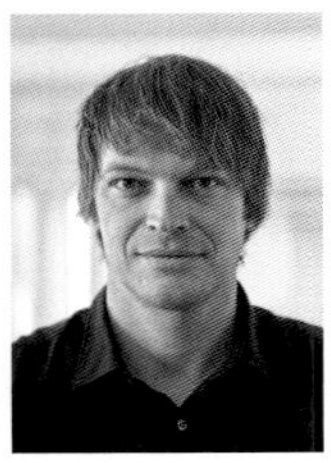

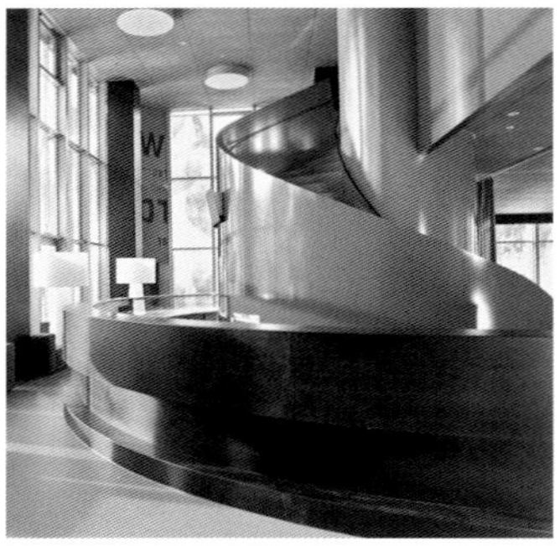

Restaurant and bar in Zurich

Operator: Compass Group, Kloten
Financing: Nicole Baumgartner, Christine Fürst
Architects: Burkhalter Sumi Architects, Zurich
Project leader: Yves Schihin
Site management: GMS Partner AG, Zurich, Ralph Eschmann
Structural engineer: Walt + Galmarini AG, Zurich
Technical services: Huwyler + Koch, Zurich
Electrical planner: Schmidiger + Rosasco AG, Zurich
Building physics: Kopitsis AG, Wohlen
Date of completion: 2006

office@burkhalter-sumi.ch
www.burkhalter-sumi.ch

Marianne Burkhalter
Born 1947 in Thalwil; 1973–1975: attended the University of Princeton; 1987 guest professor at the Southern Institute of Architecture in Los Angeles,1999 EPFL in Lausanne.

Christian Sumi
Born 1959 in Biel; 1977: degree from the ETH Zurich; 1990–1991 guest professor at the École d'Architecture in Geneva, 1994 Harvard in Boston, 1999 EPFL in Lausanne, 2003 University of Strathclyde in Glasgow.

1984: establishment of Burkhalter Sumi Architects in Zurich; professors at the Accademia di architettura in Mendrisio since 2008.

Yves Schihin
Born 1970 in Bern; 2000: degree from the EPFL in Lausanne; employee since 2000, partner since 2004 at Burkhalter Sumi Architects.

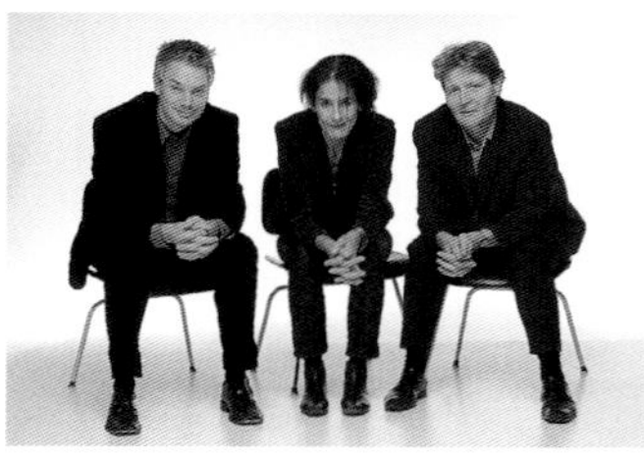

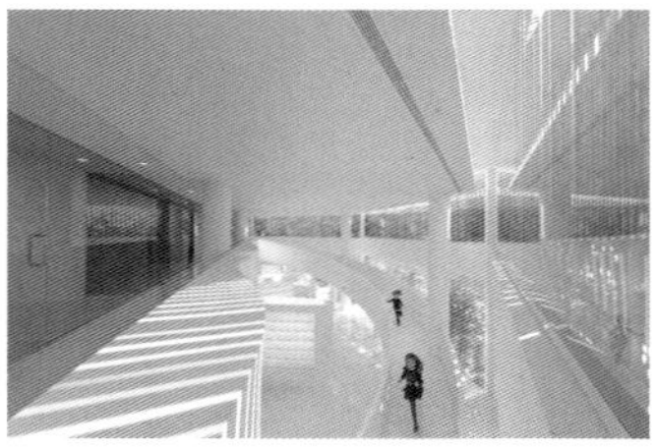

Pharmacy in Athens

Client: Morfopoulos Pharmacist, Athens
Architects: KLab architects, Athens / London
Konstantinos Labrinopoulos
Contributors: Xara Marantidou, Enrique Ramirez, Mark Chapman
Structural planning: Future Constructions, Athens
Year of completion: 2010

www.klab.gr
info@.klab.gr

Konstantinos Labrinopoulos
Born 1970 in Athens; 1994: obtained degree at the National Technical University in Athens; 1996: obtained master's degree at the SCI-Arc in Los Angeles; 2007: formation of KLab architecture.

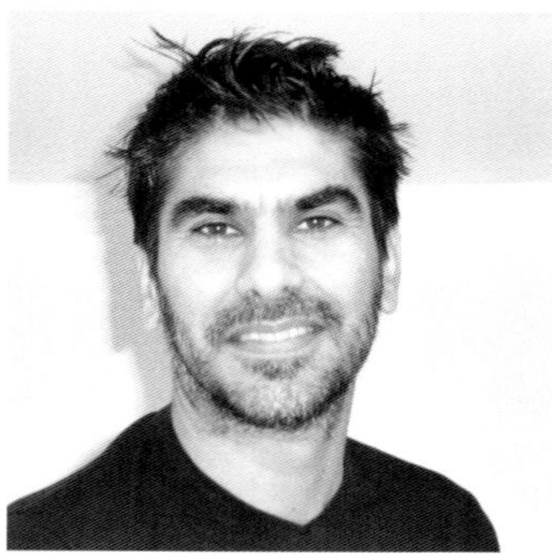

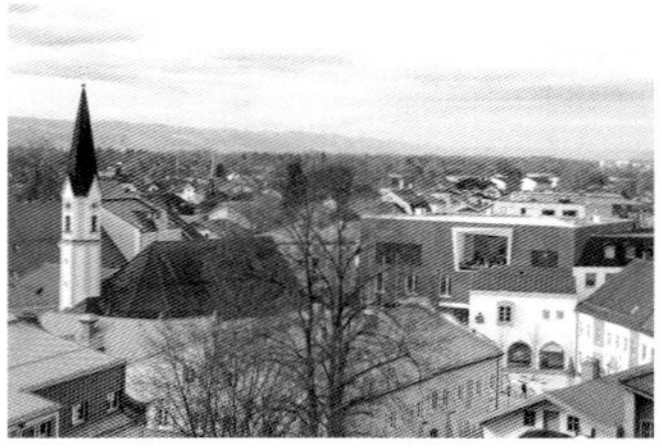

Town hall in Bad Aibling

Client: Municipality of Bad Aibling
Architects: Behnisch Architekten, Munich; Stefan Behnisch, Robert Hösle
Project management: Stephan Leissle
Supervision of works: Beratende Ingenieure Schaaf, Munich
Structural planning: Planungsgesellschaft Dittrich mbH, Munich
HVAC planning: Transplan, Munich
Electrical planning: elo plan engineering GmbH, Rosenheim
Lighting planning: Bartenbach Lichtlabor, Aldrans/Innsbruck
Fire protection: Brandschutz Consulting, Munich
Year of completion: 2012

www.behnisch.com
ba@behnisch.com

Stefan Behnisch
Born 1957 in Stuttgart; studied philosophy and economics in Munich; 1987: obtained degree in architecture at the University of Karlsruhe; lectureships in Stuttgart, Portsmouth, Nancy and Austin; visiting professor at the Yale School of Architecture in New Haven, University of Pennsylvania in Philadelphia and at the EPFL in Lausanne.

Robert Hösle
Born 1968 in Nesselwang; trained as carpenter; 1999: obtained degree in architecture at the University of Stuttgart; from 1999 worked at Behnisch & Partner; from 2004 worked at Behnisch Architekten Stuttgart; since 2009 head of Behnisch Architekten, Munich.

1989: formation of Behnisch Architekten, Stuttgart; 2009: establishment of Munich office.

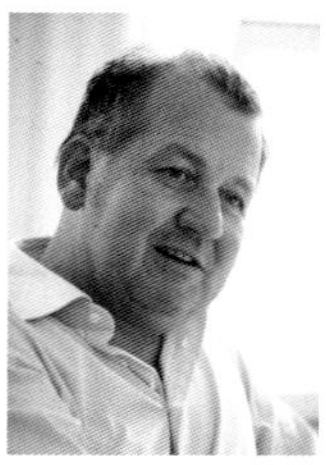

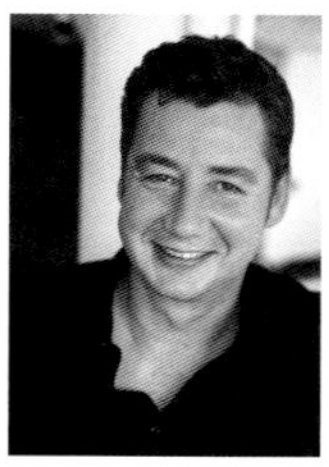

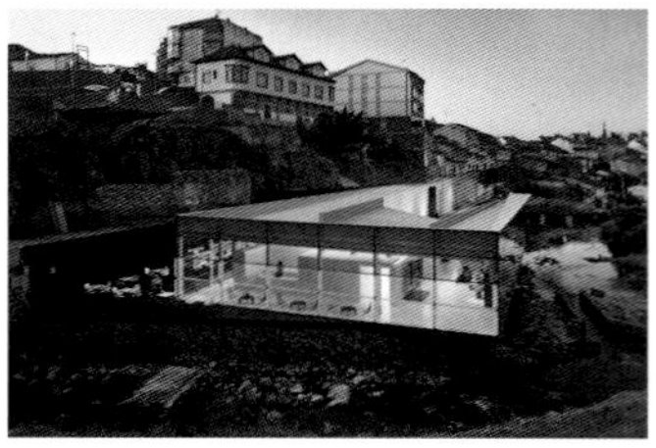

Café and exhibition spaces in Bragança

Client: Bragançapolis, Bragança
Architect: Giulia De Appolonia, Brescia
Contributors: Joana Sousa, Leonardo Paiella, Tiago Castella, Ivan Teixeira, Ruben Ferreira
Structural planning: Ara engenheiros, Fernando Rodrigues, Lisbon
Supervision of works: Santana SA
Building installations: NaturalWorks, Lisbon
Electrical planning: Ruben Sobral, Lisbon
Year of completion: 2007

www.abdarchitetti.com
info@abdarchitetti.com

Giulia De Appolonia
Born 1969 in Pordenone; 1994: obtained degree in architecture at the Politecnico di Milano; 1993–2000: worked with J. L. Carrilho da Graça in Lisbon; 2000–2004: freelance work in Lisbon; 2001–2004: professor at the Universidade Autonoma de Lisboa; 2005: established practice in Pordenone; 2008: formation of abda – de appolonia & associati with Camillo Botticini.

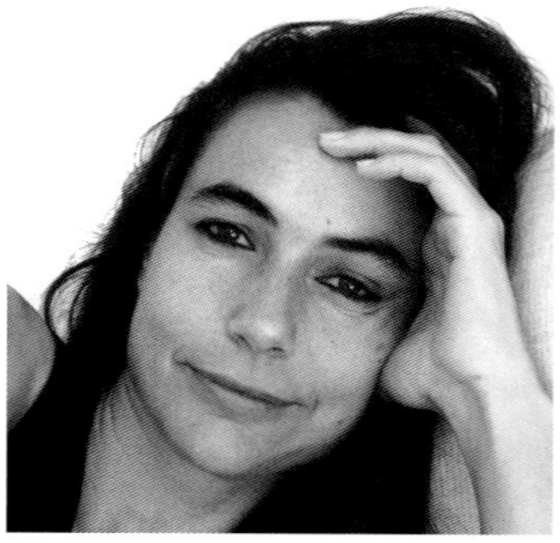

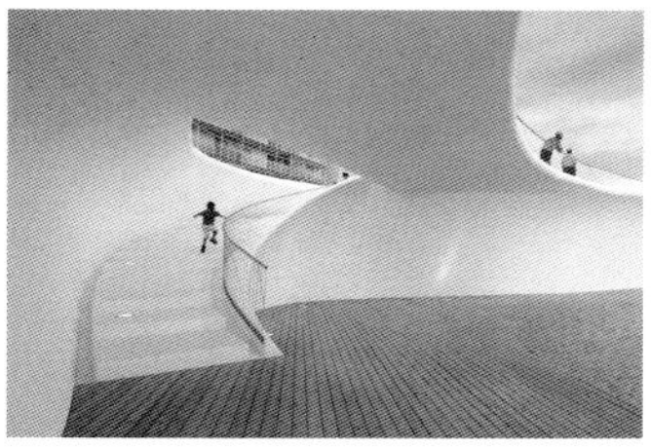

Beachfront promenade in Benidorm

Client: Generalitat Valenciana – Ajuntament De Benidorm
Architects: OAB – Office of Architecture in Barcelona
Carlos Ferrater, Xavier Martí Galí
Project management/supervision of works: Luca Cerullo
Contributors: Núria Ayala, Sofia Machado dos Santos, David Jiménez, David Abondano
Structural planning: Pondio Ingenieure, Juan Calvo
Construction: Ecisa-Dragados, Benidorm
Year of completion: 2009

www.ferrater.com
oab@ferrater.com

Carlos Ferrater
Born 1944 in Barcelona; 1971: graduated from the School of Architecture of Barcelona (ETSAB); lectureships at the ETSAB; 1973: established an architectural practice in Barcelona.

Xavier Martí Galí
Born 1969 in Barcelona; 1995: graduated from the School of Architecture of Barcelona (ETSAB); worked since 2003 at Carlos Ferrater Studio; since 2006 partner at OAB.

2006: formation of OAB – Office of Architecture in Barcelona.

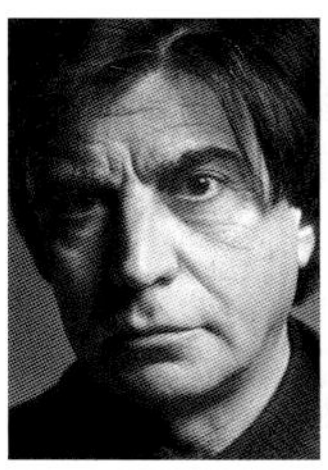
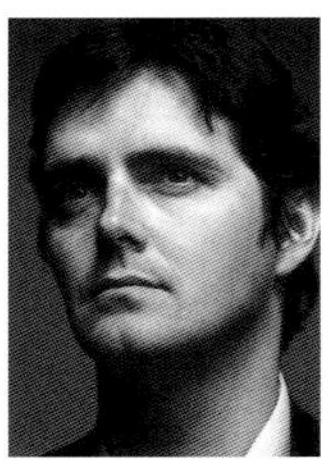

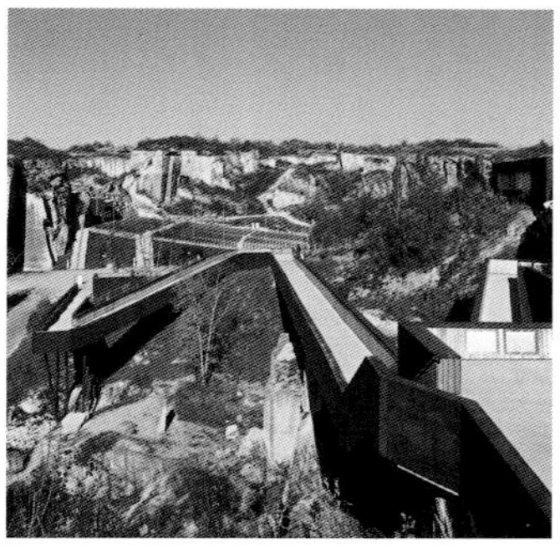

Festival arena in Roman quarry in St. Margarethen

Client: Fürst Esterházy Familienprivatstiftung
Architects: AllesWirdGut Architektur, Vienna
Contributors: Ecki Csallner, Elmir Smajic, Ferdinand Kersten, Maria Magina, Mareike Kuchenbecker, Martin Brandt, Michael Sohm
Structural planning: Gmeiner Haferl ZT GmbH, Vienna
Project management: FCP Fritsch Chiari & Partner ZT GmbH, Vienna
Service engineering, electrical planning: HPD Planungsdienst, Vösendorf
Geotechnical engineering: 3P Geotechnik, Vienna
Lighting design: Pokorny Lichtarchitektur, Vienna
Escape route planning: Büro Düh, Vienna
Local supervision of works: Werner Lacroix Bauplanungsges.m.b.H., Vienna
Year of completion: 2008

www.alleswirdgut.cc
awg@alleswirdgut.cc

Andreas Marth
Born 1969 in Fließ; studied architecture at the TU Vienna.

Friedrich Passler
Born 1969 in Bruneck; studied architecture at the TU Vienna.

Herwig Spiegl
Born 1973 in Innsbruck; studied architecture at the TU Vienna.

Christian Waldner
Born 1971 in Tscherms; studied architecture at the TU Vienna.

1999: formation of AllesWirdGut Architektur.

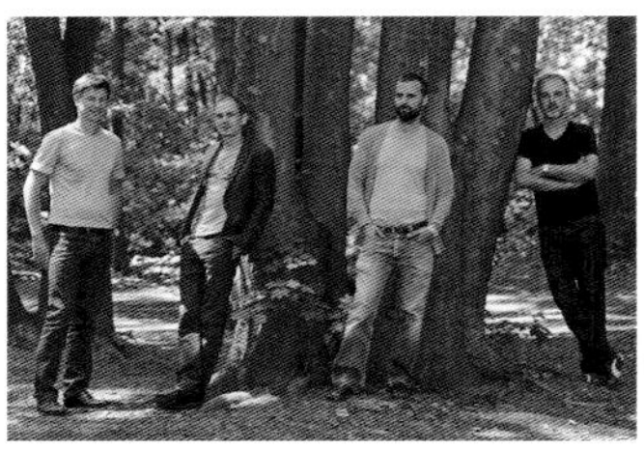

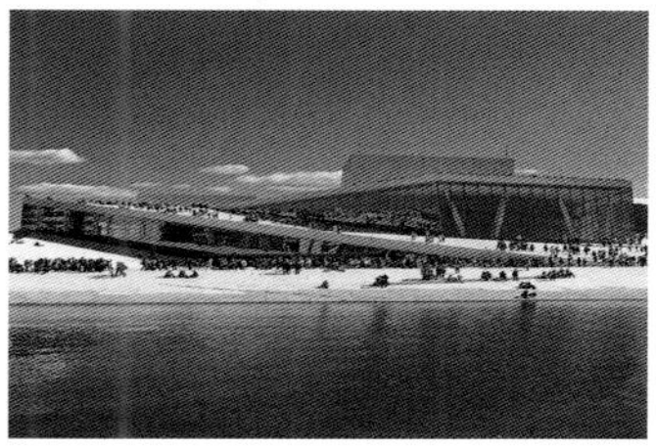

Opera house in Oslo

Client: Ministry of Church and Cultural Affairs, Statsbygg
Architects: Snøhetta A/S, Oslo
Project management: Tarald Lundevall, Sigrun Aunan, Craig Dykers, Simon Ewings, Kjetil Trædal Thorsen, Rune Grasdal, Tom Holtmann, Elaine Molinar, Kari Stensrød, Øystein Tveterk
Contributors: Anne-Cecilie Haug, Ibrahim El Hayawan, Tine Hegli, Jette Hopp, Zenul Khan, Frank Kristiansen, Cecilia Landmark, Camilla Moneta, Aase Kari Mortensen, Frank Nodland, Andreas Nygaard, Michael Pedersen, Harriet Rikheim, Margit Tidemann Ruud, Marianne Sætre, Knut Tronstad, Tae Young Yoon
Landscape architects: Snøhetta A/S, Ragnhild Momrak, Andreas Nypan
Structural planning: Reinertsen Engineering, Trondheim/Oslo
Participating artists: Kristian Blystad, Kalle Grude, Jorunn Sannes, Astrid Løvaas, Kirsten Wagle
Acoustics: Brekke Strand Akustikk, Oslo; Arup Acoustic, Winchester
Year of completion: 2007

www.snoarc.no
contact@snoarc.no

Kjetil Trædal Thorsen
Born 1958 in Haugesund; 1985: Master of Engineering at the TU Graz.

Tarald Lundevall
Born 1948 in Oslo; 1973: graduated from the Oslo School of Architecture.

Craig Edward Dykers
Born 1961 in Frankfurt; obtained bachelor's degree at the University of Texas in Austin.

1989: formation of Snøhetta AS.

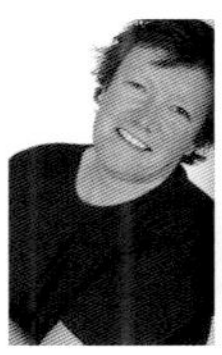
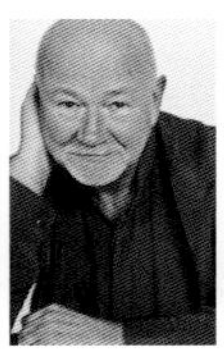
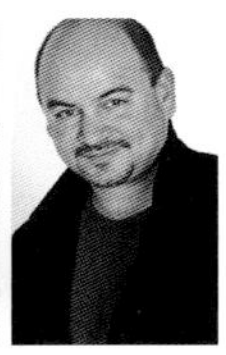

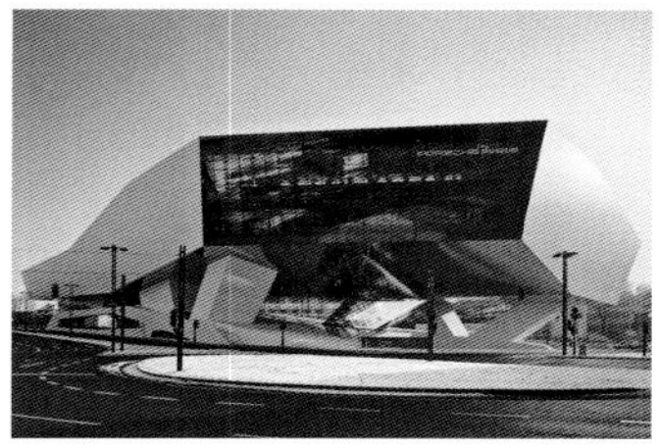

Porsche Museum in Stuttgart

Client: Porsche AG, Stuttgart
Architects: Delugan Meissl Associated Architects, Vienna
Project management: Martin Josst
Contributors: Zoltan Adorjani, Philip Beckmann, Sebastian Brunke, Gerhard Gölles, Imke Haasler, Tom Hindelang, Tapio Lassmann, Jörg Rasmussen, Torsten Sauer, Hendrik Steinigeweg
Execution planning: Wenzel + Wenzel Architekten, Stuttgart
Supervision of works: Gassmann + Grossmann Baumanagement GmbH, Stuttgart
Project management: Drees & Sommer, Stuttgart
Structural planning: Leonhardt, Andrä und Partner Beratende Ingenieure VBI GmbH, Stuttgart
Building physics: Wolfgang Sorge IB für Bauphysik GmbH, Nuremberg
Exhibition design concept: hg merz, Stuttgart
Year of completion: 2009

www.dmaa.at
office@dmaa.at

Elke Delugan-Meissl
Born in Linz; 1987: obtained degree in architecture at the TU Innsbruck.

Roman Delugan
Born in Meran; studied architecture at the University for Applied Art in Vienna.

Dietmar Feistel
Born in Bregenz; studied architecture at the TU Vienna.

Martin Josst
Born in Hamburg; studied architecture at the Muthesius University of Art and Design in Kiel.

1993: formation of Delugan Meissl.

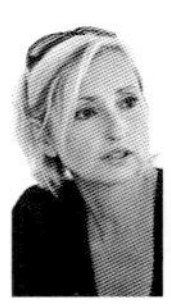
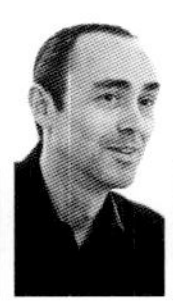
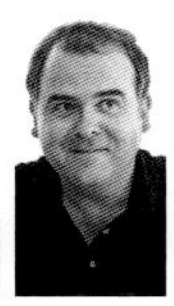
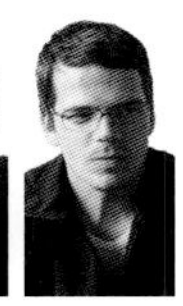

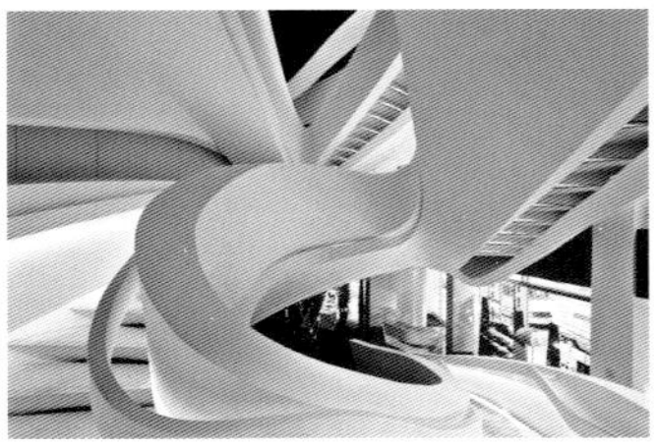

Armani Fifth Avenue in New York

Client: Gruppo Giorgio Armani
Architects: Doriana and Massimiliano Fuksas, Rome
Project management: Davide Stolfi
Interior design: Fuksas Design, Rome
Contractor: Americon Construction Inc.
Structural planning: Gilberto Sarti, Rimini
Lighting planning: Speirs & Major Associates, London
Year of completion: 2009

www.fuksas.com
office@fuksas.com

Massimiliano Fuksas
Born 1944 in Rome; 1969: graduated from the University La Sapienza in Rome; since 1967 has had his own practice in Rome, since 1989 in Paris, since 2008 in Shenzhen; visiting professor at the École Spéciale d'Architecture (ESA) in Paris, the Akademie der Bildenden Künste in Vienna, the Staatliche Akademie der Bildenden Künste in Stuttgart and Columbia University in New York.

Doriana O. Mandrelli
Born in Rome; 1979: obtained degree in the history of modern and contemporary architecture from the University La Sapienza in Rome; degree in architecture at the École Spéciale d'Architecture (ESA) in Paris; lectureship at the University La Sapienza in Rome; since 1985: practice partnership with Massimiliano Fuksas in Rome; since 1997 responsible for Fuksas Design.

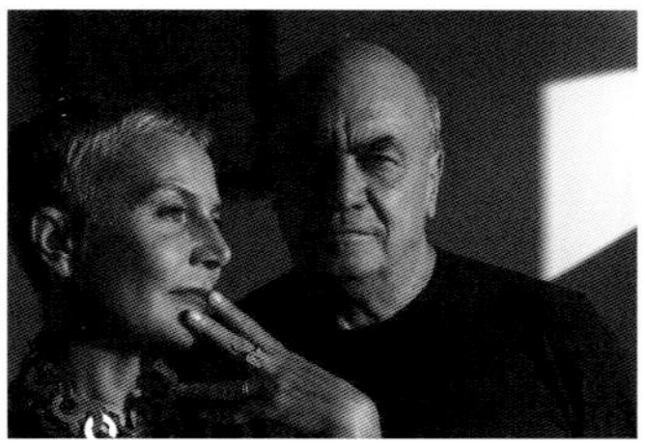

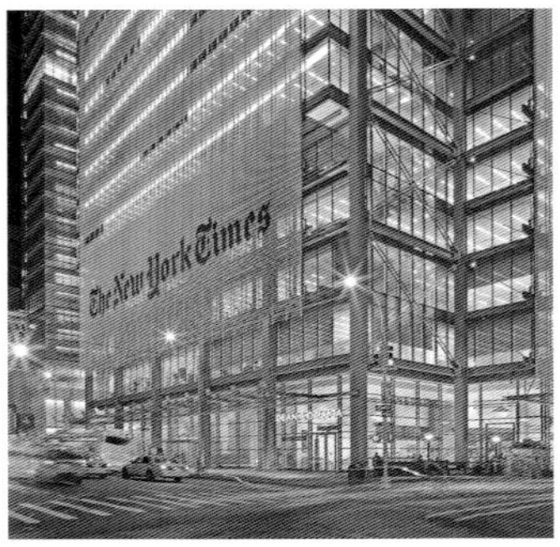

New York Times Building in New York

Client: The New York Times, New York; Forest City Ratner Companies, New York
Architects: Renzo Piano Building Workshop, Paris/Genoa
Bernhard Plattner (Project Director)
Erik Volz (Project Manager)
FXFowle Architects, New York/Dubai
Contributors FXFowle: Bruce Fowle, Daniel Kaplan, Gerald Rosenfeld, Joseph Hand, Elizabeth Finkelshteyn, Scott Wood, Ray Williams, Doug Freeman, James Adams, Jason Abbey, Nick Tocheff, Xiaotong Wu, Zheng Dai, John Secreti, Xander Redfern, Chiam Zeitz
Structural planning: Thornton Tomasetti, New York
Interior design (NY Times): Gensler, New York
Vertical access: Jenkins & Huntington Inc., New York
Security consultants: Kroll Schiff & Associates, New York
Year of completion: 2007

www.rpbw.com
italy@rpbw.com
www.fxfowle.com
info@fxfowle.com

Renzo Piano
Born 1937 in Genoa; 1964: obtained degree at the Polytechnic in Milan; 1971: formation of the practice Piano & Rogers; 1974: formation of the Renzo Piano Building Workshop.

FXFowle Architects
Founded in 1978; offices in New York and Dubai (photos: Senior Principals Bruce Fowle, Daniel Kaplan).

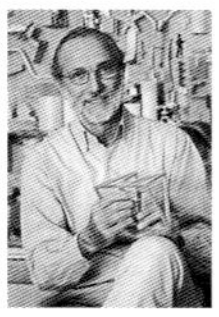
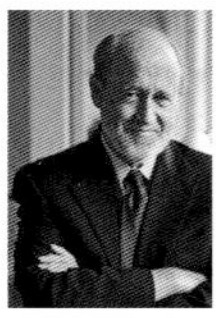
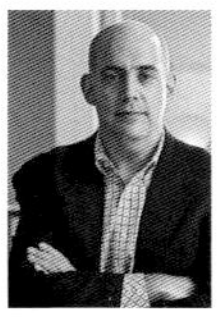

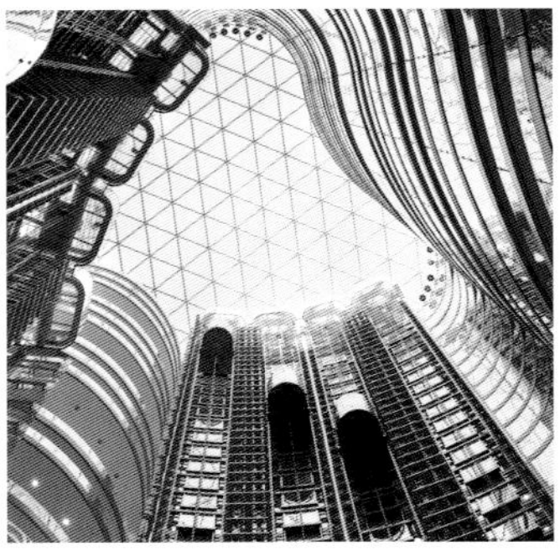

Office building in Sydney

Client: DEXUS Wholesale Property Fund; Cbus Property, Sydney
Architects: ingenhoven architects, Düsseldorf; Architectus, Sydney
Structural planning: Enstruct Group, Sydney
Facade planning: DS Plan, Stuttgart; Arup Facade with Enstruct, Sydney
Sustainability consultants: Cundall, Sydney
Lighting planning: Arup Electrical, Sydney; Tropp Lighting Design, Weilheim
Lift systems: Norman Disney and Young, Sydney
Mechanical engineering: Arup Mechanical, Sydney
Hydraulic engineering: Steve Paul & Partners, Sydney
Landscape planning: Sue Barnsley Design, Sydney
Project management: APP Corporation, Sydney
Year of completion: 2011

www.ingenhovenarchitects.com
info@ingenhovenarchitects.com
www.architectus.com.au
sydney@architectus.com.au

Christoph Ingenhoven
Born 1960 in Düsseldorf; 1978–1984: studied architecture at the RWTH in Aachen; 1985: formation of the practice ingenhoven architects.

Ray Brown
Born 1963 in Melbourne; studied architecture at the University of New South Wales; studied General Management at the Australian School of Business, University of New South Wales; from 1989 worked at Architectus, since 1998 Managing Director.

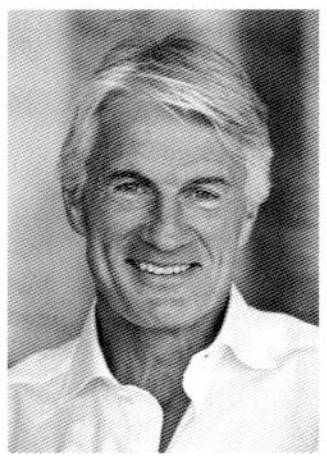
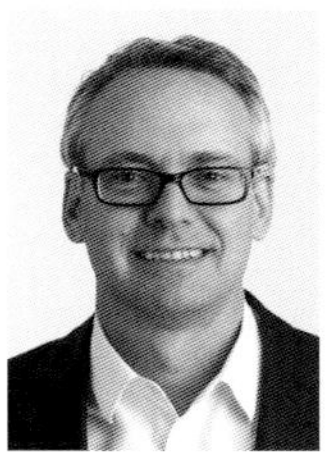

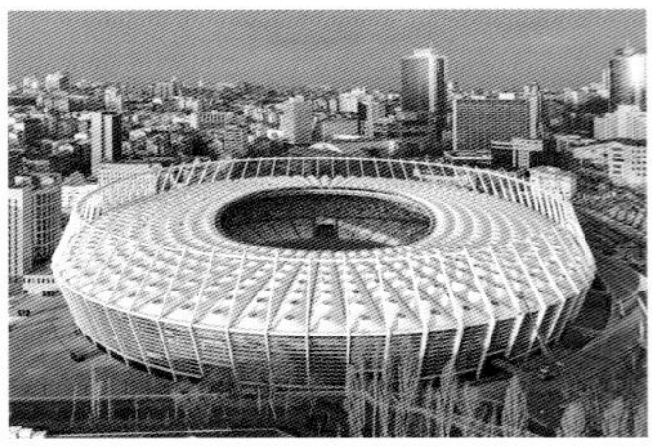

Stadium in Kiev

Client: National Sport Complex "Olimpiyskiy"
Architects: gmp • Architekten von Gerkan, Marg und Partner, Hamburg
Design: Volkwin Marg with Christian Hoffmann and Marek Nowak
Project management: Martin Bleckmann, Roman Hepp
Contributors (design): Michael König, Christoph Salentin, Olaf Peters, Heiko Faber, Sebastian Möller
Contributors (execution): Andreas Wietheger, Clemens Dost, Christiane Wermers, Jonathan Gerlach, Anke Appel, Irina Stoyanova, Franz Lensing, Jan Philipp Weber, Dominik Heizmann, Sebastian Hilke, Irina Bohlender; in consortium with Personal Creative Architectural Bureau Y. Serjogin LLC, Kiev
Supporting structure for roof: schlaich bergermann und partner, Knut Göppert, Markus Balz, Thomas Moschner, Stuttgart
Statics: Kempen Krause Ingenieurgesellschaft, Aachen
Building installations: b.i.g. Bechtold Ingenieurgesellschaft mbH, Karlsruhe
Landscape planning: ST raum a. Gesellschaft von Landschaftsarchitekten mbH, Berlin
Year of completion: 2011

www.gmp-architekten.de
hamburg-e@gmp-architekten.de

Volkwin Marg
Born 1936 in Königsberg; 1958–1964: studied architecture in Berlin and at the TU in Braunschweig; 1965: co-founder of the practice gmp • Architekten von Gerkan, Marg und Partner.

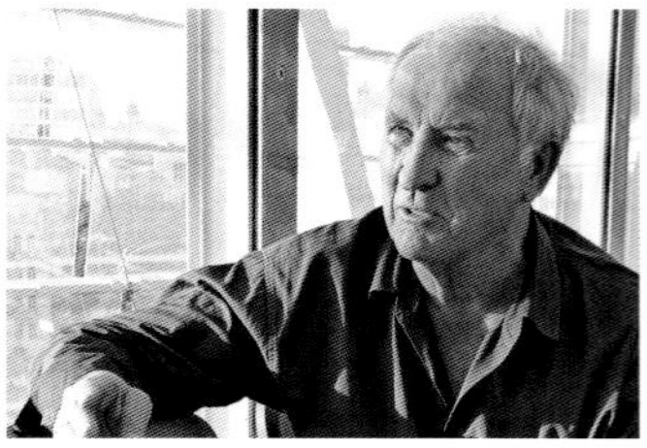

Authors

Christian Schittich (editor)
Born 1956
Studied architecture at the University of Technology in Munich; followed by seven years of practical experience, journalistic work; since 1991: editorial staff of DETAIL, magazine for architecture and architectural details;
editor since 1992, editor-in-chief since 1998;
author and publisher of numerous specialist publications and articles.

Arno Lederer
Born 1947
Studied architecture at the University of Stuttgart and the Technical University in Vienna, obtained degree in 1976;
1979: formation of the Lederer practice, since 1985 in partnership with Jórunn Ragnarsdóttir, since 1992 with Marc Oei;
1985–1990: professor of construction and design at the Hochschule für Technik, Stuttgart;
1990–1997: professor of building construction and design, 1997–2005: professor of building studies at the University of Karlsruhe;
since 2005 professor of public building and design at the University of Stuttgart.

Jeannot Simmen
Born 1946
Studied history of art, philosophy and religious studies in Zurich and at the Free University in Berlin; doctoral studies under Klaus Heinrich and Jacob Taubes; post-doctoral studies under Bazon Brock and Klaus Heinrich;
author and publisher of "The Lift. The History of Vertical Conquest" (with Uwe Drepper), Munich 1984 and "Vertical. Lift – Escalator – Paternoster. A Cultural History of Vertical Transportation", Berlin 1994;
curator of art and technology exhibitions;
1990–2002: visiting and substitute professor of art theory and design studies at the Universities of Kassel, Wuppertal and Essen.

Thomas Schielke
Born 1973
Studied architecture at the University of Technology in Darmstadt: obtained degree in 2001;
since 2001 responsible for the area of didactic communication with the lighting manufacturer ERCO;
teaching appointments at various universities;
co-author of the book "Perspectives on Lighting – Between Culture and Technology", Munich 2009.

Jimmy Schmid
Born 1961
Trained as a teacher at the Cantonal Teacher Training Seminary in Hitzkirch/Lucerne;
studied graphics at the School of Design in Lucerne and at the École nationale supérieure des Arts Décoratifs in Paris;
many years of agency work in the areas of communication design, corporate design, knowledge visualization, information logistics, exhibitions, visitor guidance, signage;
since 2001 professor of communication design at the Visual Communication department at the Bern University of the Arts, since 2008 director of the vocational master's degree course in Signage – Environmental Information Design.

Oliver Herwig
Born 1967
Studied humanities in Regensburg, Williamstown / MA, Champaign-Urbana/IL and Kiel; doctoral studies on "Wortdesign. Eugen Gomringer und die bildende Kunst";
works as a freelance journalist in Munich;
lecturer in design theory at the universities in Karlsruhe and Linz.

Illustration credits

The authors and editor wish to extend their sincere thanks to all those who helped to realize this book by making illustrations available. All drawings contained in this volume have been specially prepared in-house. Photos without credits are from the architects' own archives or the archives of "DETAIL, Review of Architecture". Despite intense efforts, it was not possible to identify the copyright owners of certain photos and illustrations. Their rights remain unaffected, however, and we request them to contact us.

From photographers, photo archives and image agencies:

- p. 8: Daniel Clements /view/arturimages
- pp. 9, 12 bottom, 141, 142: Christian Richters, Berlin
- p. 10 top: Carsten Klein/arturimages
- pp. 10 middle, bottom, 11 top, 13 top, 37 top, 106, 127: Christian Schittich, Munich
- pp. 11 bottom, 12 top: Frank Kaltenbach, Munich
- p. 13 bottom: Christian Richters /arturimages
- p. 14: Stefan Müller-Naumann, Munich
- p. 16 top: Florian Monheim /arturimages
- p. 17: Werner Huthmacher, Berlin
- p. 18 bottom: Adolf Bereuter, Lauterach
- p. 19: Jens Hauspurg, Weimar
- p. 21: Barbara Staubach /arturimages
- p. 23 top, 57–59: Walter Mair, Zurich
- p. 24 bottom: Zooey Braun/arturimages
- p. 25 bottom left: Julia Knop, Hamburg
- p. 26 bottom: Ivan Nemec, Berlin
- p. 27 bottom: Florian Holzherr, Munich
- p. 28: Christina Merkan, Berlin
- pp. 30 top left und right, 31 top: Archiv Simmen /Drepper, Berlin
- p. 30 middle: Deutsches Museum, Munich
- p. 30 bottom: according to Uwe Drepper, Berlin
- p. 31 bottom: Chevojon, Paris (Spadem)
- p. 32 top: Schindler Group, Ebikon
- p. 32 bottom: James Morris /view/arturimages
- p. 33 left, 156 top, 157 bottom: Whitney Cox, New York
- p. 34: Alexander Ring/ERCO
- p. 36 top: Simulationen: Axel Groß/ERCO
- p. 37 bottom: Thomas Mayer/ERCO
- p. 38 top: Rudi Meisel/ERCO
- p. 38 bottom: Nicholas Kane/Arcaid Images
- p. 39 top: Werner Huthmacher/ERCO
- pp. 40, 89 bottom right: Andreas Körner, Stuttgart
- p. 42 top und middle: Integral Ruedi Baur/Lisa Jacob
- p. 42 bottom: Ricardo Gonçalves, Lissabon
- p. 43: Yves André, Vaumarcus
- p. 44 top: LM communiquer, Paris
- p. 44 middle: André Baldinger, Paris
- p. 44 bottom: Kaspar Schmid, Zurich
- p. 45 middle und bottom: L2M3 Kommunikationsdesign, Stuttgart
- p. 46 top: Hi Visuelle Gestaltung, Lucerne
- p. 46 middle left: bauer – konzept & gestaltung, Vienna
- p. 46 middle right: Nik Thoenen/re-p.org
- p. 46 bottom: Stefan Wülser/ www.openmindpp.ch
- pp. 47 top left, 74–76, 89 top, bottom left, 90–92: Roger Frei, Zurich
- p. 47 top middle: Bringolf Irion Vögeli, Zurich
- p. 47 top right: Hochschule der Künste Bern, CAS Signaletik
- p. 47 bottom: Intégral Ruedi Baur, Paris
- pp. 48, 53 middle: Doppelmayr Seilbahnen Gmbh, Wolfurt
- p. 49: Kölner Seilbahn GmbH
- p. 50 top: © Transport for London
- p. 50 middle: www.flickr.com/Embassy of the Bolivian Republic of Venezuela, Photo: Abraxas Iribarren
- pp. 50 bottom, 51, 66–69, 71, 95, 146 bottom, 147, 148 top, middle bottom, bottom: Iwan Baan, Amsterdam
- p. 52: www.flickr.com/CUP Projects
- p. 53 top left und right: Eric Staudenmaier Photography, Los Angeles
- p. 53 bottom: www.flickr.com/Alamagordo, Wieland Van Dijk
- pp. 54, 81–85: Robert Leš, Zagreb
- p. 56: Georg Aerni, Zurich
- pp. 60–65: Simon Menges
- pp. 70, 72, 73: Daici Ano, Tokyo
- pp. 77, 79 bottom: JDS Architects /Felix Luong
- pp. 78, 169 3rd column bottom: Ulrik Jantzen, Copenhagen
- p. 79 top und middle: Jens Lindhe, Copenhagen
- p. 80: Jakob Boserup, Frederiksberg
- pp. 86–88, 171 4th column bottom left: Paul Ott, Graz
- pp. 93, 94 bottom, 96 top, 148 middle top: Roland Halbe, Stuttgart
- pp. 94 top, 96 bottom : Roland Halbe/arturimages
- pp. 98–103: Adam Mørk, Copenhagen
- pp. 104 top, 105: Wolfram Janzer/arturimages
- p. 104 bottom: Eun Young Yi, Cologne
- p. 107: Jana Rackwitz, Munich
- p. 108 top: Kozo Takayama, Tokyo
- p. 108 bottom: Pacia
- p. 109: Shinkenchiku-sha, Tokio
- pp. 110–113: Bruno Klomfar, Vienna
- pp. 114–119: Jens Kirchner, Düsseldorf
- pp. 120–122, 172 1st column bottom: Heinz Unger, Schlieren
- p. 123: Cordula Rau, Munich
- pp. 124–125: Panos Kokkinias, Athens
- pp. 126, 128: David Matthiessen, Stuttgart
- pp. 129–131: Fernando Guerra, Lisbon
- pp. 132, 133 right, 135: Alejo Bague, Barcelona
- p. 134 top und bottom: Office of Architecture, Barcelona
- pp. 136–139: Hertha Hurnaus, Vienna
- pp. 140, 144 Middle: Claudia Fuchs, Munich
- p. 143 left: Hélène Binet, London
- p. 143 right: Ivan Brodey, Oslo
- p. 144 top: Jens Sölvberg, Stockholm
- p. 144 bottom: Andreas Gabriel, Munich
- p. 145: Jiri Havran, Oslo
- p. 146 top: Brigida Gonzáles, Stuttgart
- pp. 150, 153: Ramon Prat, Barcelona
- p. 152: Archivio Fuksas, Rome
- p. 154: Andreas Keller/arturimages
- pp. 155, 157 top: Thomas Madlener, Munich
- pp. 156 middle, bottom: Nathan Willock /view/arturimages
- pp. 158–163: H. G. Esch, Hennef
- pp. 164, 165 right, 167: Marcus Bredt, Berlin
- p. 165 left: Jborzicchi /Dreamstime.com
- p. 168 3rd column bottom: Inga Powilleit, Haarlem
- p. 170 3rd column bottom: Reiner Zettl
- p. 170 4th column bottom: André Andersen
- p. 172 3rd column bottom: Mark Ostow, Cambridge
- p. 173 3rd column bottom: AllesWirdGut Architektur, Vienna
- p. 174 1st column bottom: Maurizio Marcato, Zevio
- p. 174 3rd column bottom left: Edgar Schoepal, Düsseldorf

From books and journals:

- p. 15: Peter Faller: Der Wohngrundriss. Munich 2002, p. 17
- p. 16 bottom: Wilfried Koch: Baustilkunde. Munich 1991, p. 316

Articles and introductory b/w photos:

- p. 8: Solomon R. Guggenheim Museum, New York (USA) 1959, Frank Lloyd Wright
- p. 12: Youth Hostel, Possenhofen (D) 2002, Hierl Architekten
- p. 28: Lift bell, Berlin-Charlottenburg (D)
- p. 40: Hochschule Osnabrück (D) 2004, Architecture: Jockers Architekten, Signage: büro uebele visuelle kommunikation
- p. 48: Thames Cable Car, London (GB) 2012
- p. 54: Hostel, Split (HR) 2010, STUDIO UP

Dust jacket:

Armani Fifth Avenue, New York (USA) 2009, Doriana and Massimiliano Fuksas, Rome
Photo: Archivio Fuksas, Rome

Project data are provided as is by the repsonsible architectural offices. The publisher is not responsible for correctness of provided data.